Laser in Environmental and Life Sciences

Springer-Verlag Berlin Heidelberg GmbH

Peter Hering • Jan Peter Lay • Sandra Stry
(Editors)

Laser in Environmental and Life Sciences

Modern Analytical Methods

With 166 Figures

Springer

EDITORS:

Professor Dr. Peter Hering
University of Düsseldorf
Institute of Laser Medicine
Universitätsstraße 1
40225 Düsseldorf
Germany
E-mail: *hering@uni-duesseldorf.de*

Dr. Sandra Stry
Sacher Lasertechnik
Hannah-Arendt-Straße 3-7
35037 Marburg
Germany
E-mail: *sandra@sacher-laser.com*

Dr. Jan Peter Lay
Deutsche Bundesstiftung Umwelt (DBU)
An der Bornau 2
49090 Osnabrück
Germany
E-mail: *jp.lay@dbu.de*

ISBN 978-3-642-07309-0 ISBN 978-3-662-08255-3 (eBook)
DOI 10.1007/978-3-662-08255-3

Library of Congress Cataloging-in-Publication Data Applied For

A catalog record for this book is available from the Library of Congress.
Bibliographic information published by Die Deutsche Bibliothek
Die Deutsche Bibliothek lists this publication in die Deutsche Nationalbibliographie; detailed bibliographic data is available in the Internet at <http://dnb.ddb.de>.

http://www.springer.de

Originally published by Springer-Verlag Berlin Heidelberg New York in 2004
Softcover reprint of the hardcover 1st edition 2004

Cover Design: Erich Kirchner, Heidelberg
Artwork: Andreas B. Broessel
Images Provided by: Andreas B. Broessel, Sacher Lasertechnik Group
With the friendly assistance of: Blackfur Media & Sacher Lasertechnik Group
Typesetting: Camera-ready by Gerd Laschinski

Printed on acid free paper 30/3141/LT – 5 4 3 2 1 0

Foreword

Studies of environmental conditions have become more and more important for our knowledge about the complex processes resulting from natural and man-made influences on our environment. During recent years new analytical techniques have been developed that are mainly based on laser spectroscopy. They are not only more sensitive and reach lower detection limits but they also give more detailed information on the species under investigation and their interactions with the environment. They therefore can considerably deepen our understanding of correlations between different processes and their causes, which is the precondition for avoiding ecological damage and for strengthening effective engagement in environmental protection.

Also in life sciences revolutionary new techniques for diagnostics have been introduced which allow a thorough investigation of biological processes in plants and their dependence on environmental conditions. In particular for medical applications these new mainly non-invasive techniques have influenced medical diagnostics and treatment considerably.

This book deals with these very interesting interdisciplinary subjects which demand experience and knowledge about both fields: the basic physics of the detection techniques and the relevant problems in environmental and life sciences. Therefore authors from a broad area of fundamental and applied research have contributed to this book, which is organized in 4 sections: Remote sensing of the atmosphere, detection techniques for liquid and solid samples, applications to gaseous substances and aerosols and finally applications in biology and medicine. Each section is opened by a review article, which hints to the relations between the following articles and surveys the literature.

The intention of the authors is to transfer their knowledge about new interdisciplinary fields to a wide section of biologists, medical people or environmentalists, which should recognize the new possibilities opened by these techniques.

Since there are not many books on these subjects, this volume may also serve to start more initiatives for a closer cooperation between physicists, chemists, biologists and medical people. One of the editors of this book, Prof. P. Hering, has a long standing experience about such a successful cooperation between a physicist and his medical colleagues in an institute of laser medicine and he wants to inspire more scientists to start or to intensify such very awarding cooperations.

I hope that this book contributes to this common goal by reaching a wide distribution. I wish the authors a critical and positive audience and a vivid discussion about the advantages and limitations and possible further applications of the new methods presented here.

Wolfgang Demtröder

Contents

List of Contributors

Prof. Dr. Angelika Anders
Institute of Biophysics
University of Hannover
Herrenhäuser Str. 2
D-30419 Hannover, Germany
anders@biophysik.uni-hannover.de

Golo von Basum
Institut für Lasermedizin
Heinrich-Heine Universität
Universitätsstraße 1
D-40225 Düsseldorf, Germany

Prof. Dr. Hans Bettermann
Institut für Physikalische Chemie und Elektrochemie I
Heinrich-Heine Universität
Universitätsstraße 1
D-40225 Düsseldorf, Germany

Prof. Dr. Ulrich Boesl-von Grafenstein
Technische Universität München
Institut für Physikalische und Theoretische Chemie
Lichtenbergstraße 4
D-85748 Garching, Germany
Phone: +49-89-289 13397, Fax: +49-89-289 14430
ulrich.boesl@ch.tum.de

Eric Crosson, Ph.D.
Informed Diagnostics, Inc.
1050 E. Duane Ave, Suite H
Sunnyvale, California, 94085 USA

Dr. Hannes Dahnke
Institut für Lasermedizin
Heinrich-Heine Universität
Universitätsstraße 1
D-40225 Düsseldorf, Germany
hannes.dahnke@uni-duesseldorf.de

F. Frischkorn
Institut für Lasertechnik an der Fachhochschule
Oldenburg / Ostfriesland / Wilhelmshaven
Constantiaplatz 4
D-26723 Emden, Germany

Prof. Dr. Peter Hering
Institut für Lasermedizin
Heinrich-Heine Universität
Universitätsstraße 1
D-40225 Düsseldorf, Germany

Prof. Dr. Karl Kleinermanns
Institut für Physikalische Chemie und Elektrochemie I
Heinrich-Heine Universität
Universitätsstraße 1
D-40225 Düsseldorf, Germany

H. Kreitlow
Institut für Lasertechnik an der Fachhochschule
Oldenburg / Ostfriesland / Wilhelmshaven
Constantiaplatz 4
D-26723 Emden, Germany

Privatdozent Dr. Frank Kühnemann
Institut für Angewandte Physik der Universität Bonn und
Max-Planck-Institut für Chemische Ökologie Jena
Institut für Angewandte Physik
Wegelerstraße 8
D-53115 Bonn, Germany
frank.kuehnemann@iap.uni-bonn.de

Dr. Joerg Lauterbach
Institut für Physikalische Chemie und Elektrochemie I und
Institut für Lasermedizin
Heinrich-Heine Universität
Universitätsstraße 1
D-40225 Düsseldorf, Germany

Dr. Matthias Lemke
Institute of Chemistry
University of Potsdam
PO Box 601 553
D-14415 Potsdam, Germany
lemke@chem.uni-potsdam.de

Dr. F. Lewitzka
Laser-Laboratorium Göttingen e.V.
Hans-Adolf-Krebs-Weg 1
D-37077 Göttingen, Germany
Phone: +49-551-503552, Fax: +49-551-503599

Prof. Dr. Hans-Gerd Löhmannsröben
Institute of Chemistry
University of Potsdam
PO Box 601 553
D-14415 Potsdam, Germany
loeh@chem.uni-potsdam.de

Prof. Dr. G. Marowsky
Laser-Laboratorium Göttingen e.V.
Hans-Adolf-Krebs-Weg 1
D-37077 Göttingen, Germany
Phone: +49-551-503530, Fax: +49-551-503599,
gmarows@gwdg.de

J. Miesner
Institut für Lasertechnik an der Fachhochschule
Oldenburg / Ostfriesland / Wilhelmshaven
Constantiaplatz 4
D-26723 Emden, Germany

Priv.-Doz. Dr. Manfred Mürtz
Institut für Lasermedizin
Heinrich-Heine Universität
Universitätsstraße 1
D-40225 Düsseldorf, Germany

M. Niederkrüger
Laser-Laboratorium Göttingen e.V.
Hans-Adolf-Krebs-Weg 1
D-37077 Göttingen, Germany
Phone: +49-551-503552, Fax: +49-551-503599
mniederk@llg.gwdg.de

Barbara Paldus, Ph.D.
Informed Diagnostics, Inc.
1050 E. Duane Ave, Suite H
Sunnyvale, California, 94085 USA

PD Dr. Ulrich Panne
Laboratory for Applied Laser Spectroscopy
Technical University Munich
Institute of Hydrochemistry
Marchioninistr. 17
D-81377 Munich, Germany
Phone: +49-89-7095 79 87, Fax: +49-89-7095 79 99
ulrich.panne@ch.tum.de

Dipl.-Phys. Ralf Pätzold
Institute of Biophysics
University of Hannover
Herrenhäuser Str. 2
D-30419 Hannover, Germany
paetzold@biophysik.uni-hannover.de

Miguel Rodriguez
Freie Universität Berlin
Institut für Experimentalphysik
Arnimallee 14
D- 14195 Berlin, Germany
Phone: +49-30-838 56119, Fax: +49-30-838 55567
miguel.rodriguez@physik.fu-berlin.de

Dr. Oliver J. Schmitz
University of Wuppertal
Department of Analytical Chemistry
Gauss-Str. 20
D-42119 Wuppertal, Germany
Phone: +49-202-4392492, Fax:+ 49-202-4393915
olivers@uni-wuppertal.de

Dr. Rainer H. Schultze
Optimare GmbH
Jadestraße 59
D-26382 Wilhelmshaven, Germany
rainer.schultze@optimare.de

B. Stark
Technische Fachhochschule Wildau
Bahnhofstraße 1
D-15745 Wildau, Germany

Dr. Roland Steinert
Institut für Physikalische Chemie und Elektrochemie I
Heinrich-Heine Universität
Universitätsstraße 1
D-40225 Düsseldorf, Germany

Dr. S. Stry
Sacher Lasertechenik
Hannah-Arendt Straße 3-7
D-35037 Marburg, Germany
Phone: +49-6421-305305, Fax : +49-6421-305299
sandra@sacher-laser.com

Dr. Karen Tönnies
Institut für Energietechnik
Technische Universität Berlin
Fasanenstr. 89
D-10623 Berlin, Germany

Matthias Ulbricht
ADLARES GmbH
Potsdamer Str. 48
D-14513 Teltow, Germany
Phone: +49-3328-33060, Fax: +49-3328-330629
ulbricht@adlares.com

Prof. Dr. Wolfgang Urban
Institut für Angewandte Physik
Wegelerstr. 8
53115 Bonn
correspondence to:
Prof. Dr. Wolfgang Urban
Adelheidstr.7
D-61462 Königstein, Germany
Phone: +49-6174-21185, Fax: +49-6174-24502
w.urban-koenigstein@t-online.de

Priv.-Doz. Dr. Christian Weickhardt
Lehrstuhl Physikalische Chemie und Analytik
Brandenburgische Technische Universität
Erich-Weinert-Straße 1
D-03044 Cottbus, Germany

Dr. Peter W. Werle
National Institute for Applied Optics
Largo E. Fermi, 6
I-50125 Florence, Italy
PWWerle@inoa.it

Prof Dr. Ludger Wöste
Freie Universität Berlin
Institut für Experimentalphysik
Arnimallee 14
D-14195 Berlin, Germany
Phone: +49-30-838 55566, Fax: +49-30-838 55567
ludger.woeste@physik.fu-berlin.de

Dr. Wiebke Zimmer
Freie Universität Berlin
Institut für Experimentalphysik
Arnimallee 14
D- 14195 Berlin, Germany
Phone: +49-30-838 56119, Fax: +49-30-838 55567
wiebke.zimmer@physik.fu-berlin.de

List of Abbreviations

2-D	two-dimensional
AAI	aristolochic acid I
AAS	atomic absorption spectroscopy
AM	amplitude modulation
AOM	acousto-optic modulator
BBO	beta-barium-borat
BEC	background equivalent concentration
BfG	German Federal Institute of Hydrology
Bodipy-EDA	4,4-difluoro-5,7-dimethyl-4-bora-3a,4a-diaza-s-indacene-3-propionyl ethylene-diamine
BTEX	benzene, toluene, ethyl-benzene, xylene
CALOS	cavity leak-out spectroscopy
CCD	charge coupled device
CE-LIF	capillary electrophoresis with laser-induced fluorescence detection
CO	carbon monoxide
CO_2	carbon dioxide
CPA	chirped pulse amplification
CR	certified reference (material)
CRDS	cavity ring-down spectroscopy
cw	continuous wave
dAMP	2'-deoxyadenosine-3'-monophosphates
DBS	doppler beam swinging
dCMP	2'-deoxycytidine-3'-monophosphates
DDT	1,1,1-trichloro-2,2-bis(p-chlorophenyl)ethane
DFB	distributed feedback lasers
DFG	difference frequency generation
dGMP	2'-deoxyguanosine-3'-monophosphates
DIAL	differential absorption Lidar
dNMP	2'-deoxynucleoside-3'-monophosphates
DOAS	differential optical absorption spectroscopy

dTMP	2'-deoxythymidine-3'-monophosphates
ECL	external cavity lasers
eCO	exhaled Carbon Monoxide
EDC	1-ethyl-3-(3′-N,N′-dimethylaminopropyl)-carbodiimide
EEM	excitation emission matrix
eNO	exhaled Nitric Oxide
EOM	electro optical modulator
EPA	Environmental Protection Agency (USA)
FM	frequency modulation
FTIR	fourier transform infrared
GC	gas chromatography
GC/MS	gas chromatography/mass spectrometry
GVD	group velocity dispersion
HEPES	N-(2-hydroxyethyl)-piperazine-N′-2-ethane sulfonic acid
HO	heme oxygenase
HPLC	high pressure liquid chromatography
HR	high reflective
IB	inverse Bremsstrahlung
ICCD	intensified charge coupled device (camera)
ICP	inductively coupled plasma
IMS	ion mobility spectrometry
IR	infrared
IRMS	isotope ratio mass spectrometry
KTP	potassium titanyl phosphate
LA	laser ablation
LA-ICP-MS	laser ablation inductively coupled plasma mass spectrometry
LA-ICP-OES	laser ablation inductively coupled optical emission spectroscopy
LAMMA	laser microprobe mass analyzer
LC-MS	liquid chromatography-mass spectrometry
LD/LI	laser desorption laser post-ionization
LDA	laser doppler anemometry
LDI	laser desorption ionization
LIBS	laser-induced breakdown spectroscopy
LIF	laser-induced fluorescence
LMR	laser magnetic resonance
LMRS	laser magnetic resonance spectroscopy
LN_2	liquid nitrogen
LOD	limit of detection

LR	laboratory reference (material)
MALDI	matrix assisted LDI
MCB	monochloro-benzene
MDA	malondialdehyde
MEKC	micellar electrokinetic chromatography
MIR	mid infrared
MOPA	master oscillator power amplifier
MOPO	master oscillator power oscillator
MPI	multiphoton ionization
MS	mass spectrometry
MT	migration time
MWD	measuring while drilling
Nd:YAG	neodymium doped yttrium aluminum garnet (laser medium)
NIR	near infrared
NO	nitric oxide
NOS	nitrous oxide synthase
NPA signal	normalised photoacoustic signal
OAP	off axis parabola
OAS	optoacoustic spectroscopy
OES	optical emission spectroscopy
OPO	optical parametric oscillator
PA	photoacoustic
PAH	polycyclic aromatic hydrocarbons
PAS	photoacoustic spectroscopy
PBL	planetary boundary layer
PCP	pentachlorophenol
PDB	Pee Dee Belemnite
PI	photoinoization
PLS	partial least squares
PM	photomultiplier tube
ppb	parts per billion
PPLN	periodically poled lithium niobate
ppm	parts per million
ppt	parts per trillion
QCL	quantum cascade laser
QY	quantum yield
RAL	relative adduct labeling
RDC	ring-down cell
REMPI	resonance enhanced multi-photon ionization

RS	Raman spectroscopy
RSD	relativ standard deviation
sccm	standard cubic centimeters per minute
SCG	supercontinuum generation
SDS	sodium dodecylsulfate
SID	surface-induced dissociation
SPM	self-phase modulation
SRG	tracer Sulfhorodamine G
TBR	three-body recombination
TCSPC	time-correlated single-photon-counting
TDL	tunable diode laser
TDLAS	tunable diode laser absorption spectroscopy
TEM	transverse electric magnetic
THF	tetra-hydrofurane
THG	third harmonic generation
Ti:Sa	titanium doped sapphire (laser medium)
TOF	time-of-flight
TOFMS	time-of-flight mass spectrometer
TOPAS	travelling-wave optical parametric amplifier
TPH	total petroleum hydrocarbon
US-EPA	US Environmental Protection Agency
UV	ultraviolet
VAD	velocity azimuth display
VDI	Verein Deutscher Ingenieure
VOC	volatile organic compounds
VUV	vacuum ultraviolet

List of Symbols

A_{nm}	Einstein coefficient
B	distance between CCD-sensor and imaging lens
c	speed of light
χ_i	ionization potential
D_1, D_2	size of particles
Δ,ξ $\Delta\psi$	pixel size of the CCD-sensor
$\Delta\zeta$	depth of reconstruction
e	electron charge
ε_0	permittivity of vacuum
E_i	energy of level i
f	focal length of imaging lens
$\Gamma(\nu,\mu)$	reconstructed plane in the digital image with the pixels x and y
g_i	statistical weight of state E_i
h	Planck's constant
I	(laser) light intensity
I	intensity
I_B	irradiance
ϑ_d	detection angle
k	Boltzmann's constant
λ	optical wave length
l	path length
λ	wavelength
$\lambda/\Delta\lambda$	spectral resolution
m	optical magnification
m_e	electron mass

n	index of refraction
ν	frequency
n_0	linear index of refraction
n_2	coefficient of the nonlinear index of refraction
n_e	electron density
n_i	density of species i
ν_L	laser frequency
ν_P	plasma frequency
ρ_0	molecule density
σ_n	n-photon-absorption cross-section
$T(k,l)$	recorded hologram - interference pattern from the CCD with its pixel coordinates k and l
T_e	electron temperature
τ_L	laser pulse width
U_n	intensity of fluorescence
z	distance where the reconstruction plane $\Gamma(m,n)$ will be spanned up
Z(Te)	partition function

Part I

Remote Sensing Methods in the Atmosphere

1 Lidar: An Overview

Matthias Ulbricht

1.1 Introduction

The understanding of atmospheric processes has become an important scientific topic and economic factor: Weather forecasting influences wide areas of modern life; if certain industries are located in unfavorable sites, residential areas may suffer from industrial pollution. A comprehensive understanding of these processes requires theoretical models as well as extensive measurements.

Optical methods have been widely applied in atmospheric sensing. They enable high sensitive remote sensing of atmospheric parameters at sites which are difficult to access with conventional methods. Furthermore, these methods measure the parameter in their natural environment reducing or even avoiding the influence of the measurement on the measured parameter.

Several laser based analytical methods have been developed. The high intensity and spectral quality of these light sources enable highly sensitive detection of various substances. Due to high light intensities some of these methods are not only based on "classical" light-matter interactions like the molecular absorption, but utilize nonlinear optical effects, which are not accessible with low-intensity light sources. The use of pulsed lasers with pulse durations of a few Nanoseconds or even shorter allow for range resolved remote sensing of atmospheric parameters with a range of several kilometers and a spatial resolution of a few meters.

Lidar (light detection and ranging) (Measures 1984) techniques extend the wide range of laser analytical methods to remote sensing. They offer range-resolved determination of different atmospheric parameters. They are monostatic, i.e. they do not require any target or retroreflector. Hence the "lidar beam" can be steered in any direction enabling the large-scale three-dimensional measurement even at sites which are not directly accessible with the instrument.

Numerous scientific groups have built lidar systems for many different applications (Ansmann 1997). Nevertheless, only a few systems have been commercialized. The main reasons for the lack of commercialization are high investment costs for the user as well as the need of scientific supervision for operation and data evaluation. Successful commercialization's had overcome these obstacles at least up to an acceptable level.

In this chapter, major variants of lidar technology (Backscatter lidar, DIAL, Raman lidar, Doppler lidar) are discussed. These techniques have been modified in many ways, which cannot be described here. The focus is on the scientific background and exemplary applications in atmospheric sensing. Fluorescence lidar is not described here, because its application to atmospheric sensing is limited. Future applications of femtosecond laser based lidar systems will be shown in chapter 2.

1.2 Backscatter Lidar (Rayleigh-Mie-Lidar)

Lidar methods are the optical equivalent of microwave based radar and sonic based sodar techniques. Its simplest version is the backscatter lidar. It consists of a pulsed laser, a receiver telescope, a fast sensitive detector, and a data acquisition system (Fig. 1.1).

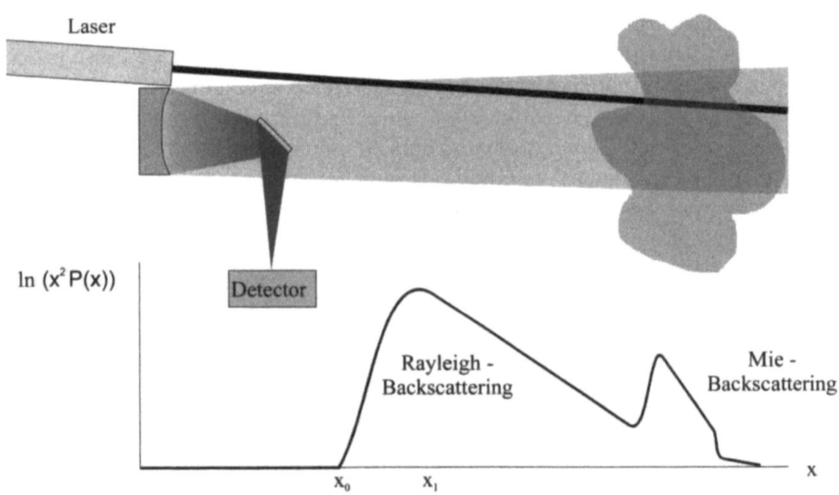

Fig. 1.1. Scheme of a backscatter lidar. The atmospheric backscatter signal is received from a minimum distance x_0 where the laser beam enters the field-of-view of the receiver telescope.

The laser pulse is transmitted into the atmosphere where it interacts with molecules and suspended particles. The most important interactions are scattering processes like Rayleigh-scattering (elastic molecular scattering) or Mie-scattering[1] (elastic scattering by particles), which causes an attenuation of the laser light de-

[1] In this chapter the term "Mie-scattering" is used for light scattering by all particles, although the classical Mie-theorie is only valid for spherical particles.

scribed by the atmospheric extinction coefficient α. A small fraction of the light is scattered back to the system, where it is collected with the receiver telescope and focused on the detector. The detector signal is acquired with high temporal resolution to obtain maximum spatial resolution: the propagation time of the light from the laser to the scattering object and back to the system corresponds to the distance of the scattering molecules or particles. Hence the spatial resolution is proportional to the temporal resolution of the system. The signal intensity contains the information about atmospheric properties.

Beside Rayleigh- and Mie-scattering the light may undergo other interaction processes like Raman scattering, fluorescence, or specific molecular absorption. Observation of these effects yield further information on the state of the atmosphere. The distinction between different lidar variants, which will be discussed later, is made depending on the investigated interaction process. For a basic backscatter lidar, only Rayleigh- and Mie-scattering is relevant.

The received backscatter lidar signal received from the distance x is given by:

$$P(x,\lambda) = P_0(\lambda)\frac{C}{x^2}O(x)\beta(x,\lambda)\tau^2(x,\lambda) \tag{1.1}$$

where $P(x,\lambda)$ is the intensity received from distance x at wavelength λ, $P_0(\lambda)$ is the emitted laser power, C is an instrument constant, $\beta(x,\lambda)$ is the backscatter coefficient, which describes the fraction of the incident light scattered back to the system, and $\tau(x,\lambda)$ is the atmospheric extinction with respect to absorption and scattering given by Beer-Lambert's law ($\alpha(\xi,\lambda)$ is the atmospheric extinction coefficient):

$$\tau(x,\lambda) = e^{-\int_0^x \alpha(\xi,\lambda)d\xi} \tag{1.2}$$

O(x) is the so called *Overlap function*, which describes the overlap of the emitted laser beam with the receiver field-of-view (FOW). As long as the laser beam divergence is smaller than the FOW, and the system is well aligned, O(x) = 1 for larger distances. In the near field, O(x) < 1 causing a reduction of the signal intensity. The determination of O(x) in the near field is difficult (Harms 1978, Velotta 1998, Wandinger 2002). Hence the evaluation of the lidar signal $P(x,\lambda)$ mostly is performed from a minimum distance x_l where the overlap function is 1. This procedure causes a blind range between the lidar system and the distance x_l, which depends strongly on the optical configuration of the system.

The lidar equation (1.1) contains one measured quantity ($P(x,\lambda)$), but two atmospheric parameters (α and β). Therefore the determination of α and β requires further knowledge about the relation between both parameters. For particles, this is difficult to access, because they vary strongly in chemical composition, shape, size and hence in its optical properties. Even methods were developed, which allow a determination of the aerosol extinction profile under certain conditions (Klett 1981), no general mathematical solution for this problem exists. To determine the aerosol extinction, Raman lidar is used, which will be discussed later.

Further aerosol parameters like size distribution can be determined with the extension of a backscatter lidar from one to multiple wavelengths (Müller 1998). This method was successfully applied for the determination of stratospheric aerosols.

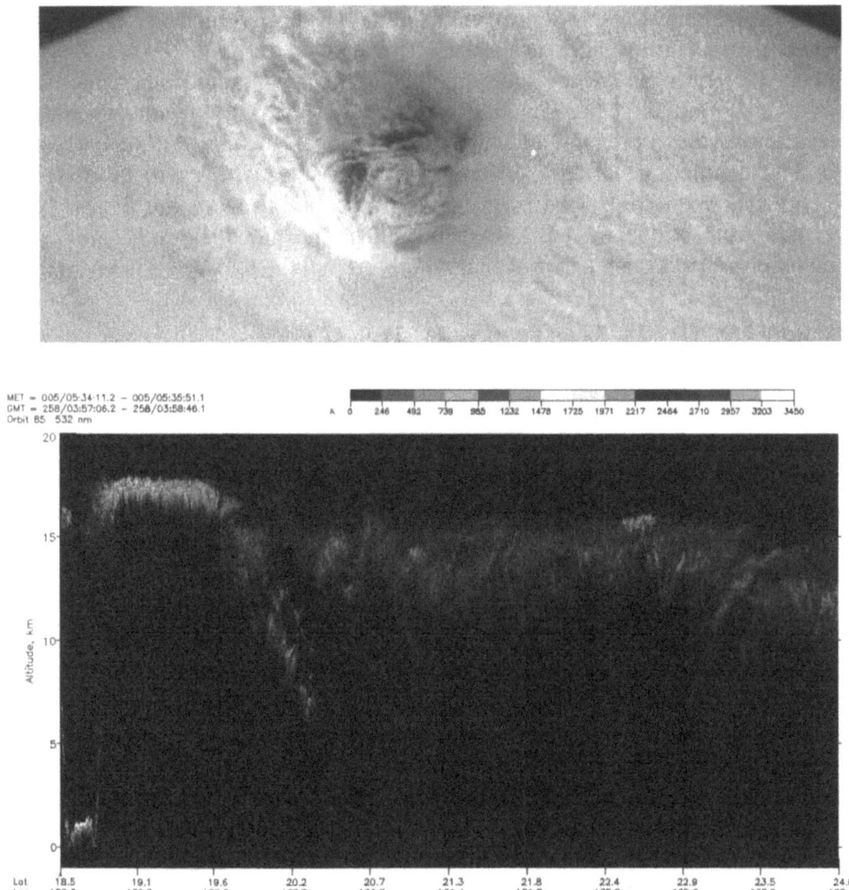

Fig. 1.2. Photo and backscatter lidar measurement of super typhoon Melissa detected during orbit 85 of LITE. The eye of Melissa had a diameter of about 30 km at the surface. The backscatter intensity is color-coded. The lidar measurements taken with the 532 nm channel show the eye at the left edge of the diagram (from http://www-lite.larc.nasa.gov)

The measurement of the depolarization of the scattered light yields information about the shape of particles. These measurements e.g. help to distinguish between different states (ice, liquid) of aerosols.

Even if a quantitative determination of extinction profiles from a single-wavelength backscatter lidar is difficult, the method is a powerful tool for vertical profiling of the atmosphere. The observed relative particle distribution shows the stratification of the atmosphere which helps to understand exchange processes of different air masses.

Backscatter lidar systems were installed on different platforms. An important milestone in scientific lidar application was the Lidar In-space Technology Experiment (LITE), where a three-wavelengths backscatter lidar was flown on the Space Shuttle Discovery. The aim was the measurement of clouds and aerosols in the stratosphere and the troposphere, the height of the planetary boundary layer (PBL), and the atmospheric temperature and density in the stratosphere. For an overview of LITE see Ansmann 1997, several results have been published in Winkler 1996.

Numerous scientific backscatter lidar have been build. A commercial compact eye-safe backscatter lidar is manufactured by SESI, Burtonsville, MD, USA (www.sesi-md.com). This so-called *Micro Pulse Lidar (MPL)* emits 10 μJ laser pulses with a repetition rate of 2.5 kHz at 523 nm. It is a stand-alone system for continuous vertical profiling of the atmosphere. More than 20 systems have been sold worldwide.

1.3 DIAL

The remote measurement of gas concentrations is one of the major applications of lidar techniques. For this, the DIAL-method (differential absorption lidar) is applied (Measures 1984, VDI 1999). This technique is based on the specific light absorption of pollutant molecules. Two laser pulses with different wavelengths are emitted into the atmosphere. One wavelength (λ_{on}) is tuned to a specific absorption line of the molecule of interest. The second wavelength (λ_{off}) is a reference wavelength with no specific absorption of the pollutant molecule. If both laser beams are well aligned (i.e. $O(x,\lambda_{on})=O(x,\lambda_{off})$), the ratio of both backscatter signals is given by:

$$\frac{P(x,\lambda_{on})}{P(x,\lambda_{off})}=\frac{P_0(\lambda_{on})C(\lambda_{on})\beta(x,\lambda_{on})e^{-2\int\alpha(\xi,\lambda_{on})d\xi}}{P_0(\lambda_{off})C(\lambda_{off})\beta(x,\lambda_{off})e^{-2\int\alpha(\xi,\lambda_{off})d\xi}} \tag{1.3}$$

α can be separated into three parts :

$$\alpha(\xi,\lambda_i)=\alpha_{Rayleigh}(\xi,\lambda_i)+\alpha_{Mie}(\xi,\lambda_i)+\alpha_{Gas}(\xi,\lambda_i) \tag{1.4}$$

If $\lambda_{on}\approx\lambda_{off}$, the backscattering as well as the Rayleigh- and Mie extinction coefficients for both wavelengths are approximately equal, because they vary slow with the wavelength:

$$\alpha_{Rayleigh}(\lambda_{on}) = \alpha_{Rayleigh}(\lambda_{off})$$
$$\alpha_{Mie}(\lambda_{on}) = \alpha_{Mie}(\lambda_{off}) \quad (1.5)$$
$$\beta(\lambda_{on}) = \beta(\lambda_{off})$$

The absorption of the molecule of interest is described by:

$$\alpha_{Gas}(x,\lambda) = N_{Gas}(x)\sigma_{Gas}(\lambda) \quad (1.6)$$

where N(x) is the number density and $\sigma(\lambda)$ the molecular absorption cross section of the molecule of interest. If the absorption spectrum of the relevant gas is known, the gas concentration can be directly calculated from the lidar signals:

$$N_{Gas}(x) = \frac{1}{2\Delta\sigma_{Gas}} \frac{d}{dx} \ln \frac{P(x,\lambda_{off})}{P(x,\lambda_{on})} \quad (1.7)$$

with

$$\Delta\sigma_{Gas} = \sigma_{Gas}(\lambda_{on}) - \sigma_{Gas}(\lambda_{off}) \quad (1.8)$$

DIAL has been widely applied for the remote detection of many gases like SO_2 (Frederiksson 1979, Fuji 2001, Edner 1994), NO (Kölsch 1992, Aldén 1982), NO_2 (Kölsch 1989, Galle 1988) , CH_4 (Menyuk 1987), NH_3 (Force 1985), Cl_2 (Edner 1987), HCl (Menyuk 1987), Hg(g) (Edner 1989), and different volatile organic compounds (VOCs) (Menyuk 1982, Milton 1992). The applicability of DIAL requires well separated absorption lines of the molecules of interest. Furthermore tunable lasers or lasers with natural coincidence of its wavelength with the molecular absorption line are needed. Many DIAL systems work in the ultraviolet spectral range, where gases like SO_2, Ozone, NO, NO_2, and some aromatic hydrocarbons show specific absorptions. These systems are based on Dye-lasers, Ti:Sapphire-lasers or OPOs. Ozone, which has a broad unstructured absorption band, can be detected with Excimer- and Nd:YAG-lasers with harmonic generation. Other VOCs, HCl, and NH_3 are detected in the infrared either between 9 µm and 11 µm using CO_2-lasers or in the range of 3 µm to 5 µm (Dye-lasers with difference frequency generation, OPOs, chemical lasers). Gas detection in the IR require the presence of a target or aerosols in the air, because here Rayleigh scattering is too weak to generate a sufficient lidar signal. Furthermore the sensitivity of fast IR-detectors is smaller than equivalent detectors in the UV. Some gases like water vapor can be detected in the visible and near infrared range. Here special care has to be taken to ensure eye-safety.

DIAL is a powerful tool for emission monitoring. By scanning the lidar beam over the region of interest, emission sources can be pinpointed (Fig 1.3). The large spatial coverage of the DIAL measurement is especially important in the case of diffuse emissions (e.g. from larger industrial pants, waste disposal sites). Here conventional point monitors are difficult to apply, because the gas concentration show strong spatial variations, and many measurements have to be taken to determine the emission.

Air pollution is not a local phenomena of industrial sites or residential areas. Hence pollution transport plays a dominant role in urban air pollution. Pollution reduction concepts always have to consider pollution transport on local and regional scale. Regional pollution transmission (few hundred to one thousand kilometers range) occurs in altitudes of a few hundred meters up to a few kilometers. Continuous monitoring of pollution transport is difficult to obtain with classical methods like balloon sounding or airborne measurements. DIAL enables the observation of pollution transport even across borders (Fig 1.4).

DIAL is also applied in ambient air monitoring. Here, because of its role in summer smog, the detection of tropospheric Ozone and its precursors is of major interest (Weidauer 1998).

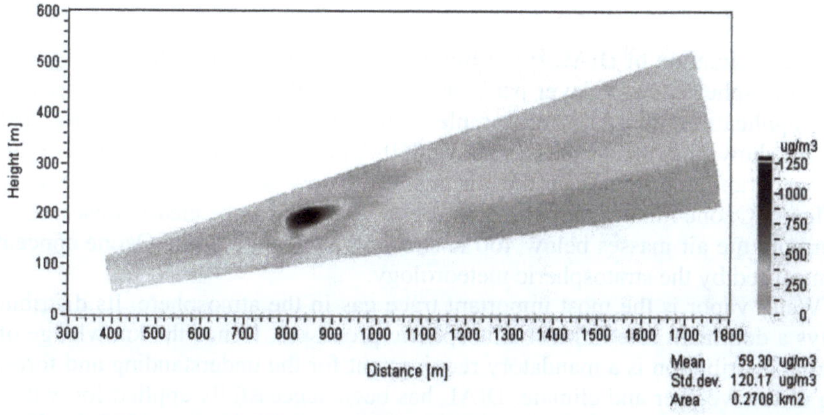

Fig. 1.3. SO_2 emission from a refinery stack. A vertical cut through the plume perpendicular to the wind field was taken with a mobile DIAL system of Elight Laser Systems, Teltow, Germany. By integrating the concentration profile over the area of the plume, and by multiplying with the wind vector, the emission can be determined.

Even if DIAL systems are commercially available, its recent use in air quality monitoring is mostly limited to research applications. A major reason is the lack of regulations, which rule the application of DIAL in air pollution monitoring. A first step was taken in Germany, where a VDI guideline (VDI 1999) was passed, which describes the usage of DIAL in air pollution control.

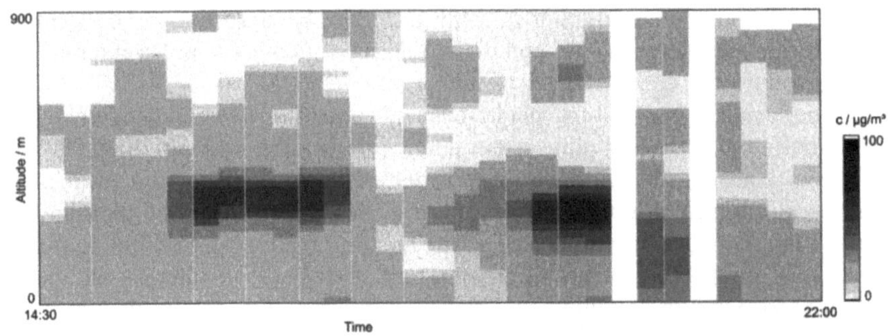

Fig. 1.4. Observation of pollution transport from Dunkerque (France) to De Panne (Belgium). The diagram shows the temporal evolution of the vertical distribution of SO_2. The measurement was taken with a mobile DIAL system of Elight Laser Systems, Teltow, Germany.

The application of DIAL is not limited to the troposphere. The investigation of the stratospheric Ozone layer particularly in the polar regions has become a standard application of DIAL. An example is shown in Fig. 1.5. Back trajectory calculations show, that the air masses above 460 K potential temperature have a different geographic origin than the air masses below 460 K. Hence one expects different Ozone mixing ratios in both regions, which can be clearly observed. Furthermore the air masses below 460 K show a fast change of the Ozone concentration caused by the stratospheric meteorology.

Water vapor is the most important trace gas in the atmosphere. Its distribution plays a dominant role in most atmospheric processes. Hence the knowledge of its spatial distribution is a mandatory requirement for the understanding and forecasting of the weather and climate. DIAL has been successfully applied for water vapor profiling. A comprehensive overview is given in Wulfmeyer 1998, Wulfmeyer 1998/2 and Bösenberg 1998. The detection of water vapor using DIAL requires very narrow bandwidth lasers like injection-seeded Ti:Sapphire- (Moore 1997), Alexandrite- (Wulfmeyer 1998), Nd-YAG-lasers (Lehmann 1997), and OPOs (Fix 1997).

For quantitative gas measurements it is important, that the absorption cross sections have no or only a negligible temperature dependence. On the other hand, the temperature dependence of some absorption can be utilized to determine the atmospheric temperature profile using DIAL (Wulfmeyer 1998, Bösenberg 1998, Theopold 1993). Here the gas mixing ratio must be known. Preferable gases are Oxygen, which has a constant mixing ratio in the atmosphere, or water vapor, where its concentration must be measured in parallel.

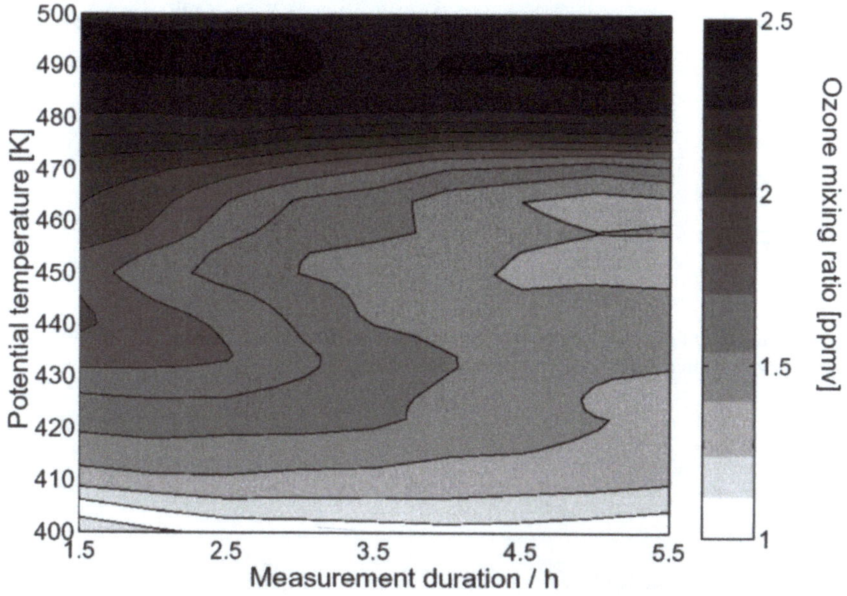

Fig. 1.5. Temporal evolution of stratospheric Ozone in Ny-Ålesund (Svalbard, Norway) (Wahl 2002). The potential temperature between 400 K and 500 K corresponds to an altitude of 15.5 km to 21 km.

DIAL systems have been successfully commercialized. Elight Laser Systems, Teltow, Germany (www.elight.de), has developed a mobile scanning UV-DIAL system based on a flashlamp-pumped Ti:Sapphire laser with harmonic generation. The systems operates with a repetition rate of 20 Hz and can be tuned on the absorption lines of SO_2, NO_2, O_3, Toluene and Benzene. More than 10 systems have been sold in Europe. Furthermore Elight Laser Systems offers the *Ozone Profiler* based on a Nd:YAG-laser pumped OPO for automatic continuous vertical ozone concentration measurements. NPL, Teddington, UK (www.npl.co.uk), has built IR-DIAL systems for industrial emission monitoring, especially for the petrochemical industry. This system is based on dye-lasers and OPOs (Gardiner 1997).

1.4 Raman Lidar

Beside Rayleigh- and Mie-scattering, the inelastic Raman-scattering can be utilized for lidar measurements.

Raman-scattering causes a shift of the scattered wavelength (Fig. 1.6). The wavelength shift depends only on the scattering molecule, hence the wavelength selection for a Raman lidar is less critical than e.g. for a DIAL system. Usually Nd:YAG- or Excimer-lasers are used.

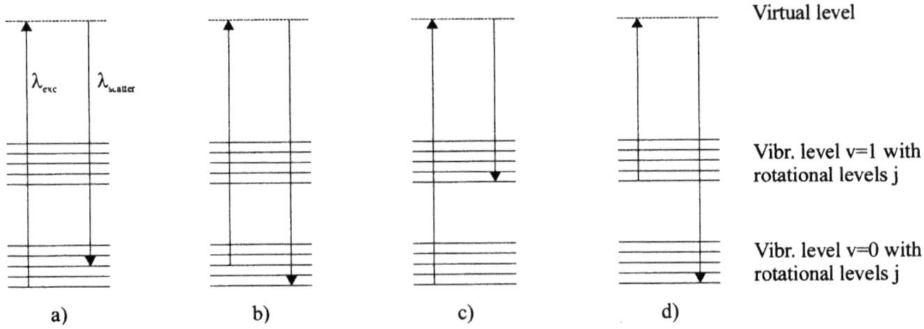

Fig. 1.6. Different Raman scattering processes : a) rotational Stokes, b) rotational anti-stokes, c) vibrational stokes, d) vibrational anti-stokes

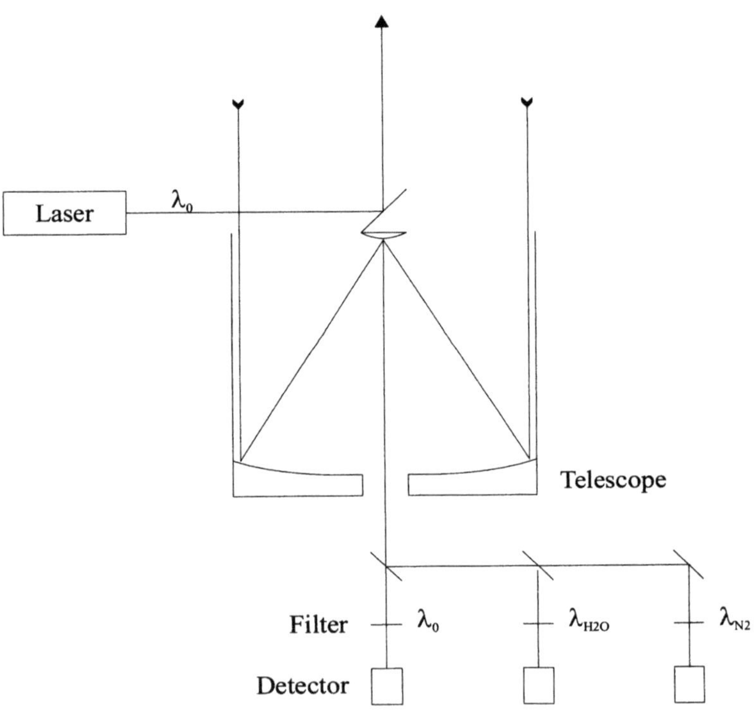

Fig. 1.7. Scheme of a basic Raman lidar for water vapor detection

The atmospheric backscatter signal contains both elastic and Raman scattered fraction. Since the Raman-scattering cross section is about three orders of magnitude smaller than the Rayleigh scattering cross section, the Raman detection channel requires a strong suppression of the Rayleigh-Mie-line, which can be obtained with narrow-band filters or gratings (Fig. 1.7).

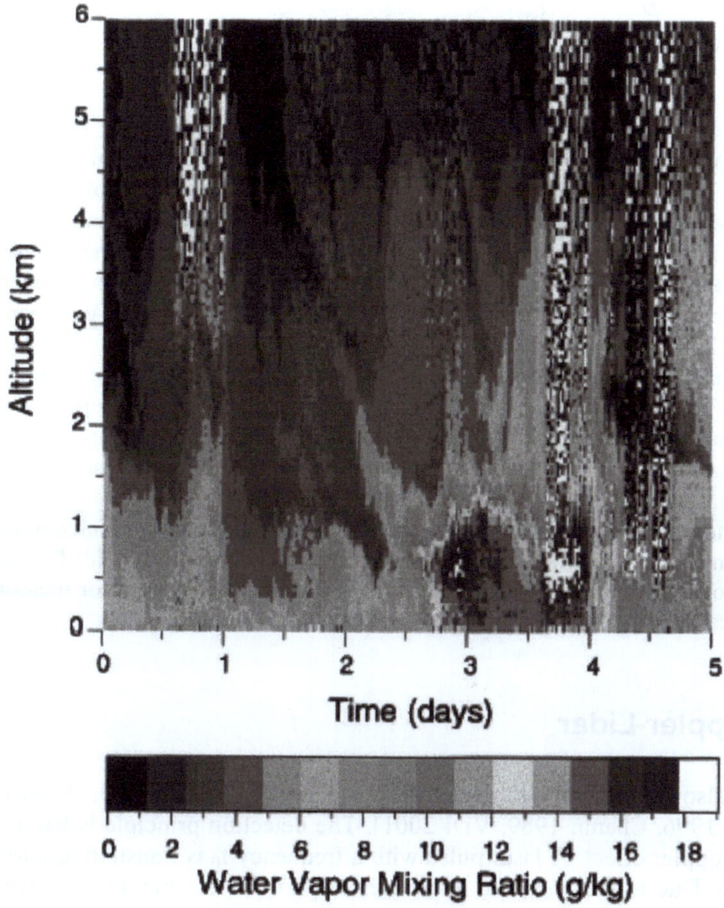

Fig. 1.8. Water vapor mixing ratio measured with a Raman lidar system built by Sandia National Laboratories, Livermore, CA. The temporal resolution is 10 min. The grainier portion of the image represents the daytime measurements. The speckled vertical strips are caused by attenuation of the laser beam in the presence of clouds (Goldsmith 1998).

During daytime many Raman lidar systems suffer strongly from daylight. To allow for daytime measurements, advanced optical design of the system is needed (Goldsmith 1998). For the lower atmosphere, operation in the solar-blind region

($\lambda < 300$ nm) is possible (Renaut 1988). Strong Rayleigh extinction and Ozone absorption will limit the range of those systems.

An important application of Raman lidar is the measurement of the vertical water vapor distribution (Fig. 1.8). Compared with DIAL, the required lasers are less sophisticated. On the other hand, daytime performance may be limited.

The water vapor mixing ratio can be derived from Nitrogen and water vapor Raman backscattering:

$$w(x) \propto \frac{P_{Raman,H_2O}(x)}{P_{Raman,N_2}(x)} \tag{1.9}$$

The measurement of the temperature profile of the atmosphere is another important application of Raman lidar. The population of the molecules in different rotational states depends on the temperature and is given by the Boltzmann distribution. To measure the temperature, the anti-stokes rotational Raman scattering is detected. The intensity of the Raman scattering signal is proportional to the population of the rotational levels, hence the temperature profile can be derived from the relative intensities of the different Raman scattering signals (Vaughan 1993, Arshinov 2001, Nedeljkovic 1993).

The main difficulty for rotational Raman lidar is the suppression of the Rayleigh line, because the wavelength separation between Rayleigh- and rotational Raman-lines is small (1-2 nm), and the anti-stokes lines are approx. 10^{-4} less intense than the Rayleigh line. To overcome this difficulty, combined vibrational-rotational Raman spectroscopy can be used (Heaps 1997).

Until now, Raman lidar has been a scientific domain, commercialization is still outstanding. When routine daytime measurements become possible, Raman lidar has the potential to become a powerful routine measurement tool for meteorological applications.

1.5 Doppler-Lidar

Lidar is also used to remote detect the wind vector (Bilbro 1980, Werner 2001, Huffaker 1996, Chanin 1989, VDI 2001). The detection principle is based on the optical Doppler effect. A laser pulse with a frequency f_0 is transmitted into the atmosphere. Due to the wind the molecules or particles are moving relative to the light source with the wind velocity component in direction of the laser beam (line of sight) v_{LOS}. The movement causes a shift Δf of the frequency f of the laser light received by the detection system.

$$\Delta f = f' - f_0 = 2\frac{v_{LOS}}{c} f_0 \tag{1.10}$$

For detecting the small frequency shift in the order of 10^{-6} usually a heterodyne detection is applied. Here, light from a second laser working as local oscillator

(LO) is mixed with the light received from the atmosphere. This yields a detector current which can be splitted into three parts :

$$i_{Signal} = i_{DC} + \rho\sqrt{2I_{x,\lambda}I_{LO}}\begin{Bmatrix}\cos\left[2\pi\left(f_{LO} - \left(f_0 \pm \Delta f\right)\right)t\right] \\ + \cos\left[2\pi\left(f_{LO} + \left(f_0 \pm \Delta f\right)\right)t\right]\end{Bmatrix} \tag{1.11}$$

The detector cannot follow the high frequencies $f_{LO}+(f_0\pm\Delta f)$, hence only the beat frequency $f_{LO}-(f_0\pm\Delta f)$ is detected.

Due to the small frequency shift, a very narrow laser bandwidth combined with high frequency stability of the laser is required. Usually CO_2-laser or Tm:YAG-laser are used. In case of a MOPO design, the master oscillator can be used as seeder for the power oscillator and as local oscillator, hence the beat frequency is the frequency shift to be detected.

Using the above method, only the wind velocity in beam direction is measured. To obtain the wind vector, the measurement must be done in several non-coplanar directions. This can be either done with sequential scanning of a few beam directions (DBS: Doppler Beam Swinging) or by applying the Velocity Azimuth Display (VAD) scan technique. In the case of the VAD scan technique, a conical scan of the laser beam is performed. The result is a sinusoidal curve of v_{LOS}, from which all components of the wind vector can be derived.

Instead of using a pulsed laser, as well a cw-laser can be applied. To obtain spatial resolution, the laser beam is focused at a certain distance. By variation of the focal distance, the wind vector is determined as a function of range. This method is called Laser Doppler Anemometry (LDA) (Werner 1995).

Direct detection, as an alternative to heterodyne detection, was also demonstrated (Abreu 1992, Irrgang 2002). Here a spectroscopic element must be implemented to resolve the spectroscopic shift. It must be either scanned in wavelength (etalon) (McGill 1997) or have an extremely well-known spectral response (narrowband filter) (Korb 1992).

An eyesafe wind lidar operating at 2 µm has been commercialized by Coherent Technologies Inc., Lafayette, CO, USA (www.ctilidar.com). Beside for meteorological applications it can be used for the detection of wind-shears and turbulences affecting air traffic.

1.6 Outlook

Lidar techniques have demonstrated their usefulness for many different tasks in atmospheric diagnostics. Certain tasks will further remain of mainly scientific interest, while dedicated applications in meteorology and air pollution monitoring have the potential for use outside the scientific community. Here highly reliable laser sources and user-friendly software will allow for the design of fully automatic lidar systems for continuous monitoring of the atmosphere.

References

(Abreu 1992) VJ Abreu, JE Barnes, PB Hays: "Observations of winds with an incoherent lidar detector", Appl. Opt. **31**, 4509-4514 (1992)

(Aldén 1982) M Aldén, H Edner, S Svanberg: "Laser monitoring of atmospheric NO using ultraviolet differential absorption techniques", Opt. Lett. **7**, 543-545 (1982)

(Ansmann 1997) A Ansmann, R Neuber, P Rairoux, U Wandinger (eds.): "Advances in Atmospheric Remote Sensing with Lidar", Springer, New York, Heidelberg, Berlin (1997)

(Arshinov 2001) Arshinov, S. Bobrovnikov, I. Serikov, A. Ansmann, D. Althausen, I. Mattis, and U. Wandinger: "Spectrally Absolute Instrumental Approach to Isolate Pure Rotational Raman Lidar Returns from Nitrogen Molecules of the Atmosphere",in Dabas, Loth, Pelon (eds.), Advances in Laser Remote Sensing, Selected Papers Presented at the 20th International Laser Radar Conference (ILRC), Vichy, France 10-14 July 2000. Edition del'Ecole Polytechnique, 121-124, (2001)

(Bilbro 1980) J.W. Bilbro: "Atmospheric laser Doppler velocimetry : an overview", Opt. Eng. **19**, 533-542 (1980)

(Bösenberg 1998) J. Bösenberg: "Ground-based differential absorption lidar for water-vapor and temperature profiling: methodology", Appl. Opt. **37**, 3845-3860 (1998)

(Chanin 1989) M.L. Chanin, A. Garnier, A. Hauchecorne, J. Porteneuve: "A Doppler lidar for measuring winds in the middle atmosphere"; Geophys. Res. Lett. **16**, 1273-1276 (1989)

(Edner 1987) H. Edner, K. Fredriksson, A. Sunesson, W. Wendt: "Monitoring Cl_2 using a differential absorption lidar system", Appl. Opt. **26**, 3183-3185 (1987)

(Edner 1989) H. Edner, G.W. Faris, A. Sunesson, S. Svanberg: "Atmospheric atomic mercury monitoring using differential absorption lidar techniques", Appl. Opt. **28**, 921-930 (1989)

(Edner 1994) H. Edner, P. Ragnarson, S. Svanberg, E. Wallinder, R. Ferrara, R. Cioni, B. Raco, G. Taddeucci: "Total fluxes of sulfur dioxide from the Italian volcanoes Etna, Stromboli, and Vulcano measured by differential absorption lidar and passive differential absorption spectroscopy", J. Geophys. Res. **99**, 18827-18838 (1994)

(Fix 1997) A. Fix and G. Ehret: "Injection Seeded Optical Parametric Oscillator System for Water Vapor DIAL Measurements", in A. Ansmann, R. Neuber, P. Rairoux, U. Wandinger (eds.): "Advances in Atmospheric Remote Sensing with Lidar", Springer, New York, Heidelberg, Berlin (1997)

(Force 1985) A.P. Force, D.K. Killinger, W.E. DeFeo, N. Menyuk: "Laser remote sensing of atmospheric ammonia using a CO_2 lidar system", Appl. Opt. **24**, 2837-2841 (1985)

(Frederiksson 1979) K. Frederiksson, B. Galle, K. Nystrom, S. Svanberg: "Lidar System Applied in Atmospheric Pollution Monitoring", Appl. Opt. **18**, 2998-3003 (1979)

(Fuji 2001) T. Fuji, T. Fukuchi, N. Goto, K. Nemoto, N. Takeuchi: "Dual differential absorption lidar for the measurement of atmospheric SO_2 of the order of parts in 10^9", Appl. Opt. **40**, 949-956 (2001)

(Galle 1988) B. Galle, A. Sunesson, W. Wendt: "NO_2-Mapping using laser radar techniques", Atmosph. Environ. **22**, 569-573 (1988)

(Gardiner 1997) T.D. Gardiner, M.J.T. Milton, F. Molero, P.T. Woods: "Infrared DIAL Measurements with an Injection-Seeded OPO", in A. Ansmann, R. Neuber, P. Rairoux, U. Wandinger (eds.): "Advances in Atmospheric Remote Sensing with Lidar", Springer, New York, Heidelberg, Berlin (1997)

(Goldsmith 1998) J.E.M. Goldsmith, F.H. Blair, S.E. Bisson, D.D. Turner: "Turn-key Raman lidar for profiling atmospheric water vapor, clouds, and aerosols", Appl. Opt. **37**, 4979-4990 (1998)

(Harms 1992) J. Harms, W. Lahmann, and C. Weitkamp: "Geometrical compression of lidar return signal", Appl. Opt. **17**, 1131-1135 (1978)

(Heaps 1997) W.S. Heaps, J. Burris, J.A. French: "Lidar technique for remote measurement of temperature by use of vibrational-rotational Raman spectroscopy", Appl. Opt. **36**, 9402-9405 (1997)

(Huffaker 1996) R. M. Huffaker, R. M. Hardesty: "Remotes sensing of atmospheric wind velocities using solid-state and CO_2 coherent laser systems", Proc. IEEE **84**, 181-204 (1996)

(Irrgang 2002) T.D. Irrgang, P.B. Hays, W.R. Skinner: „Two-channel direct-detection Doppler lidar employing a charge-coupled device as a detector", Appl. Opt. **41**, 1145-1155 (2002)

(Klett 1981) J.D. Klett: "Stable analytical inversion solution for processing lidar returns", Appl. Opt. **20**, 211-220 (1981)

(Kölsch 1989) H.J. Kölsch, P. Rairoux, J.P. Wolf, L. Wöste: "Simultaneous NO and NO_2 DIAL Measurement using BBO Crystals", Appl. Opt. **28**, 2052-2056 (1989)

(Kölsch 1992) H.J. Kölsch, P. Rairoux, J.P. Wolf, L. Wöste: "Comparative Study of Nitric Oxide Immission in the Cities of Lyon, Geneva, and Stuttgart Using a Mobile Differential Absorption LIDAR System", Appl. Phys. B **54**, 89-94 (1992)

(Korb 1992) C.L. Korb, B.M. Gentry, C.Y. Weng: "Edge technique, theory and application to the lidar measurement of atmospheric wind", Appl. Opt. **31**, 4202-4213 (1992)

(Lehmann 1997) S. Lehmann and J. Bösenberg: "A Water Vapor DIAL System Using Diode Pumped Nd:YAG Lasers" in A. Ansmann, R. Neuber, P. Rairoux, U. Wandinger (eds.): "Advances in Atmospheric Remote Sensing with Lidar", Springer, New York, Heidelberg, Berlin (1997)

(McGill 1997) M.J. McGill, W.R. Skinner, T.D. Irrgang: "Analysis techniques for the recovery of winds and backscatter coefficients from a multiple-channel incoherent Doppler lidar", Appl. Opt. **36**, 1253-1268 (1997)

(Measures 1984) R.M. Measures: "Laser Remote Sensing", Wiley, New York (1984)

(Menyuk 1982) N. Menyuk, D.K. Killinger, W.E. DeFeo: "Laser remote sensing of hydrazine, MMH, and UDMH using a differential-absorption CO_2 lidar", Appl. Opt. **21**, 2275-2286 (1982)

(Menyuk 1987) N. Menyuk, D.K. Killinger: "Atmospheric Remote Sensing of Water Vapor, HCl and CH_4 Using a Continuously Tunable Co:MgF_2 Laser", Appl. Opt. **26**, 3061-3065 (1987)

(Milton 1992) M.J.T. Milton, P.T. Woods, B.W. Jolliffe, N.R.W. Swann, T.J. McIlveen: "Measurements of Toluene and Other Hydrocarbons by Differential-Absorption LIDAR in the Near-Ultraviolet", Appl. Phys. B **55**, 41-45 (1992)

(Moore 1997) A.S. Moore, Jr., K.E. Brown, W.M. Hall, J.C. Barnes, W.C. Edwards, L.B. Petway, A.D. Little, W.S. Luck, Jr., I.W. Jones, C.W. Antill, Jr., E.V. Browell, and S. Ismail: "Development of the Lidar Atmospheric Sensing Experiment (LASE) – An Advanced Airborne DIAL Instrument", in A. Ansmann, R. Neuber, P. Rairoux, U. Wandinger (eds.): "Advances in Atmospheric Remote Sensing with Lidar", Springer, New York, Heidelberg, Berlin (1997)

(Müller 1998) D. Müller, U. Wandinger, D. Althausen, I. Mattis, A. Ansmann: "Retrieval of physical particle properties from lidar observations of extinction and backscatter from multiple wavelengths", Appl. Phys. **37**, 2260-2263 (1998)

(Nedeljkovic 1993) D. Nedeljkovic, A. Hauchecorne, M.L. Chanin: "Rotational Raman lidar to measure the atmospheric temperature from the ground to 30 km", IEEE Trans. Geosci. Remote Sens. **31**, 90-101 (1993)

(Renaut 1988) D. Renaut and D. Capitini: "Boundary-layer water vapor probing with a solar blind Raman lidar: validations, meteorological observations and prospects", J. Atmos. Oceanic Technol. **5**, 585-601 (1988)

(Theopold 1993) F.A. Theopold, J. Bösenberg: "Differential absorption lidar measurements of atmospheric temperature profiles: theory and experiment", J. Atmos. Ocean. Tech. **10**, 165-179 (1993)

(Vaughan 1993) G. Vaughan, D.P. Wareing, S.J. Pepler, L. Thomas, V. Mitev: "Atmospheric temperature measurements by rotational Raman scattering", Appl. Opt. **32**, 2758-2764 (1993)

(VDI 1999) VDI guideline 4210, part 1: "Remote sensing, Atmospheric measurements with LIDAR, Measuring gaseous air pollution with DAS LIDAR", Beuth, Berlin, 1999

(VDI 2001) VDI guideline 3786, part 14: "2001-12 Environmental meteorology, ground-based remote sensing of the wind vector, Doppler wind lidar", Beuth, Berlin, 2001

(Velotta 1998) R. Velotta, B. Bartoli, R. Capobianco, L. Fiorani, and N. Spinelli: "Analysis of the receiver response in lidar measurements", Appl. Opt. **37**, 6999-7007 (1998)

(Wahl 2002) P. Wahl, Observation and Characterization of laminated ozone structures in the polar stratosphere, in print, Ber. Polarforsch. Meeresforsch. (2002)

(Wandinger 2002) U. Wandinger and A. Ansmann: "Experimental determination of the lidar overlap profile with Raman lidar", Appl. Opt. **41**, 511-514 (2002)

(Weidauer 1998) D. Weidauer, H.D. Kambezidis, P. Rairoux, D. Melas, M. Ulbricht, Atmosph. Environ. **32**, 2173-2183 (1998)

(Werner 1995) Ch. Werner, F. Köpp, R.L. Schwiesow: "Remote measurements of boundary-layer wind profiles using a cw Doppler lidar", J. Clim. and Appl. Meteor. **34**, 2055-2067 (1995)

(Werner 2001) C. Werner, P.H. Flamant, O. Reitebuch, F. Köpp, J. Streicher, S. Rahm, E. Nagel, M. Klier, H. Herrmann, C. Loth, P. Delville, Ph. Drobinski, B. Romand, Ch. Boitel, D. Oh, M. Lopez, M. Meissonnier, D. Bruneau, A. Dabas: "Wind infrared Doppler lidar instrument", Opt. Eng. **40**, 115-125 (2001)

(Winkler 1996) D.M. Winkler, R.H. Couch, and M.P. McCormick: "An overview of LITE: NASA's Lidar In-space Technology Experiment", Proc. IEEE **84,2**, 164-180 (1996)

(Wulfmeyer 1998) V. Wulfmeyer: "Ground-based differential absorption lidar for water-vapor and temperature profiling: development and specifications of a high-performance laser transmitter", Appl. Opt. **37**, 3804-3824 (1998)

(Wulfmeyer 1998/2) V. Wulfmeyer, J. Bösenberg: "Ground-based differential absorption lidar for water-vapor profiling: assessment of accuracy, resolution, and meteorological applications", Appl. Opt. **37**, 3825-3844 (1998)

2 Application Perspectives of Intense Laser Pulses in Atmospheric Diagnostics

Wiebke Zimmer, Miguel Rodriguez and Ludger Wöste

2.1 Introduction

Since many years lasers have been used in atmospheric diagnostics, especially in remote sensing. In particular the lidar method (described in Chap. 1) has become a powerful technique to monitor atmospheric parameters and has helped to understand a variety of atmospheric phenomena.

In the field of laser research, on the other hand, the generation of ultra-short laser pulses made the decisive step with the development of the chirped pulse amplification (CPA) technique in 1985 (Strickland and Mourou 1985; Backus et al. 1998). It provides light pulses with peak power values exceeding 10^{12} W.

Both aspects together suggest to evaluate, if the application of high-intensity lasers can improve existing techniques or create new methods in atmospheric diagnostic, i.e. for remote sensing. The outstanding new aspect thereby is that the air, which is matter composed of gases and particles, i.e. aerosols, plays a double role. It is not only the object of analysis, but also a medium on which the possible measuring methods are based. Extremely non-classic light-matter interactions promise new possibilities for atmospheric science, but, to be able to interpret the measurements correctly, they also require intensive investigations which belong to the current basic research in quantum optics.

The state of these investigations, including first results of laboratory experiments as well as measurements in the atmosphere, will be presented in the following Sects. Beforehand, a short introduction in the high-power laser technique and the basic effects of nonlinear optics is given.

Generation of ultra-short high-intensity laser pulses

The active laser medium commonly used for the generation of short pulses is the titanium-doped sapphire crystal which has a very broad gain bandwidth (~ 700 to 1000 nm). Its optimal activation frequency lies in the green spectral region, therefore powerful Nd:YAG lasers, emitting at a wavelength of 532 nm (frequency doubled), are applied as pump lasers. In the CPA technique an initial short pulse, e.g. with a duration of some tens of femtoseconds (~ 10^{-14} s) generated by a mode-locked Ti:Sa oscillator, is first temporally stretched in order to reduce its peak power. In this way it can be amplified without damaging the optics, passing through a series of amplifier stages containing the same active medium as the oscillator. The amplified pulse is finally recompressed to nearly its original duration. In both, stretcher and compressor, gratings are used to control the pulse duration. The temporal expansion is therefore a consequence of the separation of the different colors in the pulse spectrum, the so-called *chirping* of the pulse. This is possible because of the fact that the initial pulse, due to its temporal sharpness, has a broad spectrum, typically of the order of 30 nm full width at half maximum.

Today. commercial CPA systems with output energies of several hundreds of millijoules are available. With pulse durations of about 100 fs these systems provide a peak power of several terawatts (10^{12} W). For laboratory experiments as described in Sect. 2.3 also femtosecond laser systems of lower pulse energies (in the order of 1 mJ) can serve to create the required light intensities. Laser systems of that kind are used by many research groups, e.g. in chemical physics.

Nonlinear optics

When a high power laser pulse propagates through a transparent medium it undergoes nonlinear self-action effects which lead to strong changes of the spectral, temporal and spatial characteristics of the pulse. Those phenomena have been extensively studied since the early 1970s (Alfano and Shapiro 1970).

A basic principal of nonlinear optics is the variation of the refractive index of matter in the presence of strong electro-magnetic fields. For a given radiation frequency the refractive index n, which is no more a constant of the medium, depends in the form

$$n = n_0 + n_2 I \quad (+\text{ higher terms}), \tag{2.1}$$

on the light intensity I, where n_0 is the linear refractive index and n_2 is the coefficient of the nonlinear refractive index. In the case of a high-power laser beam the transverse intensity profile leads to a higher refractive index in the centre of the beam than at the edge, inducing a so-called Kerr lens. The result is a self-focusing (Shen 1984; Strickland and Corkum 1994) of the beam, causing a further increase of intensity. The variation of the refractive index on the longitudinal axis, due to the trailing and the falling edge of the very short pulse, is also important. The resulting self-phase modulation (Alfano and Shapiro 1970; Manassah 1989) changes, i.e. broadens, the pulse spectrum, while temporal effects like self-steepening and pulse splitting (Ranka et al. 1996) may occur. In addition to the n_2 effects, multiphoton processes, as the generation of high harmonics, four-wave mixing or multiphoton ionization (MPI), take place at very high intensities.

Nonlinear propagation of laser beams in matter is known as long as the laser itself and it could first be observed in solid media, i.e. crystals. Some well known nonlinear effects like frequency doubling or tripling are specifically related to the anisotropic structure of crystals. But only the development of ultra-short pulse lasers permitted to observe nonlinear effects even in gases at atmospheric pressure.

2.2 Interaction of Intense Laser Pulses With Air – Detection of Gases

Propagation and white-light generation

The n_2 coefficient (see eq. 2.1) of air is quite difficult to retrieve experimentally. We assume a value of 5×10^{-19} cm^2/W at 800 nm which leads to a self-focusing threshold power of 2 GW for a diffraction limited beam. In this section we deal with terawatt laser pulses, exceeding by far this threshold. As a typical example, the self-focusing length for a 3-TW pulse at 800 nm with a beam diameter of 5 cm is, in theory, approximately 60 m.

When the intensity of the self-focusing beam reaches a certain threshold, the light starts to generate a plasma because of MPI. From that point on the presence of free charges lowers the refractive index which, in contrast to the Kerr lens effect, causes a defocusing of the beam. A balance of both effects leads to an outstanding phenomenon of nonlinear propagation called self-guiding or self-channelling (Braun et al. 1995). Generally, TW-pulses are not focused to one plasma channel but are converted into a bundle of filaments (see Fig. 2.1). Single-filament guiding is achieved with lower laser power and/or very "clean" beam profiles. The guiding can be stable over very long paths compared to the length of a "classic" focal region. Filaments of up to 200 m in length have been reported (La Fontaine et al. 1999). The diameter of such filaments is about 100 μm and the critical intensity of self-guided filaments lies between 10^{13} and 10^{14} W/cm^2 for 800-nm pulses, slightly above the threshold for multiphoton ionisation (MPI).

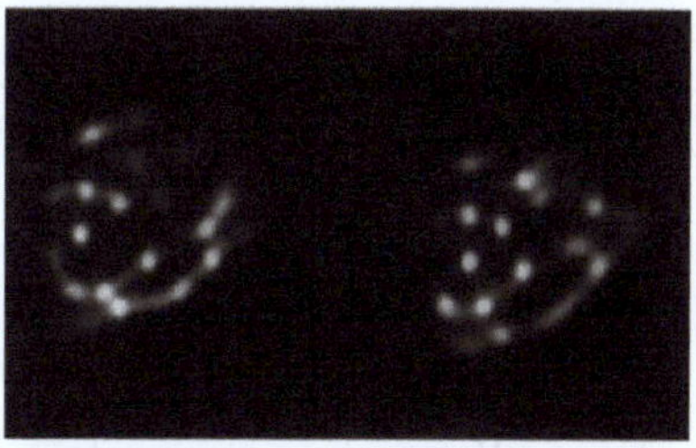

Fig. 2.1. Photograph of filament patterns of two consecutive self-focusing TW-pulses after 50 m of propagation in air (each whole-beam diameter is approx. 5 cm).

Inside those filaments the spectral shape of the pulse is strongly modified. The most striking change is visible to the naked eye: the infrared beam converts into white light (note that Fig. 2.1 is a real photograph of a scattering screen). This phenomenon is called supercontinuum generation (SCG) (Alfano 1989). Recently such spectra, which maintain their maximum at the laser wavelength (~ 800 nm), have been measured between the UV and the mid-IR. White light could be detected from 150 nm to 4.5 μm (Kasparian et al. 2000). Self-phase modulation (SPM) is believed to be the main source of the continuum, but it is influenced by other effects, mainly MPI and the group velocity dispersion (GVD), which is a linear effect but has special relevance for ultra-short pulses. Emission of the low-density plasma produced in gases by filamentation plays a minor role. This may be different when the gas, e.g. air, contains aerosols (see Sect 2.3). In contrast to plasma emission, the white light generated in coherent processes, like SPM or four-wave mixing, is forward directed (Chin et al. 1999).

There are different approaches to successful modelling of filamentation and SCG, but a complete theory is still lacking. Numerical simulations are still limited by enormous computing times and have only been performed for short distances. Therefore long-distance propagation experiments (see the "Outlook" Sect. 2.4) are needed, in general for an understanding of the physics, but also for the application of high-power femtosecond lasers to analyse the atmosphere.

Femtosecond white-light lidar: remote atmospheric spectroscopy

The main advantage of lidar over other remote sensing techniques, as differential optical absorption spectroscopy DOAS or fourier transform infrared spectroscopy FTIR (both using natural or artificial white-light sources), is the high spatial resolution over long distances, which is achieved by the use of short laser pulses and fast electronics to record the signal of the backscattered light. A terawatt laser can serve as a fast-pulsed highly collimated white-light source for lidar. In first experiments to evaluate this possibility (Wöste et al. 1997) the laser-like divergence of the white light was verified by detecting the backscattered light from 12-km high cirrus clouds. Therefore a self-focusing 2-TW 790-nm laser beam was used measuring only the light between 350 and 650 nm (BG39 filter).

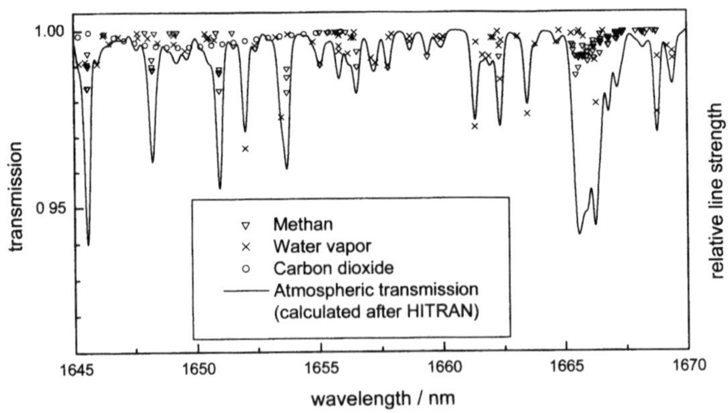

Fig. 2.2. Simulated absorption spectrum of 3 km atmosphere under normal conditions. The line-by-line calculation of the spectrum (solid line) implies an experimental resolution of 0.25 nm. The symbols indicate the relative oscillator strength (a.u.) of the single absorption lines (disregarding weak lines). Line data have been taken from HITRAN 2000 (Rothman et al. 1996).

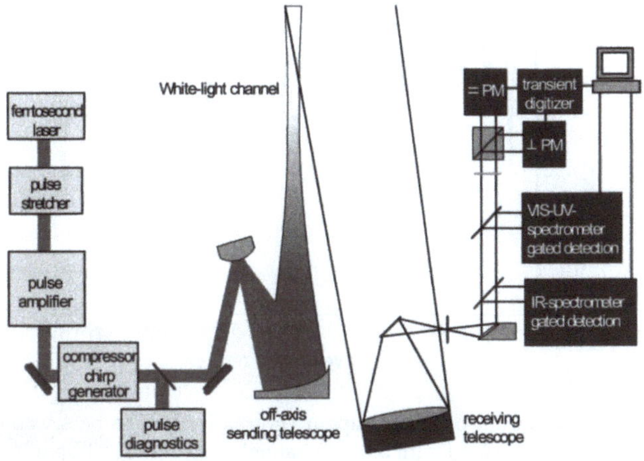

Fig. 2.3. Schematic setup of the femtosecond white-light lidar.

The near to mid-IR is of great interest in atmospheric science, because many pollutant and climate relevant gases, e.g. the volatile organic compounds (VOC) have strong absorption bands in this spectral region. Great efforts are undertaken in the development of (tunable) lasers which cover this wavelength range. But taking into account that, in contrast to typical UV absorption bands, in the IR the molecules have narrow absorption lines in bands which strongly overlap with those of other molecules (Fig. 2.2), it is easier to measure precisely the absorption with a light continuum than with monochromatic lasers. Even if the spectrum can not be measured in full resolution (the natural line-widths determined by atmospheric pressure and Doppler broadening) exact fitting of the data is possible, while using monochromatic lasers the spectral shape and exact position of the laser line for the measurement is very critical.

A highly collimated pulsed white-light source, as terawatt lasers now provide, does not only offer the mentioned advantages for the retrieval of gas concentrations, but could also become a real multicomponent lidar for the simultaneous detection of various gases. This would be very much acclaimed by the modellist. Although a limitation still lies in the availability of suitable detectors for the infrared, particularly multichannel devices. If a single-point detector has to be used in a scanning process, the measurement is very time-consuming, as it would be using a continuously tunable laser.

The schematic setup of the femtosecond white-light lidar with a multispectral detection is shown in Fig. 2.3. An example of a measured water vapour spectrum and the fitting of the calculated absorption is shown in Fig. 2.4. In this experiment, where a 2-TW 790-nm laser was used, the H_2O mixing ratio could be obtained with an estimated accuracy of 15% (Rairoux et al. 2000). The uncertainty was mainly due to a lack of knowledge about the distance at which the white-light is created. In this regard further experiments dedicated to the study of SCG are needed.

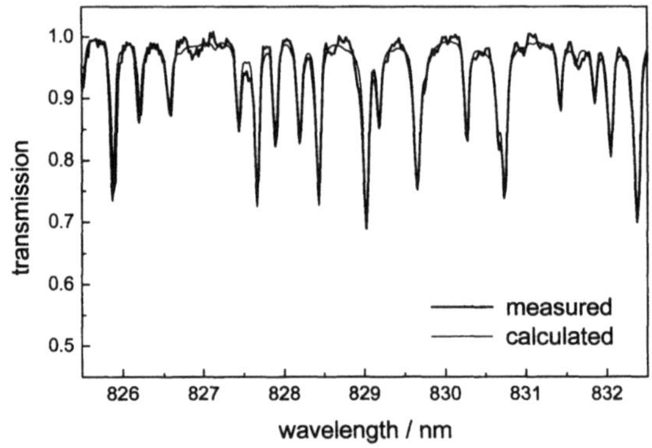

Fig. 2.4. Absorption spectrum of water vapour (normalized). The backscatter signal from 600-1000 m height was measured with a multichannel analyzer. The humidity (mixing ratio of H_2O) is retrieved through a fit to the calculated spectrum (based on HITRAN).

The laser parameters can be set to optimise the SCG in filaments, e.g. through a slight initial focusing of the beam with an adjustable sending telescope. A special feature is the so called negative pre-chirp: in order to compensate the GVD the faster components of the spectrum are set to the back of the initial pulse by tuning the pulse compressor (see Sect. 2.1). In this way the intensity can be enhanced at a certain distance and the position of the filaments can be shifted. This fact might have a huge impact on lidar, because an enhanced backscatter of light by the air in the filament region was recently demonstrated in laboratory experiments (Yu et al. 2001).

2.3 Interaction of Intense Laser Pulses with Microdroplets – Characterization of Aerosols

A further step in the evaluation of new lidar techniques, beyond the ones described in the previous section, lies in the interaction of intense laser pulses with atmospheric aerosols. In this context the question arises, if there are nonlinear optical effects which can provide information about the composition of aerosols. To address this issue in detail it is important to study at first the interaction of femtosecond laser pulses with microdroplets in laboratory experiments. It has to be evaluated which nonlinear optical effects can be generated in droplets and if their size and their chemical composition are determinable. Particularly, with regard to a possible lidar application, the angular dependence of the emitted light is of major interest. It would be a great advantage, if the angular intensity distribution favored the backscatter direction due to anisotropic scattering. First an experimental arrangement is presented which conforms to the described requirements. Afterwards the most important nonlinear optical effects of the interaction of intense laser pulses with microdroplets will be discussed.

Experimental setup

The schematic experimental setup is given in Fig. 2.5. An amplified Ti:Sa femtosecond laser system generates light pulses with a repetition rate of 1 kHz, a pulse duration of 80 fs and a central wavelength of 810 nm. A TOPAS (travelling-wave optical parametric amplifier) allows to access a spectral range from 800 nm to 1600 nm. A piezo driven nozzle produces single droplets with a range in diameter from 30 to 50 μm. The droplet diameter can be modified by the amplitude and duration of the driving pulse of the nozzle and is continuously monitored by a combination of a microscope and a CCD-Camera. Under fixed conditions the droplet diameter is found to be constant with a drop to drop variation of less then 2%. The droplet generation is synchronized to the laser system and the femtosecond pulses are focused on the droplets using a 20 mm-focal length lens. The setup permits to vary the intensity between 10^{11} and 10^{13} W/cm². To observe the angular dependence of the emitted light, an optical fiber is mounted on a goniometer which has its centre directly under the point of interaction. This arrangement allows an angular range from –60° to 178° to be used. The 0°-position is defined to be the forward direction of the laser beam. The emitted light is guided to a grating-spectrometer and is detected by a photomultiplier tube (PM). To reduce the noise, the PM-signal passes a boxcar averager module with an integration window of about 30 ns duration and an average over 100 laser shots. This arrangement allows to perform two types of measurements:

- the angular dependence of the scattered light intensity at one particular wavelength
- the spectrum of the emitted light at one defined angle.

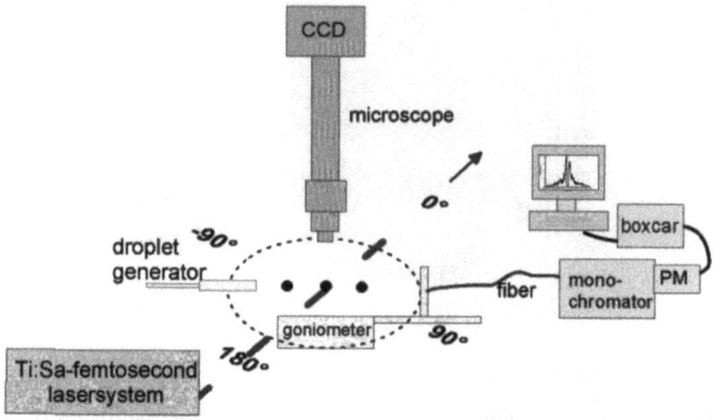

Fig. 2.5. Schematic setup for the measurement of the spectral and angular distribution of the light emitted by microdroplets, due to the interaction with femtosecond laser pulses.

Nonlinear optical effects in microdroplets

Microdroplets are attractive systems for the study of several nonlinear optical effects. In microspheres the intensity required for such processes is much lower than in the macroscopic volume due to their strongly curved liquid-air interface. It acts as a lens and focuses the incident radiation on some small regions inside the droplet. At these areas of high intensity the efficiency for nonlinear optical processes is strongly enhanced.

With Lorenz-Mie-theory the scattering of a planar electromagnetic wave at an homogeneous spherical particle can be described (Barber and Hill 1990; Barton 1995). The scattered wave and the internal intensity distribution depend only on the refractive index of the droplet medium and on the size parameter, which is the ratio of the droplet diameter to the wavelength of the incident light. In case of the interaction of femtosecond laser pulses with microdroplets the properties of ultra short laser pulses have to be taken into account. An ultra short laser pulse has a large spectral bandwidth. Therefore a large size parameter range has to be considered calculating the internal intensity distribution of the incident light in the droplet by Mie-theory. From the 3-dimensional illustration of this intensity distribution in the equatorial plane of a sphere it is clearly visible, that the droplet concentrates the internal field at some regions, the so called hot spots (Fig. 2.6). At these points the efficiency for nonlinear optical processes is enhanced and depends strongly on the order of the process.

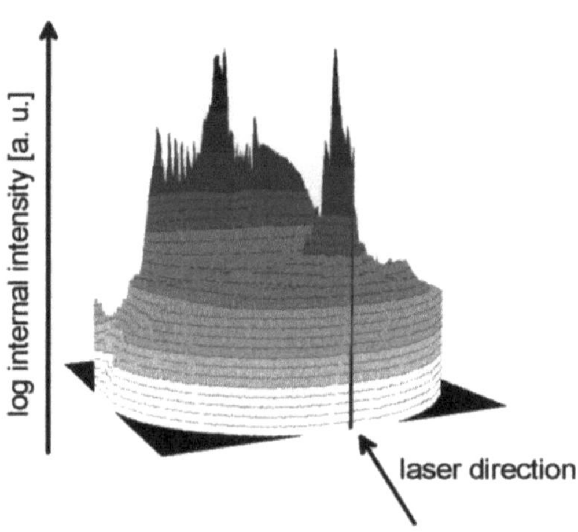

Fig. 2.6. 3-dimensional logarithmic illustration of the relative internal intensity in the equatorial plane of a droplet with a diameter of 55 μm and a refractive index of 1.329 excited by a femtosecond laser pulse.

High harmonic generation

Among the nonlinear optical effects coherent and incoherent processes have to be distinguished. Coherent nonlinear optical processes like third harmonic generation (THG) and sum frequency generation in microdroplets are well studied using picosecond laser pulses (Hill et al. 1993; Leach et al. 1993). THG was also observed to be the most efficient nonlinear optical effect induced by femtosecond laser pulses in water droplets (Kasparian et al. 1997). Spectral analysis of the light emitted by water droplets show no higher harmonic of even order, as expected due to the inversion symmetry of liquids. Only at the surface of the droplet this symmetry is broken enabling the oriented molecules at the interface liquid-air to generate an even order polarisation. In pure water droplets the intensity of the second harmonic generated at the surface is very low (Goh et al. 1988), but it was observed in water droplets coated with surfactant molecules (Hartings et al. 1997). Odd harmonics however were measured up to the fifth harmonic generated by femtosecond infrared laser pulses (Zimmer et al. 2002). The inset in Fig. 2.7 (left) shows the VIS-spectrum of the light emitted by a water droplet with a diameter of 55 μm excited at 810 nm and a pulse intensity of $3.5 \cdot 10^{11}$ W/cm². The intense THG signal at 270 nm is clearly visible. The angular distribution of the third harmonic radiation is given in Fig. 2.7 (left). It shows a maximum in forward direction between 0° and 42° which is strongly modulated. Backward scattering is hardly detectable. By decreasing the size parameters the pattern simplifies. For smaller size parameter a model was developed using the standard Green function (Carrol and Zheng 1998). The spatial and angular distribution of THG can be calculated assuming that the THG results from the surface and the interior of the droplet. In Fig. 2.7 (right) the experimental result and the calculation for a size parameter of 80 are compared and show a remarkable agreement. The fifth harmonic indicates nearly the same angular behaviour (Zimmer et al. 2002).

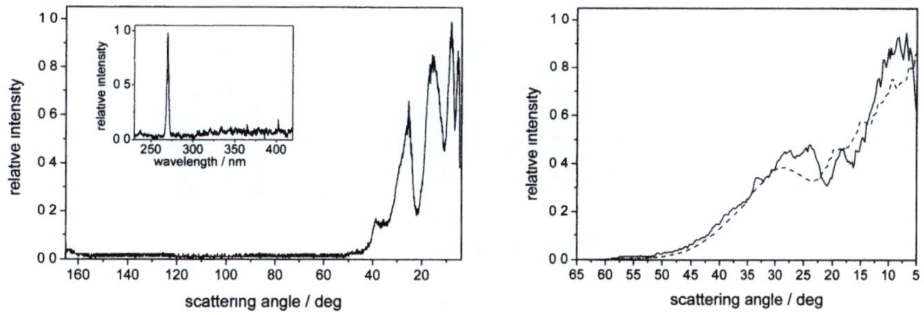

Fig 2.7. Left: Angular distribution of the THG-intensity generated in a water droplet with a diameter of 55 μm, excited by a pump wavelength of 810 nm. Inset: Generated light in the UV-VIS region at a scattering angle of 30°. Right: Angular distribution of the THG-intensity of the experiment (solid) and theory (dashed) for a size parameter of 80.

Multiphoton fluorescence

In contrast, incoherent nonlinear optical effects can have an angular intensity distribution enhanced in the backscatter direction. The fluorescence of an ensemble of freely rotating molecules for example is isotropic. But when these molecules are homogeneously embedded in a microdroplet, the emission can become anisotropic. Since the sphere concentrates the incident radiation at certain regions an angular dependent re-emission is initiated. Due to the reciprocity principle the fluorescence radiation returns of regions of high intensity to the light source (Hill et al. 1997). The intensity of the fluorescence $U_n(\vartheta_d)$ which is detected at an angle ϑ_d is the sum of the contributions from all molecules in the droplet of volume V

$$U_n(\vartheta_d) = \int_V \sigma_n I_{\text{int}}{}^n(\vec{r})\rho_0\left[F(\vec{r},\vartheta_n)\right]dV \tag{2.2}$$

where σ_n is the n-photon-absorption cross-section, $I_{\text{int}}(\vec{r}) = \vec{E}_{\text{int}}(\vec{r})\vec{E}_{\text{int}}{}^*(\vec{r})$ is the internal intensity of the incident wave, ρ_0 is the molecule density and $F(\vec{r},\vartheta_n)$ the probability that a photon emitted from a molecule at the position $\vec{r}$ is detected (Hill et al. 1996). $F(\vec{r},\vartheta_n)$ and $I_{\text{int}}(\vec{r})$ have to be integrated both, the spectral bandwidth of the laser pulse and the spectral bandwidth of the fluorescence. In the measurements reported, ethanol and methanol droplets containing the dye coumarin 510 were used. The central wavelength of the pump was 400 nm, 850 nm and 1200 nm. This corresponds to a 1-, 2- and 3-photon-excitation. As an example Fig. 2.8 shows a 3-photon-fluorescence. Both the calculation and the experimental results show an enhanced backscatter. These enhancement increases with the number of photons (Hill et al. 2000).

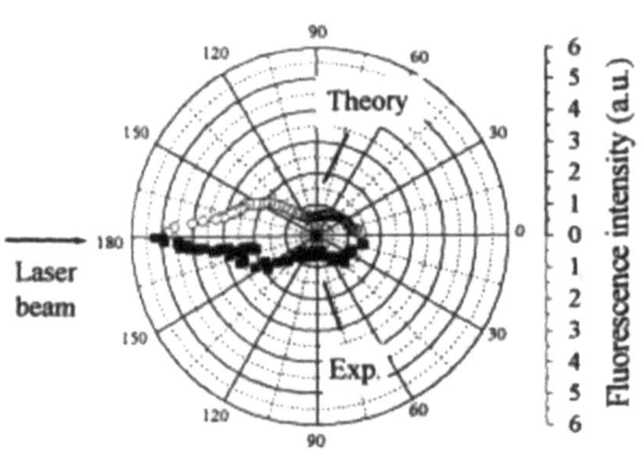

Fig. 2.8. Experimental and theoretical angular distribution of the emission of a 3-photon-fluorescence process induced in a microdroplet containing coumarin 510 (Hill et al. 2000).

Laser-induced breakdown

By the interaction of femtosecond laser pulses with water droplets the intensity is high enough to initiate a laser induced breakdown. Due to MPI a plasma is generated in the regions of high intensity inside the droplet (Hammer et al. 1996). The plasma creation within a nominally transparent droplet transforms it into a highly absorbing droplet. Once breakdown has occurred and plasma has been generated in this volume the remaining portion of the laser pulse is absorbed raising the temperature and the pressure of the volume to very high values. Plasma temperatures of 30,000 K and pressures up to 500,000 atm in droplets were observed and calculated (Carls et al. 1991). Since the internal-field distribution and the laser-induced plasma distribution are nonuniform the explosive vaporisation is asymmetrical relative to the droplet centre. This behaviour is illustrated in Fig. 2.9 for two different laser intensities. The images are taken with a CCD camera illuminated with an LED during an exposure time of 400 ns. By varying the delay between the time of laser-droplet-interaction and the illumination the distorted droplets are imaged with time resolution. The internal heating is greatest just within the shadow face and it is clearly visible that the local high pressure causes expansion. The droplet breaks open at the shadow side and a jet stream is ejected away from the droplet. At higher intensities this process is much more pronounced, the shock wave traverses the droplet also in the opposite direction and vapour is streaming additionally from the illuminated side. In less than 70 μs the droplet is completely vaporized.

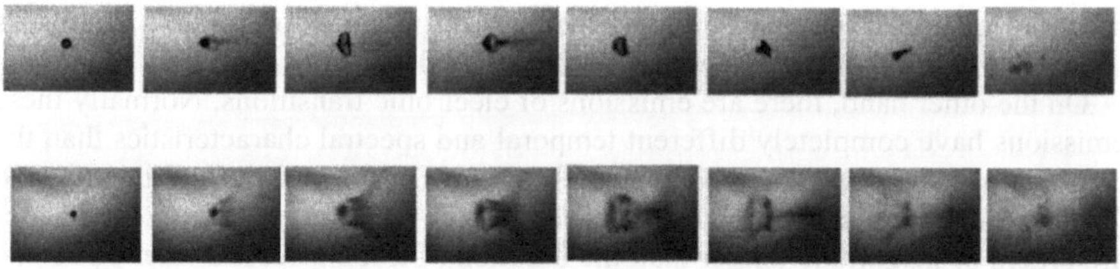

Fig. 2.9. Photographical sequences of the interaction of femtosecond laser pulses with microdroplets. The droplet diameter is 30 μm, and the laser propagates from the left to the right side. First sequence with a laser intensity of 10^{12} W/cm² represents the droplet with a delay of 0 ns, 700 ns, 1400 ns, 1800 ns, 8700 ns, 12000 ns, 20700 ns and 69700 ns. Second sequence with a laser intensity of $9{,}7 \cdot 10^{12}$ W/cm² represents the droplet with a delay of 0 ns, 500 ns, 1100 ns, 1700 ns, 3000 ns, 5900 ns and 7100 ns.

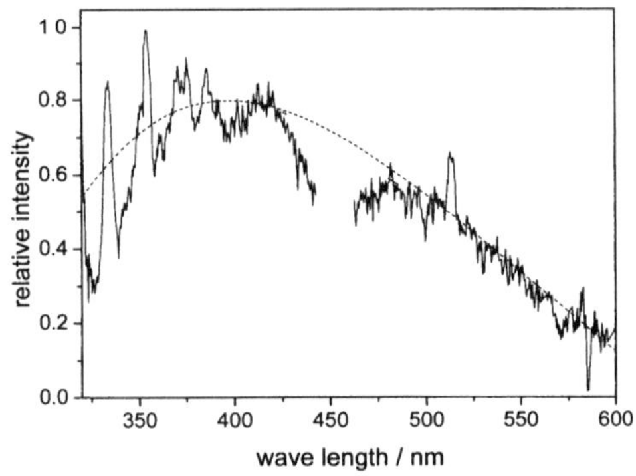

Fig. 2.10. Spectrum of the light emitted by a water droplet at a scattering angle of 12 ° by an excitation intensity of $1.7 \cdot 10^{15}$ W/cm². Due to the high intensity of the laser pulses the droplet generation was irregular. Therefore the spectrum shows intensity fluctuation. The dashed line represents the black body radiation at a temperature of 7300 K.

Plasma emission

The heated plasma produces a luminescence spectrum, which include two types of radiation. On one hand it acts as a source of a continuum, the emission of bremsstrahlung and electron-ion-recombination. Using the Planck equation the plasma temperature can be estimated (Barnes and Rieckhoff 1968). For the white-light spectrum generated in water droplets with a laser intensity of $1.7 \cdot 10^{15}$ W/cm², as shown in Fig 2.10, the estimated temperature is 7300 K.

On the other hand, there are emissions of electronic transitions. Normally these emissions have completely different temporal and spectral characteristics than the continuum emission. They have spectrally narrow peaks and their frequency and lifetime are significant for the excited species. The lifetime of these excited states are orders of magnitude longer than the duration of a femtosecond laser pulse and the emission does not start before the background radiation has decayed (Cremers et al. 1984). The weak α- and β- Balmer lines of hydrogen could be observed in the ns-regime (Eickmans et al. 1987a). The H_α-lines were used to determine the electron density and the plasma temperature was calculated by the intensity ratio of the H_α- and H_β-lines (Griem 1964; Chang et al. 1988). The observation of these plasma lines - also known as laser induced breakdown spectroscopy (LIBS) - represents an attractive method in the research of atmospheric aerosols, since these aerosols mainly consist of water containing additional substances (Poulain and Alexander 1995), which can be identified by their plasma lines. For example, the sea salt aerosol contains sodium chloride. Sodium has two strong emission bands around 589 nm. If water droplets containing 5-M sodium chloride solution are excited by laser pulses starting at an intensity of $4 \cdot 10^{14}$ W/cm², an orange flash is even visible with the naked eye. The spectrum of the emitted light, detected at a scattering angle of 165°, is displayed in Fig. 2.11. The sodium signal is so intense that the white light continuum disappears in the background noise (Zimmer 2001).

Due to the Stark effect the doublet is not resolvable (Eickmans et al. 1987b). The angular distribution of the 589 nm peak shows a strong enhancement in backscatter direction, a finding which opens fascinating prospects for atmospheric remote sensing.

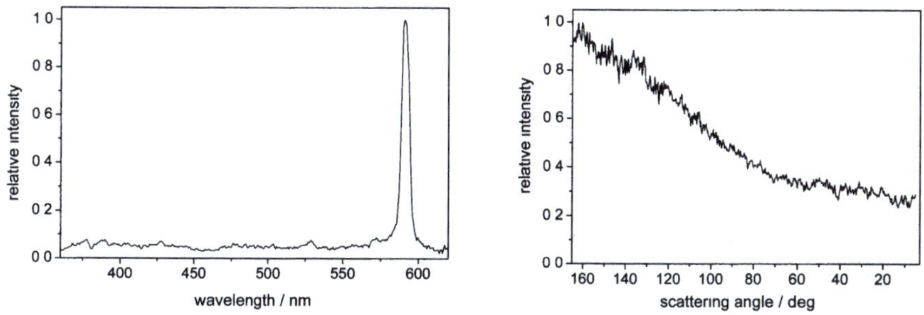

Fig. 2.11. Left: Spectrum of the light emitted in 165° direction by a microdroplet containing a 5 molar NaCl solution. Right: Angular distribution of the 598 nm signal. Laser intensity: $7.2{\cdot}10^{14}$ W/cm².

2.4 Outlook

Until now all the described results have been obtained using lasers that are stationed in laboratories. Recently, the first mobile terawatt laser has been designed and built in the framework of a French-German research project called Teramobile (Wille et al. 2002). A 5-TW laser was integrated in a standard freight container, together with detection optics and electronics for lidar. The Teramobile only requires a power supply, which can be a generator. It contains air condition and an autonomous cooling system for the laser and the electrical devices. This mobile laboratory will allow to fulfil the mentioned requirements and to investigate out the possibilities for in field analysis of the atmosphere:

- Long-distance propagation experiments will be performed on free plane sites, e.g. former air fields. In that way horizontal scanning of the laser beam over ranges of several kilometres will be possible.
- The system can be brought to nearly any place, so that the new lidar techniques can be performed under different atmospheric conditions.
- Experiments with aerosols will be done under real and defined artificial conditions. Field measurements can, on one hand, be dedicated to the study of the propagation of TW pulses in a simulated cloud or haze of water droplets, to observe how the presence of particles modifies the filament formation. On the other hand, experiments will evaluate the potential of measuring aerosol compositions based on the results described in Sect. 2.3.

References

Alfano RR (ed, 1989) The Supercontinuum Laser Source. Spinger, Berlin Heidelberg New York

Alfano RR, Shapiro SL (1970). Phys Rev Lett 24: pp 584, 592, 1217

Backus S, Durfee CG III, Murnane MM, Kapteyn HC (1998) High power ultrafast lasers. Rev Sci Instrum 69: 1207-1223

Barber PW, Hill SC (1990) Light Scattering by Particles. In: Computational Method. World Scientific, Singapore

Barnes PA, Rieckhoff KE (1968) Laser induced underwater sparks. Appl Phys Lett 13: 282-284

Barton (1995) Internal and near surface electromagnetic fields for spheroidal particle with arbitrary illumination. Appl Opt 34: 5542-5551

Braun A, Korn G, Liu X, Du D, Squier J, Mourou G (1995) Self-channeling of high-peak-power femtosecond laser pulses in air. Opt Lett 20: 73

Carls JC, Seo Y, Brock JR (1991) Laser-induced breakout and detonation waves in droplets. II Model. J Opt Soc Am B 8: 329-336

Carrol D, Zheng XH (1998) Modelling third harmonic generation from microdroplets. Pure Appl Opt 7: L49-L55

Chang RK, Eickmanns JH, Hsieh WF, Wood CF, Zhang JZ, Zheng J (1988) Laser-induced breakdown in large transparent water droplets. Appl Opt 27: 2377-2385

Chin SL, Petit S, Borne F, Miyazaki K (1999) The white light supercontinuum is indeed an ultrafast white light laser. Jpn J Appl Phys 38: L126-L128

Cremers DA, Radziemski LJ, Loree TR (1984) Spectrochemical Analysis of Liquids Using the Laser Spark. Appl Spectrosc 38: 721-729

Eickmans HE, Hsieh WF, Chang RK (1987a) Laser induced explosion of H2O droplets: spatially resolved spectra. J Opt Soc Am B 12: 22-24

Eickmans HE, Hsieh WF, Chang RK (1987b) Plasma spectroscopy of H, Li, Na in plumes resulting from laser-induced droplet explosion. Appl Opt 26: 3721-3725

Goh MC, Hicks JM, Pinto RG, Bhattacharyya K, Eisenthal KB, Heinz TF (1988) Absolute orientation of water molecules at the near water surface. J Phys Chem 92: 5074-5079

Griem HG (1964) Plasma Spectroscopy. McGraw-Hill Book Company, New York

Hammer DX, Thomas RJ, Noojin GD, Rockwell BA, Kennedy PK, Roach WP (1996) Experimental Investigation of Ultrashort Pulse Laser-induced Breakdown Thresholds in Aqueous Media. IEEE J Quantum Elektron 32: 670-677

Hartings JM, Poon A, Pu X, Chang RK, Leslie TM (1997) Second harmonic generation and fluorescence images from surfactants on hanging droplets. Chem Phys Lett 281: 389-393

Hill SC, Leach DH, Chang RK (1993) Third-order sum-frequency generation in droplets: model with numerical results for third-harmonic generation. J Opt Soc Am B 10: 16-33

Hill SC, Saleheen HI, Barnes MD, Whitten WB, Ramsey JM (1996) Modeling fluorescence collection from single molecules in microspheres: effects of position, orientation, and frequency. Appl Opt 35: 6278-6288

Hill SC, Videen G, Pendlton JD (1997) Reciprocity method for obtaining the far fields generated by a source inside or near a microparticle. J Opt Soc Am B 10: 2522-2529

Hill SC, Boutou V, Yu J, Ramstein S, Wolf JP, Pan Y, Holler S, Chang RK (2000) Enhanced Backward-Directed Multi-Photon-Excited Fluorescence from Dielectric Microcavities. Phys Rev Lett 85: 54-57

Kasparian J, Krämer B, Dewitz JP, Vajda S, Rairoux P, Vezin B, Boutou V, Leisner T, Hübner W, Wolf JP, Wöste L, Bennemann KH (1997) Angular Dependences of Third Harmonic Generation from Microdroplets. Phys Rev Lett 78: 2952-2955

Kasparian J, Sauerbrey R, Mondelain D, Niedermeier S, Yu J, Wolf JP, André YB, Franco M, Prade B, Mysyrowicz A, Tzortzakis S, Rodriguez M, Wille H, Wöste L (2000) Infrared extension of the supercontinuum generated by fs-TW-laser pulses propagating in the atmosphere. Opt Lett 25: 1397-1399

La Fontaine B, Vidal F, Jiang Z, Chien CY, Comtois D, Desparois A, Johnson TW, Kieffer JC, Pépin H (1999) Filamentation of ultrashort pulse laser beams resulting from their propagation over long distances in air, Phys Plasmas 6: 1615-1621

Leach DH, Chang RK, Acker WP, Hill SC (1993) Third-order sum-frequency generation in droplets: experimental results. J Opt Soc Am B 10: 34-45

Manassah JT (1989) Simple Models of Self-Phase and Induced-Phase Modulation, in Alfano (ed, 1989)

Poulain DE, Alexander DR (1995) Influences on Concentration Measurements of Liquid Aerosol by Laser-Induced Breakdown Spectroscopy. Appl Spectrosc 49: 569-579

Rairoux P, Schillinger H, Niedermeier S, Rodriguez M, Ronneberger F, Sauerbrey R, Stein B, Waite D, Wedekind C, Wille H, Wöste L (2000) Remote sensing of the atmosphere using ultrashort laser pulses. Appl Phys B 71: 573-580

Ranka JK, Schirmer RW, Gaeta AL (1996) Observation of Pulse Splitting in Nonlinear Dispersive Media. Phys Rev Lett 77: 3783

Rothman LS et al. (1998) The HITRAN molecular spectroscopic database and HAWKS (HITRAN atmospheric workstation), 1996 edition. J Quant Spectrosc Radiat Transfer 60: 665-710 (see www.hitran.com for HITRAN 2000)

Shen Y (1984) The principles of nonlinear optics. Wiley, New York

Strickland D, Corkum PB (1994) Resistance of short pulses to self-focusing. J Opt Soc Am B 11: 492

Strickland D, Mourou G (1985) Compression of amplified chirped optical pulses. Opt Commun 56: 219

Wille H, Rodriguez M, Kasparian J, Mondelain D, Yu J, Mysyrowicz A, Sauerbrey R, Wolf JP, Wöste L (2002) Teramobile: a mobile femtosecond-terawatt laser and detection system. Eur Phys J AP 20: 183-190

Wöste L, Wedekind C, Wille H, Rairoux P, Stein B, Nikolov S, Werner C, Niedermeier S, Ronneberger F, Schillinger H, Sauerbrey R (1997) Femtosecond atmospheric lamp. Laser und Optoelektronik 29: 51

Yu J, Mondelain D, Ange G, Volk R, Niedermeier S, Wolf JP, Kasparian J, Sauerbrey R (2001) Backward supercontinuum emission from a filament generated by ultrashort laser pulses in air. Opt Lett 26: 533-535

Zimmer W (2001) Nichtlineare optische Effekte bei der Wechselwirkung von Femtosekundenlaserpulsen mit Mikrotropfen. Ph.D. thesis, FU Berlin

Zimmer W, Krenz M, Wöste L, Leisner T (2002) High harmonic generation in microdroplets. in preparation

3 Analysis of Three Dimensional Aerosol Distributions by Means of Digital Holography

H. Kreitlow, F. Frischkorn, J. Miesner and B. Stark

3.1 Abstact

Digital holography presents a modern method for micro particle diagnostics, generating three dimensional snapshots of particle distributions in volumes in a rapid sequence.

Holographic techniques for micro particle analysis were already introduced by Silverman et. al. in 1964 (Silverman et. al. 1964), (Vikram 1992). Scientists pushed that technology forward to analyze atmospheric aerosol distributions in clouds. In 1994 the researchers Borrman and Jaenicke made a comparison concerning the performances of a conventional holographic aerosol analysis system with the standard FSSP-100 and PVM-100 devices (Borrmann et al. 1994). The particle range in that investigation was in the range of 3-15µm diameter. With their new holographic method they could store 450 liters (0,45m^3) of cloud volume in a hologram using one exposure with a recognizable particle size of 5-500µm (Uhlig 1995). However, the technical requirements for hologram exposure (ruby laser), hologram development (dark room), droplet reconstruction (reconstruction set-up in the lab) were enormous. Furthermore, analyzing the reconstructed cloud volume is extremely time consuming (Vössig et. al. 1997).

To overcome these disadvantages, a mobile system based on the method of digital holography has been developed and tested in a field campaign.

This innovative digital holographic measurement system uses the principle of Fraunhofer inline holography and provides a series of advantages compared to conventional holographic systems regarding time consumption and system design for both holographic storage and hologram reconstruction or hologram analysis.

3.2 Introduction

Aerosol particles influence strongly the global climate because they scatter and absorb the sun's radiation. Theoretical climate modeling requires a good understanding of the generation of aerosols, their size distribution and their optical properties. This should result in reasonable forecasts about climate development.

Using the holographic method, the droplet distribution can be stored in a hologram and can be analyzed simply by quantifying the diameter of the droplets and counting the number out of the reconstructed volume. Thus, the humidity of the air can be measured very precisely for example.

3.3 Digital Holography

The equipment requirements for holographic film exposure and the subsequent analysis of the reconstructed volume following the classical approach are enormous and realization is extremely time consuming (Vössig et. al. 1997). By contrast, the procedure of digital holography is favorable because of its short process times for digital hologram storage by a CCD-sensor and a computer as well as numeric reconstruction of the digital hologram. The analysis of the reconstructed sample volume may also be performed by computer which finally results in digital data output (Kreis et. al. 1998).

3.4 Storage of the Digital Hologram

The fundamental difference between the digital and the conventional holographic method is the storage medium. Instead of a holographic film, an optoelectronic CCD-sensor is used for hologram recording: The CCD-sensor receives the interference pattern of the superimposed object wave and reference wave electronically and transfers it via computer to a digital mass storage (**Fig. 3.1**).

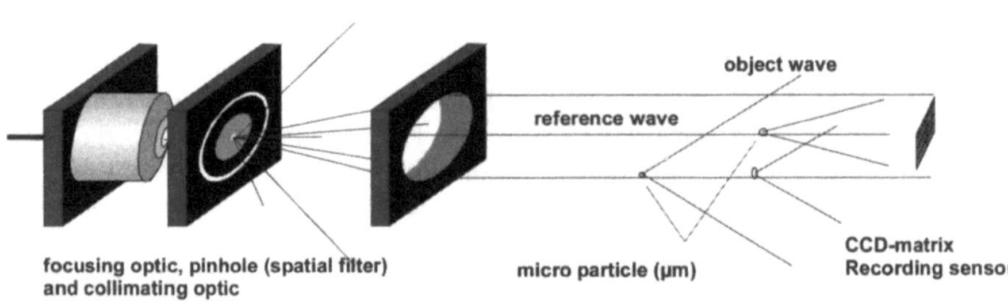

Fig. 3.1. Digital holography – the inline method

The resolving power of CCD-sensors is only one tenth that of holographic films. Therefore, the interference pattern has to be scaled such that spatial sampling with the CCD-sensor is possible and the Nyquist criterion is fulfilled avoiding aliasing effects. This adaptation is realized using an optical imaging system.

Fig. 3.2 points out that in consequence of the optical magnification the recorded volume and therefore also the information density minimizes in the same way.

The optical magnification m is a function of the focal length f and its distance B to the CCD-sensor:

$$m(f,B) = \frac{B-f}{f} \tag{3.1}$$

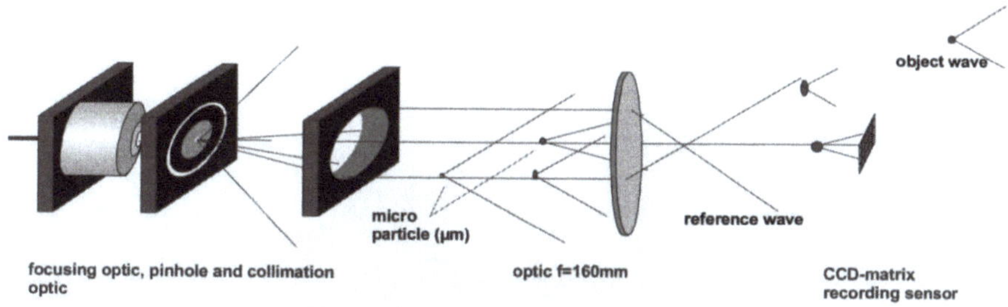

Fig. 3.2. Digital holography – adaptation of the spatial interference pattern frequency to the resolving power of the CCD-sensor

3.5 Digital Reconstruction of the Stored Volume

The descriptions of holographic processes are mathematically defined by the so-called diffraction integrals, where either the generation of holograms by the superposition of two coherent optical waves or the reconstruction of the object scene can be specified. The reconstruction takes place by diffraction of a reference wave at the holographic interference pattern. This can be done alphanumerically by discrete Fourier transform algorithm (Schnars 1994) also:

$$\Gamma(m,n) = \frac{ia}{\lambda z}\exp\left[-i\pi\lambda z\left(\frac{m^2}{N^2\Delta x^2}+\frac{n^2}{N^2\Delta y^2}\right)\right] * \sum_{k=0}^{N-1}\sum_{l=0}^{N-1} T(k,l)\exp\left[-i\frac{\pi}{\lambda z}\left(k^2\Delta x^2+l^2\Delta y^2\right)\right]\exp\left[+i2\pi\left(\frac{km}{N}+\frac{lm}{N}\right)\right] \tag{3.2}$$

$\Gamma(m,n)$	:	Reconstructed plane in the digital image with the pixels x and y
$T(k,l)$	:	Recorded hologram - interference pattern from the CCD with its pixel coordinates k and l
λ	:	Optical wave length
z	:	Distance where the reconstruction plane $\Gamma(m,n)$ will be spanned up
Δx, Δy	:	Pixel size of the CCD-sensor

The information retrieval, that is the analysis of the three dimensional reconstructed volume, takes place in the computer. However, the two dimensional display of monitors must be considered where three dimensional volumes cannot be presented directly. Through planewise reconstruction of the volume along the optical axis, each image can be analyzed first and then all planes can be put together to illustrate the volume.

The numerical plane to plane distance $\Delta z = z_1 - z_2$ (**Fig. 3.3**) can be made as small as desired. The reconstructed volume can be shown to the operator either in a stationary form or as a movie.

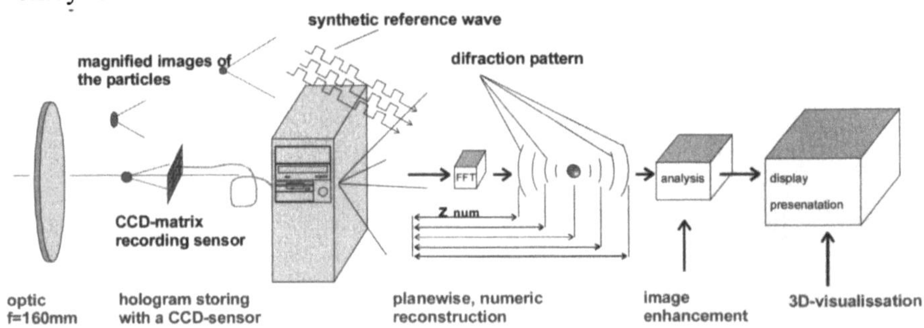

Fig. 3.3. Process chain of digital holography, see also **Fig. 3.2**

3.6 Development of an Aerosol Analysis Aystem on the Basis of Digital Holography

Because of the advantages of aerosol analysis on the basis of digital holography, an experimental holographic laboratory set-up has been developed further to a rigid system for use in field applications.

Fig. 3.4. Laser head: laser resonator top, collimating unit middle, energy and cooling supply bottom

The adverse circumstances in natural environments (wind, weather and transport) pose great demands on the system. All the housings of the holographic measurement system have to be waterproof and temperature stabilized. Furthermore, the measurement volume has to be illuminated within the shortest possible time because the droplets move in gusts at high speed in front of the holographic storage system.

Therefore, a mobile short pulse ruby laser system has been developed, which is connected directly to the beam shaping and collimating unit of the illumination system; the so-called laser head (**Fig. 3.4**). The laser pulse illuminates the cloud volume within 30ns (694nm, approx. 10mJ). The power supply, the water cooling and the control unit are installed in portable boxes.

The recording system with the optoelectronic CCD-sensor is placed opposite the laser system and orientated to be perpendicular to the propagation of the laser beam. Both systems have to be aligned carefully to be exactly on the same optical axis. In this way aberrations of the fine holographic structures can be prevented. To support the alignment, a diode laser is installed in the system.

The camera for hologram recording was chosen with a 1024x1024 pixel à 6,7x6,7μm^2 CCD chip. It has a dynamic range of 60dB, digital output and a pixel clock option for maximum image quality. To protect the sensor against humidity, cold and stray light, it is housed together with the magnification system in a mobile case (**Fig. 3.6** and **Fig. 3.7**).

Fig. 3.5. Top, control unit of laser system. Right, power supply and water cooling for the laser head

Fig. 3.6. Holographic camera: CCD-camera with magnification system in waterproof housing

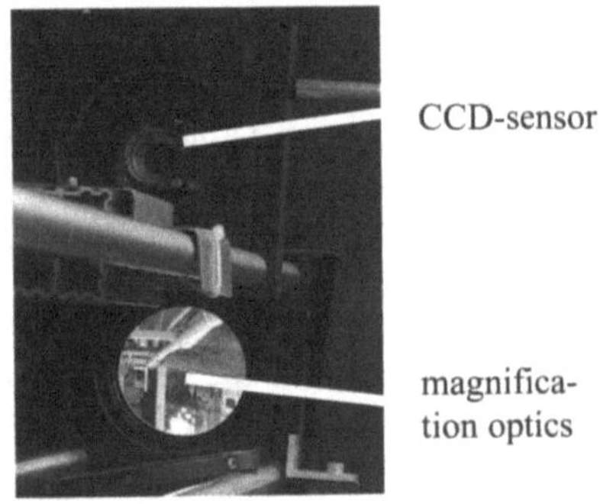

Fig. 3.7. View to the magnification optics and the CCD-sensor inside the holographic camera

Using test particles located on the surface of a glass substrate, the resolving power of the system could be determined and optimized. In this way, particles with diameters down to 4μm could be holographically stored and reconstructed (**Fig. 3.8 - Fig. 3.11**). However, the depth of field is adversely affected by the optical image. At a magnification $m \approx 13$ it reduces to less than 2mm depth which results in a reconstructable volume of only 0,077mm³ (0,22mm x 0,17mm x 2mm). Assuming a possible droplet density of $N=350\ mm^{-3}$, circa 25 droplets could be counted in a hologram.

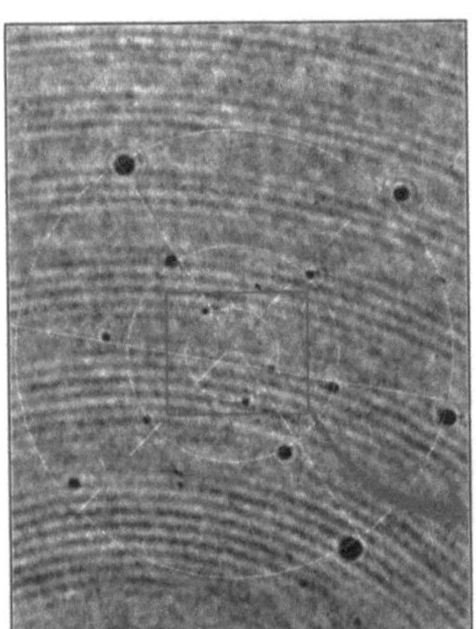

Fig. 3.9. Screenshot of imaging plane of test particles

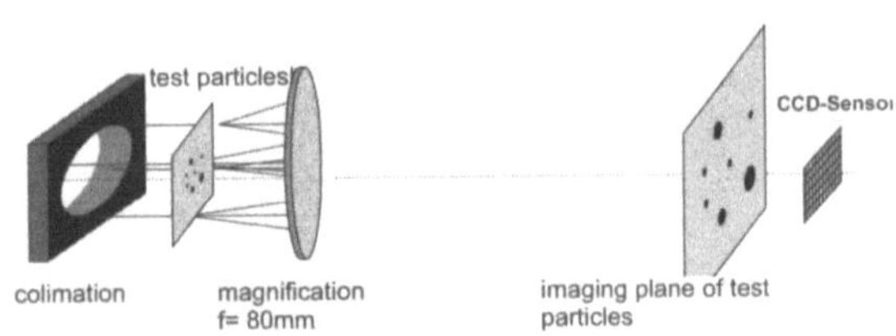

Fig. 3.8. Experimental set-up

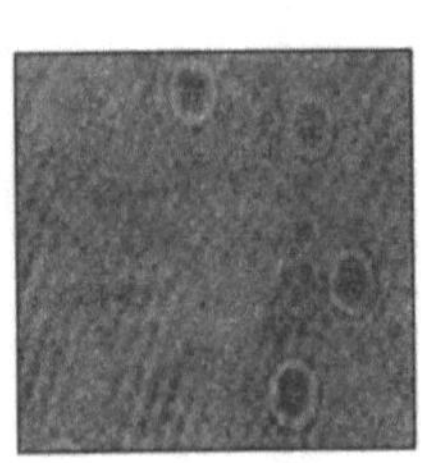

Fig. 3.10. Digital hologram of innermost circle in **Fig. 3.9**

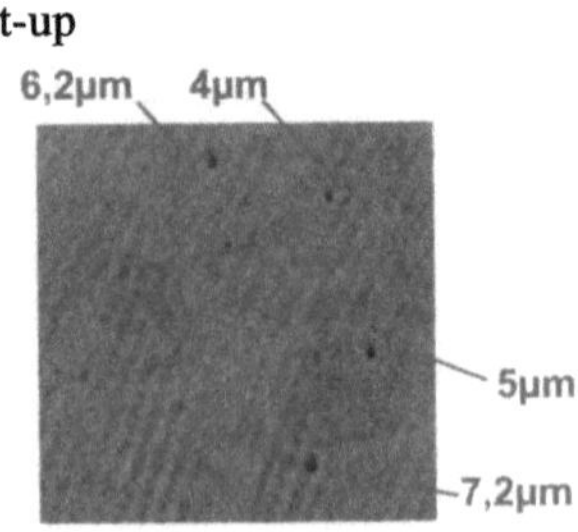

Fig. 3.11. Reconstruction of hologram of **Fig. 3.10**

3.7 Analysis of Digitally Reconstructed Droplets

The system calibration was done using a precision normal. The fully automated volume analysis requires proper image processing algorithms. Three essential routines have been developed to extract the droplets out of the numerically reconstructed measurement volume.

1) Recognition of droplet and its localization in every reconstruction plane (x- and y-component)
2) Determination of the plane of maximal contrast (z-component)
3) Determination of droplet size

3.8 Recognition of Droplets, Determination of Localization and Size

Because of the noisy and speckled background which even differs with the depth of reconstruction, filter algorithms on the basis of threshold operations or contrast enhancement are not suitable. The combination of structure filters with sequence comparison turned out to be much more promising.

Structure filters use so-called templates to search digital images for regions with similar shape. A selection of templates is shown in **Fig. 3.12,** they are abstract structures very similar to the shape of the expected object. Here the templates are basically circular or ring shaped and they investigate the reconstructed planes for locations of the droplets.

Fig. 3.12. Selection of templates

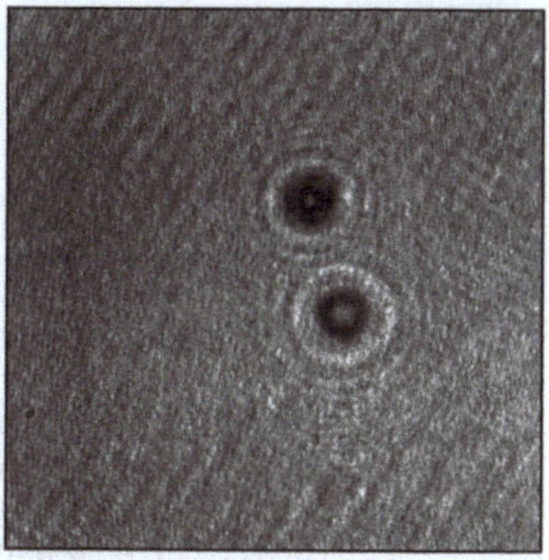

Fig. 3.13. Hologram of two tracer-particles $D_1 = D_2 = 50\mu m$

In the analysis process, a template is shifted pixel by pixel across the image plane and for every step the similarity of the image section with the template becomes quantified by the correlation coefficient. Where templates match best to the image section, the value of the correlation coefficient reaches a maximum.

The results from the template matching analysis of a plane are compared afterwards to the results from the planes next and before. Since the appearance of the centres of the droplets in x, y are the same in every plane, it is possible to distinguish between noise and droplets. The final result of this method shows the local positions of the centres of the droplets in (x, y).

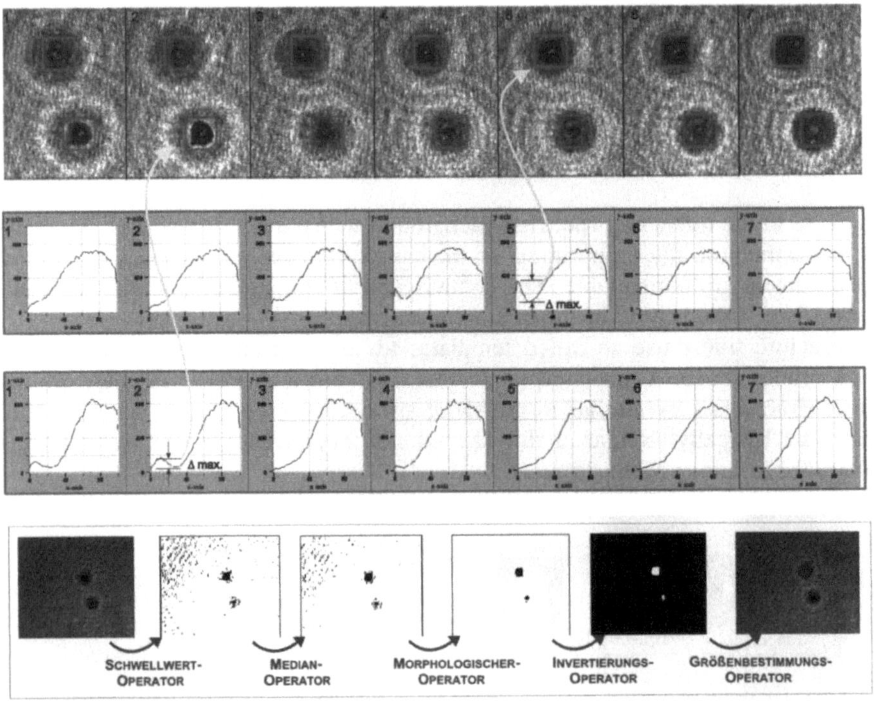

Fig. 3.14. Automated analysis of a numerically reconstructed digital hologram containing two tracer particles (droplets)

The next step is to determine the position of the droplets along the z-direction represented by the location of maximal sharpness. The prior results from the first analyzing routine are taken to specify that location. **Fig. 3.13** shows a hologram from tracer particles, stuck on both sides of a glass sample of 1,7mm thickness. A series of plane-wise reconstructed volume slides from that hologram is illustrated in the top row of **Fig. 3.14**. The images are cut to the region of interest. It can be seen clearly that there are both images with low and high sharpness of the two droplets and that they are always at the same spatial position. A certain investigation zone is now spanned over the center of area of the droplet. These zones are presented in the top row of **Fig. 3.14** by rectangles. Rows two and three display the corresponding partial histograms of the region of interest. Analysis of the histograms shows that when the sharpness of the reconstruction is maximal the difference between the first relative maximum and the following relative minimum is also maximal. Finally the so found image is used to determine the droplet size. This is performed by the following sequence of image processing operations (see bottom row). First the region of interest is binarized by applying the greylevel indicated by the minimum of the partial histogram as threshold. Following this the disturbing background noise as well as stains are removed by applying a median filter and a dilation process respectively. For convenience the resulting region of interest is inverted. Now a distinct area which is representing the size of the droplet is remaining. The absolute value of droplet size may be calculated by multiplying the number of covering pixels in horizontal and vertical direction by the scaling coefficients respectively.

3.9 Results

Center of area of droplets:	$P_1(x,y) = (203,359)$ pixel; $P_2(x,y) = (145,142)$ pixel
Depth of reconstruction:	$\Delta z = 100$mm value of num. reconstruction depth
Size of particles:	$D_1 = 72$ pixel; $D_2 = 73$ pixel

Interpreting these values by the calibration the results are:

scale x, y	:	1 pixel	⇔	0,7µm
scale z	:	1mm (num.)	⇔	17µm
$r_{1,2}(x,y,z)$	=	(0,0,0) µm, (40,150,1700) µm		
D_1 , D_2	=	50µm		

Fig. 3.16. View from the ruby laser to the holography camera.

Fig. 3.15. FELDEX 2000 – Set-up for digital holographic aerosol analysis (sunshine).

Fig. 3.17. Set-up in a cloud

3.10 Field Campaign FELDEX 2000

During FELDEX 2000, a field campaign held in autumn 2000 on top of the "Kleiner Feldberg" in the Taunus mountains close to Frankfurt (Main) / Germany, the new digital holographic analysis system has been tested for the first time under real conditions (**Fig. 3.15** - **Fig. 3.17**).

The calibration of the measuring window was performed as described earlier and showed the scale to be 0,45µm/pixel in x- and y-direction for that certain set-up.

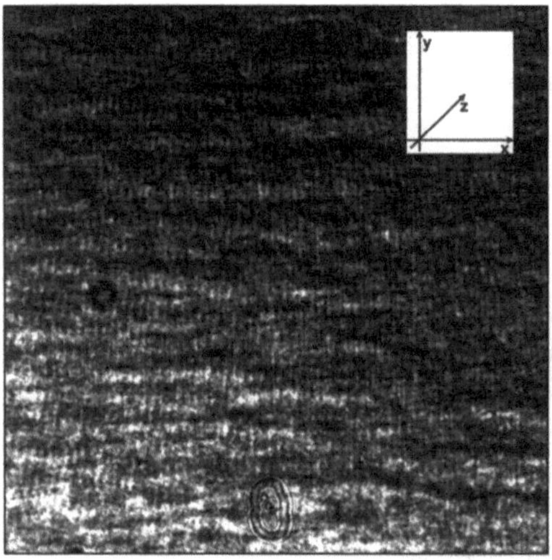

Fig. 3.19. Digital hologram of a cloud segment 1024x1024Pixel)

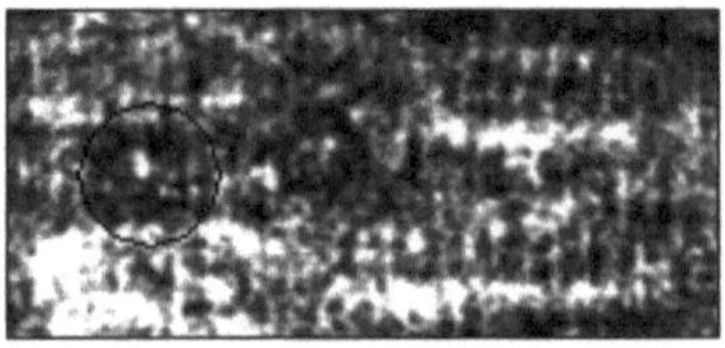

Fig. 3.18. Section of reconstruction of first particle d=18pixel or 8µm, plane A

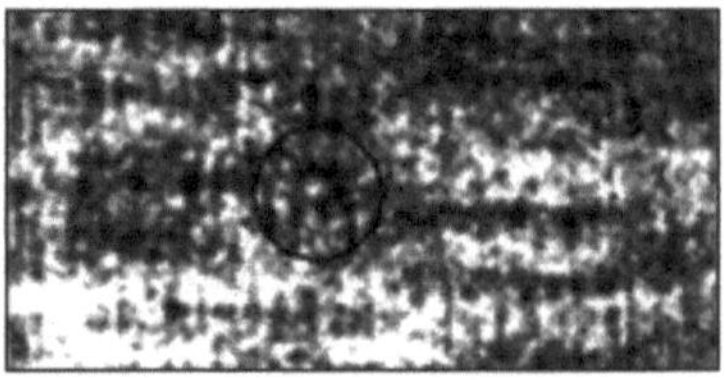

Fig. 3.20. Section of reconstruction of second particle d=16 pixel or 7µm, plane B

Holograms were stored digitally and natural cloud aerosols could be reconstructed and analyzed (**Fig. 3.18** - **Fig. 3.20**). In this example, the positions of the particles relative to each other are:

$\mathbf{r}(x,y,z)=(49, 22, 520)$pixel or $\mathbf{r}(x,y,z)= (22,10,234)\mu m$
which gives a distance of: $|\mathbf{r}(x,y,z)|=235\mu m$

A new presentation software "zebra" (Hoffmann 1999) used these data to display all droplets contained in the measurement volume in their natural location at the storage time, **Fig. 3.21**. The simulation in the right scale gives a view of the behavior of the micro structure of a cloud.

3.11 Discussion

The development of the digital holographic aerosol analysis system pointed out great advantages compared with conventional holographic systems. The time consumption for holographic droplet analysis could be reduced by a factor of 20. However, the performance of the digital system is limited by the comparatively poor resolving power of today's CCD-sensors. Using magnification optics for adaptation of the interference pattern to the low resolving power, this could be partially corrected. However, due to the optical magnification, the effective measurement volume that could be stored holographically decreases in size. Applying state-of-the-art CCD-cameras it was possible to raise the detection volume up to 1mm^3 in size, with the hologram storage time, reduced to less than one second.

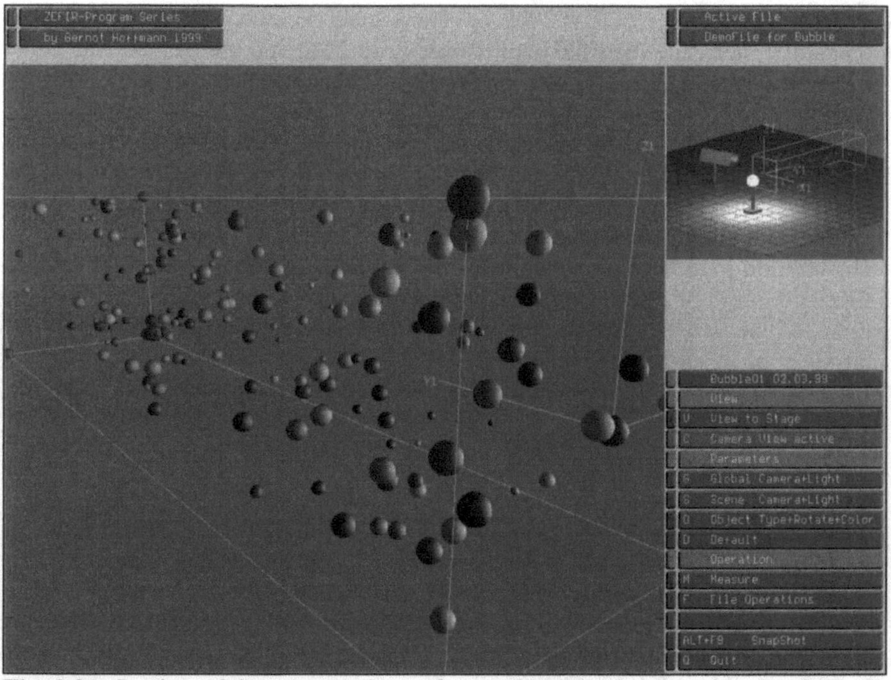

Fig. 3.21. Section of the presentation software ZEBRA, developed by Prof. Dr. G. Hoffman, Emden

Established aerosol monitoring systems or aerosol particle counters use mostly Mie scattering to determine mass concentration (Jaenicke et. al. 1993). They may even generate data in “real-time” but cannot record the spatial distribution of the particles. Methods on the basis of PIV (particle image velocimetry), LDV (laser doppler velocimetry) or LDA (laser doppler anemometry) normally apply monodispers particles as tracers to characterize three dimensional information of flow behaviour. They are not used for particle size monitoring and are not able to record spatial distributions. Also the holographical particle measuring method “light in flight” (Hinrichs et. al. 1997) can only be used for flow characterization.

The next generation of CCD-sensors with a higher number of pixels combined with smaller pixel sizes will be important, increasing both the measurement volume and the resolving power.

The alignment of the two main systems, the laser head and the holography camera, is a stringent condition which turned out to be critical especially during the field campaign FELDEX 2000 on the Kleiner Feldberg hill in the Taunus mountains. There, the distance between the laser head and the camera was about 8m whereby smallest deviations from the optical axis resulted in distortions of the hologram.

The reconstruction of the measurement volume out of the hologram was performed automatically by computer. Special numeric algorithms later analyzed the reconstructed image planes towards droplets and extracted the spatial aerosol distribution with all information such as droplet localization and size.

In addition to atmospheric diagnosis, the digital holographic system developed here can be used in a series of other applications such as pollen analysis, investigations of industrial processes e.g. varnishing systems, injection nozzles or the analysis of black carbon in exhausts.

Acknowledgement

We would like to thank the "Deutsche Bundesstiftung Umwelt (DBU)" who financed this work as part of the research project "Atmosphärische Diagnostik". The management of this sub-project was conducted by the "Technische Fachhochschule Wildau, Germany" in co-operation with the "Institut für Lasertechnik der Fachhochschule Ostfriesland, Emden, Germany" and the company "deka Sensor & Technologie, Teltow, Germany".

References

Borrmann S, Jaenicke R; Instrument Intercomparison Study on Cloud Droplet Size Distribution Measurements: Holography vs. Laser Optical Particle Counter; J. of Atmosph. Chem. 19: 253-258, 1994

Hinrichs H, Hinsch K, Kickstein J, Böhmer M; Light-in-flight holography for visualisation and velocimetry in three dimensional flows; Optics Letters Vol. 22 No. 11, p. 828-830; June 1997

Hoffmann G; Zebra; Fachhochschule Ostfriesland; 1999

Jaenicke R, Hanusch T; Simulation of the optical particle counter forward scattering spectrometer probe 100 (FSSP 100), Aerosol Sc. and Technology Vol. 18, p.309-322; 1993

Kreis T M, Jüptner W P O, Geldmacher J; Digital Holography: Methods and Applications; SPIE Vol. 3407, p.169-177, 1997

Schnars U; Digitale Aufzeichnung und mathematische Rekonstruktion von Hologrammen in der Interferometrie; VDI-Forschungsberichte; VDI-Press, Germany Reihe 8, Nr. 378, 1994

Silverman B A, Thompson B J, Ward J H; A Laser Fog Disdrometer; J. of Appl. Metrology, Vol 3, 1964

Uhlig E.; Holographische Untersuchung der Wolkenmikrostruktur unter Anwendung eines automatisierten Bildanalysesystems; Diss. Univ. Mainz; 1995.

Vikram Ch S; Particle Field Holography; Cambridge University Press, 1992

Vössig H, Borrmann S, Uhlig E; HODAR holography applied to raindrop and snowflake in-situ measurements; J. Aerosol Sci., Vol. 28, Suppl. 1, pp. S375-376, 1997

Part II

Applications in Liquid and Solid States

4 Laser-Based Analysis of Solids with Environmental Impact

Ralf Pätzold and Angelika Anders

4.1 Introduction

Within the last years, the laser-based analysis of solids has developed so far that it can be applied generally and not only in the laboratory. The increasing number of applications prove this. In the following, projects shall be presented which make use of lasers concerning the analysis of solids as well as pollutants on solid surfaces in the field of basic and mainly applied research. Of course, this can only be a specific selection of the actual applications and research programmes due to the various utilized techniques and the numerous investigated substances.

Table 4. 1. Analytical techniques based on optical detection methods

Method	Principle
Laser-induced fluorescence spectroscopy (LIF)	Fluorescence measurement after excitation by photons
Laser-induced breakdown spectroscopy (LIBS)	Detection of the atomic emission spectra after plasma generation
IR spectroscopy, Raman spectroscopy	Examination of molecular vibrations
Laser ablation inductively coupled plasma optical emission spectroscopy (LA-ICP-OES)	like LIBS, but with a carrier gas and an inductively coupled plasma

As a key technology of the 21st century, nowadays the laser can also be applied in fields which previously could only be treated conventionally with standard methods in the laboratory. Whereas in the nineties of the last century the emphasis was laid on the investigation of the various spectroscopic methods themselves, nowadays the practically orientated use and the applicability of laser-based analysis methods are of major interest.

Analytical techniques for trace detection of different kinds of solids as well as pollutants on solid surfaces are commonly used in the laboratory. Laser-based analytical techniques are not (yet) able to replace the standard analytical methods. Lasers rather are able to give access to other areas due to their special properties. The main strengths of laser-based methods are characterized by the following key words: **in-situ, rapid, on-site, on-line, remote sensing, cost-effective, safe**. Especially in the modern environmental analysis these aspects play an outstanding role. This is also reflected by the increasing number of mobile measuring apparatus which are already available.

For laser-based analyses of solids there are different kinds of processes available due to the variety of interactions of laser light and solids (Omenetto 1998; Winefordner et al. 2000): Absorption, emission (fluorescence), ionization, ablation, desorption, vaporization, and photothermal processes. The various laser-based techniques for the analysis of solids can roughly be divided into two groups corresponding to the principle of detection.

The first group includes the optical spectroscopic methods [Table 4. 1]. The laser beam interacts with the examined object and the yielded characteristic light is analysed. Here, the laser with its unique characteristics of being a special light source can be used for the direct detection of substances. With modern detection systems, analyses are possible which could previously be done only with expensive laboratory equipment. This group is characterized by rapid in-situ and on-site measurements.

Table 4. 2. Analytical techniques based on non-optical detection methods

Method	Principle
Laser-based ion mobility spectrometry (IMS)	Measurement of the mobility of ions (K_0) in a gas flow
Resonance enhanced multi-photon ionization time-of-flight mass spectrometry (REMPI-TOFMS)	Ionization of molecules by multiple absorption of photons and mass-to-charge selective detection
Laser ablation induced coupled plasma mass spectrometry (LA-ICP-MS)	Mass spectrometry with an inductively coupled plasma as ionization source
Photoacoustics / Optoacoustics	Detection of acoustic waves generated by absorption of laser radiation via localized transient heating and expansion

In the second group [Table 4. 2], the high power densities which can easily be achieved with lasers mainly serve for a specific sample preparation. Especially for the investigation of solids, the laser power is used for ablation and desorption processes in order to convert the solid material which has to be investigated into the gas phase (Howe et al. 2001). The vaporized material can be analysed by different kinds of non-optical detection systems, such as the very sensitive mass spectrometry. In case of mass spectrometry, ionization of the gaseous materials is necessary which can be done by using laser radiation, too. Multi-photon ionization includes a further specific parameter for detection because of the selective excitation with a laser.

At the moment, a universal method does not exist. For example, some analytical techniques are only suitable for analysis of elements and not for compounds; some methods are destructive and some are non-destructive. The existing general difficulties of laser-based analyses of solids (e.g. matrix effects, multi component analysis, quantification) are also shown by the number of the various utilized methods. Therefore, only methods whose development has already achieved a high standard shall be presented in the following. Furthermore, promising techniques with a high potential for modern rapid analyses (see above) are briefly explained as well. Under these conditions, the usable techniques are primarily the optical ones: LIF, LIBS, and laser-based IMS (the latter two also in combination with LIF), Raman techniques, and optoacoustic methods. In addition, also techniques shall be presented, which are very accurate, capable for quantification and suitable for trace or ultra-trace analysis. These are especially the mass spectrometric techniques. References to more comprehensive articles and reviews are given at the end of this article (see "Other Techniques and Outlook").

4.2 Laser-Induced Fluorescence Spectroscopy

Fluorescence spectroscopy is one of the most used optical method for analysis due to its low demand on the system – also concerning solids. After the absorption of a photon, the molecule emits a characteristic fluorescence spectrum (typ. 10^{-9}–10^{-8} s). This allows – like a fingerprint – a differentiated statement about the fluorescing components. Another advantage are the numerous measurands which are accessible for analysis such as: fluorescence decay, excitation wavelength, anisotropy, polarization, quantum yield, Stokes' shift, etc.

This amount of further parameters for the detection of substances allows to compensate for the difficulties which may occur when applying fluorescence spectroscopy (e.g. matrix and quenching effects). Of course, there are substances which show an extremely low quantum yield or even no fluorescence at all due to other fast decay pathways (e.g. intersystem crossing). In this case other spectroscopic methods have to be applied in order to detect such substances (see laser-based IMS or LIBS respectively).

When using LIF spectroscopy, the extraordinary characteristics of lasers are specifically utilized. So the high spectral intensity of the laser radiation allows a selective excitation of the sample. By using tunable lasers (dye laser, optical parametric oscillator, or diode laser respectively), one is able to record the so-called excitation emission matrix spectra (EEM) and thus the excitation wavelength as an additional parameter. Furthermore, the short laser pulses (10^{-14}–10^{-9} s) permit the determination of the fluorescence lifetimes of excited states. When performing the so-called wavelength time matrix spectrometry (WTM, see Fig. 4.1), one has an additional measurand – the decay time – besides the fluorescence spectra.

The main strengths of LIF are simplicity, speed, sensitivity, non-destructiveness, and the ability to analyse both organic and inorganic materials. The LIF technique as such a simple but very efficient tool for rapid analysis shall be explained in more detail looking at the detection of polycyclic aromatic hydrocarbons (PAHs) in soils and the detection of a wood preservative as examples.

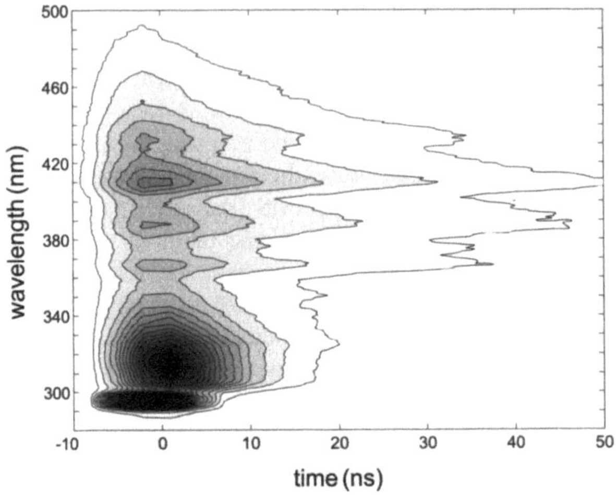

Fig. 4.1. Time-resolved fluorescence spectrum of 16 PAHs (EPA-610 standard[1], reprint with permission from Karlitschek et al. 1998)

4.2.1 Polycyclic Aromatic Hydrocarbons and Mineral Oils in Soils

LIF-based detection systems are widely applied for the analysis of soils which is proven by the dedicated literature. Besides heavy metals (which are detectable by plasma spectroscopy, see LIBS), there are mainly PAHs present in soils. PAHs are a part of many mineral oils and are regarded (because of their carcinogenic effect) as the substances in contaminated soils from which the greatest danger originates (Harvey 1991). As the present energy supply is mainly based on mineral oils, numerous areas are polluted with these substances (e.g. the soils around petrol stations) in such a way that they have to be cleaned-up. The consequence is the increasing demand for suitable fast analysis methods for screening the soils in order to be able to make clean-up decisions when starting remediation works. The demands for such an in-situ analysis are fulfilled ideally by the LIF technique. For some years, commercial LIF systems are available (e.g.: LLG, Göttingen, Germany; LLA Instruments GmbH, Berlin, Germany; Fugro Geosciences, Inc., Rookin, Texas, USA) which are able to do a reliable work in this matter. In numerous research groups, the scientific preconditions were evolved hereunto. It

[1] U.S. Environmental Protection Agency, United States of America

could be shown that PAHs, which have good up to very good fluorescence characteristics, could be detected by static as well as by time-resolved measurements in the required range: The limit of detection (LOD) for petroleum products in soil is about 100 ppm, whereas the concentration for a surveillance is about 5000 ppm (Löhmannsröben et al. 2000).

For the differentiation between the various PAHs molecules, the above mentioned methods WTM and EEM can be utilized. So, in the course of an examination (Karlitschek et al. 1998) with a portable system, it was possible to distinguish between the PAHs and mono-aromatics, such as benzene, toluene, xylene, ethylbenzene (BTXE). The system consisted of a battery-powered, frequency tripled (355 nm) and quadrupled (266 nm) Nd:YAG laser which is pumped by a diode laser. In order to record time-resolved spectra, it is equipped with an intensified CCD camera. The results of this work showed that the wavelength of 266 nm is well suited for the detection of BTXE and of small PAHs. The 355 nm radiation is preferably applied for larger PAHs. A multi-component analysis of PAH could be realized by using time-resolved measurements and a chemometrical approach.

The EEM spectroscopy is another multi-dimensional option to identify the different PAHs in soils (Hart et al. 1997). The presented system was capable of producing multiple excitation wavelengths from 250 nm up to 400 nm. This was performed by using a hydrogen-methane Raman shifter which was pumped by the 4th harmonic of a Nd:YAG laser. This system was placed in a vehicle and demonstrated successfully an area-wide screening at a contaminated air force base.

A problem is the exact quantification of the pollutants when investigating soils by LIF and thus the need for external calibration standards of the investigated object (of course, this also applies for most of the other methods in this article). The essential calibration is rather due to the lack of a standard. The properties of LIF considering a quantified determination of PAHs were examined in detail (Löhmannsröben and Roch 2000). It could be shown that after an appropriate calibration, the obtained results are quite similar to the ones attained with mobile gas chromatography/mass spectroscopy (GC/MS) and IR analysis after the extraction of the soil samples. Furthermore, optical properties of soil are taken into consideration in order to improve and to simplify the calibration process (Löhmannsröben and Schober 1999). For this purpose, diffuse reflectance spectroscopy was utilized.

Fibre-optic sensor heads play an outstanding role for the applicability of a mobile device. Of course, this applies to LIF spectroscopy as well as to other optical techniques. In the course of LIF-based trace detection of pollutants in soil, the different parameters of the fibre-optic sensor head (geometry, material) were examined by using a mathematical model which was implemented by a computer program (Bünting et al. 1999).

4.2.2 DDT on Wood

The increasing importance of the recycling of used materials is very often assimilated by the legislation of the appropriate countries. For example, a law which is valid in Germany (law for recycling management and waste material) shall prevent that raw materials are wasted and it shall also restrict the amount of waste. First of all, the used materials shall be substantially recycled and re-used. If this is not reasonable or not possible due to a high concentration of pollutants, the polluted materials have to be energetically disposed, e.g. in a combined heat and power plant. Last, if the amount of pollutants is too high (according to the legal threshold values), the used materials have to be depolluted as hazardous waste. In this connection, especially wood is an ideal substance for being recycled. Moreover, the amount of used wood is an important economical factor. The current analytical laboratory methods which are used at recycling factories so far are not suitable from an economical point of view because of their expensive, time-consuming, and spot testing characteristics.

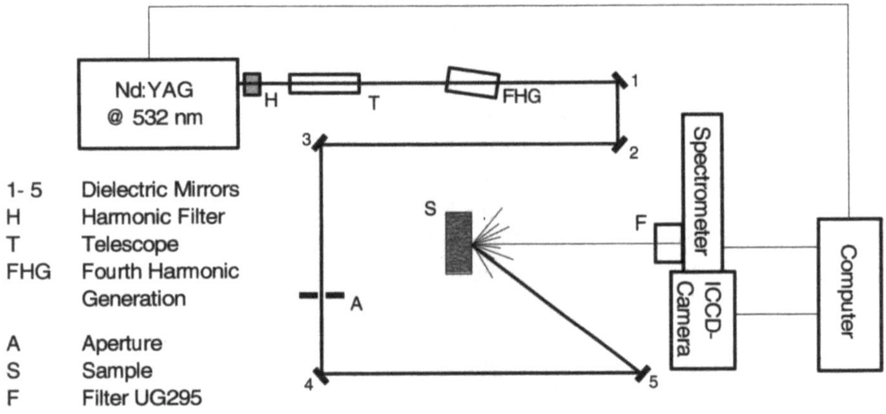

Fig. 4.2. LIF-based experimental set-up

The insecticide DDT and the wood preservative Hylotox (effective components 3.5% DDT, 0.5% Lindan) were examined and detected by LIF (Paetzold and Anders 2001). Although the use of DDT is prohibited in many countries, it can still be detected everywhere in the world due to its persistence (even in the Antarctic ice). The wood preservative Hylotox 59 was extensively used in the former German Democratic Republic until the year 1989. Thus, high DDT concentrations of up to 1000 µg/g can still be found in roof trusses for example. Wood which is polluted to such a high extent can be detected by using rapid analysis systems and then disposed on a waste site.

The experimental set-up (Paetzold and Anders 2001) consisted of a Nd:YAG laser as excitation source and a spectrometer as well as a gated intensified CCD camera (ICCD) as detection unit (Fig. 4.2). This apparatus which represents a typical experimental set-up for LIF was centrally controlled by a computer. If a fixed frequency laser is used, the excitation wavelength must match with the absorption behaviour of the sample which is examined. Therefore, a frequency quadrupled Nd:YAG laser with an output beam at 266 nm was used in these experiments for stimulating the DDT fluorescence. It has to be taken into consideration that the fluorophores are decomposed when stimulated with such highly energetic photons. For a mobile on-site analysis, the system can be advanced to a portable system if using a diode-pumped Nd:YAG laser and fibre optical components.

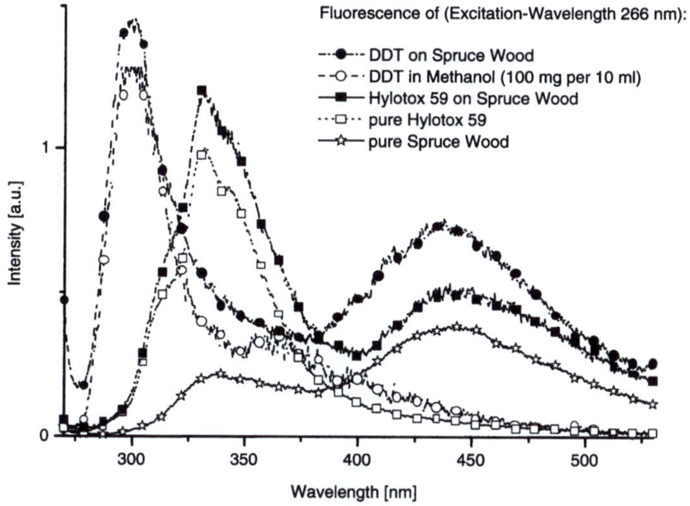

Fig. 4.3. Fluorescence spectra of spruce and wood preservatives; excitation wavelength: 266 nm

When detecting pollutants on surfaces, the influence of the background matrix on the fluorescence signal has to be determined first of all. Thus, the fluorescence spectra of various kinds of unpolluted wood were determined. Also, the influence of damp wood – as it occurs at outdoor places in every recycling firm – was examined. The spectra partially show relatively strong deviations. Hylotox 59 and a DDT-methanol solution (100 mg DDT dissolved in 10 ml Methanol) were painted on the wood samples in normally applied concentrations without any dilution (DDT: 0.04–0.4 mg/cm^2, Hylotox 59: 0.1–1 mg/cm^2).

Comparing the spectra of the pure substances (DDT in Methanol, Hylotox 59) and the untreated wood with the spectra of the contaminated wood samples, it is noticeable that the latter result from a superposition of the single spectra (Fig. 4.3). Whereas DDT in methanol and also on spruce shows a distinct maximum at 296 nm, pure Hylotox 59 as well as Hylotox 59 on spruce has its maximum towards longer wavelengths at about 333 nm. Therefore, it is easy to differentiate the spectra and assign them to the investigated substances. Thus the pollutants DDT and Hylotox 59 respectively are easy to determine because of their emission maxima which lies at lower wavelength when comparing with the spectra of wood samples.

Another parameter for the analysis of the wood preservatives was determined: the decay time of the fluorescence by means of time resolved measurements (Fig. 4.4). The very short exposure time of the gated CCD camera (5 ns) was used for time resolution (boxcar method for taking the temporally shifted spectra). A cut along the time axis at a given wavelength shows the time course of the fluorescence curve. Of course, the measured curves still had to be deconvoluted (Lakowicz 1999) because the laser pulse was in the range of the fluorescence life time. The fluorescence of spruce has a fast decay time (below the time resolution of the used system of 1 ns), whereas for pure Hylotox 59 and for Hylotox on spruce wood distinctly prolonged decay times (approx. 5 ns) were determined, which – in addition to the emission spectra – makes an identification easier.

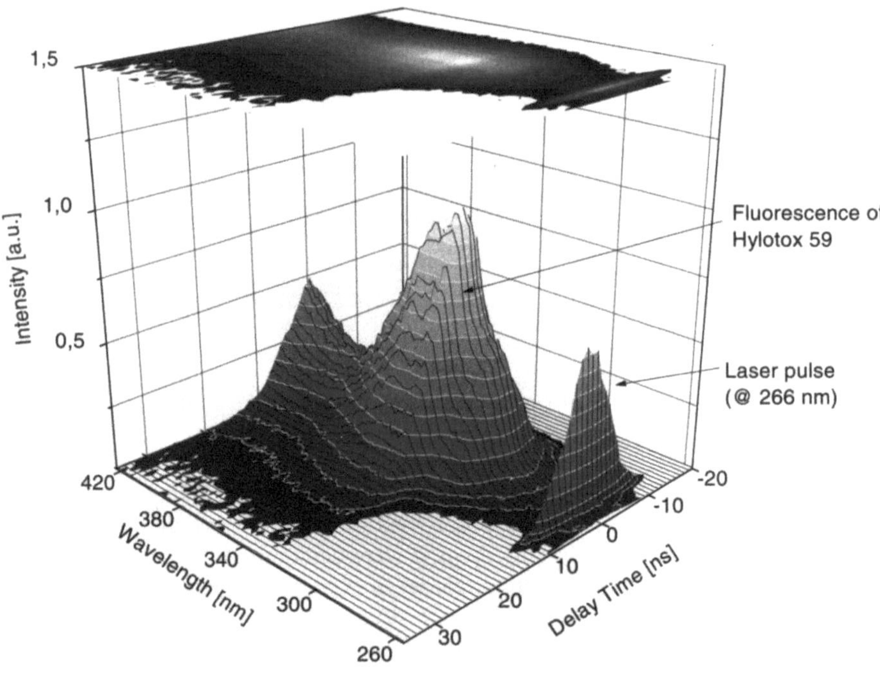

Fig. 4.4. Time resolved LIF spectrum of Hylotox 59 on spruce; excitation wavelength: 266 nm

For an additional analysis of the static and dynamic fluorescence spectra, various computer methods can be applied. By using databases, various reference spectra with variable parameters (wood type, dampness, varnish and so on) can be saved centrally and called later on for evaluation. Moreover, neural networks can be used profitably for an interpretation of the recorded spectra (Hennig et al. 2001).

As shown, LIF is a suitable method for the detection of DDT or Hylotox 59 respectively. For other protective agents such as Pentachlorophenol (PCP), Lindan or inorganic protective varnishes that do not fluoresce, further laser-based rapid analytical methods are available. Numerous inorganic salts can be determined by using laser-induced breakdown spectroscopy (see LIBS). Organic wood preservatives such as PCP or Lindan can be detected performing the laser-based ion mobility spectroscopy or with the photoacoustic technique (Beck et al. 2000).

4.2.3 Pesticides on Leaves

In order to protect the consumer from polluted imported products, systematic controls of fruits and vegetables shall be carried out, e.g. at the borders of the EU. The potential of LIF spectroscopy as a fast tool for the detection of pesticides directly on plants was tested (Jantos et al. 2000). For this purpose, different kinds of pesticides (two herbicides, two fungicides and one insecticide) were sprayed on the leaves of lettuce (*Lactuca sativa var. Capitata*) in ready-made concentrations (0.5 g/l). On this matrix, the fluorescence spectra as well as the reflectance spectra were measured by using a mobile spectrometer system. This device consisted of a nitrogen laser (which could pump a dye laser), a tungsten lamp, a sensor head coupled with fibre optics, a spectrometer and a CCD camera (Hennig et al. 2000). All components were chosen for maximal applicability in field analysis and with regard to cost-effectiveness.

The fluorescence of the five pesticides was excited by the nitrogen laser (337 nm) or by the dye laser respectively. The influence of pesticides and environmental parameters on the chlorophyll fluorescence was determined. Although two pesticides showed a specific fluorescence spectrum when dissolved in water, the direct detection on the lettuce was not possible because of the strong background fluorescence. Nevertheless, it was possible to detect the pesticides by measuring the fluorescence signal of chlorophyll that was altered due to the influence of the pesticides.

4.2.4 Rock Identification

A LIF-based system for the identification of different rocks is commercially available (LIF GmbH, Wedel, Germany). It is based on a Nd:YAG laser which emits the 3rd and 4th harmonics. The cost-effective solution for the detection unit consists of photomultiplier tubes (PMT) and spectral filters. The variable compositions of rocks cause different fluorescence signatures. Therefore, an identification and a rapid sorting (e.g. on a belt conveyor) is possible.

4.2.5 Conclusion

The laser-induced fluorescence spectroscopy is a well suited tool for modern rapid analysis based on optical measurements. In numerous research groups, the pros and cons were investigated extensively. The main problems of LIF are the limitation of quantified statements, fluorescence quenching effects, partially low quantum yields. Nevertheless, the commercial availability proves its suitability as an analytical tool.

4.3 Laser-Induced Breakdown Spectroscopy

The laser-induced breakdown spectroscopy (LIBS), also known as LIPS (laser-induced plasma spectroscopy) or LPS (laser-based plasma spectroscopy), LAS (laser ablation spectroscopy), LA-OES (laser ablation optical emission spectroscopy) or LSS (laser spark spectroscopy) is a method where an intense laser pulse (less than 10^{-8} s) is properly focused on the target (Schechter 1997). By irradiating solids with such high power densities (10^9–10^{12} GW/cm^2) a dielectric breakdown can be caused and a plasma is generated. The temperature of the plasma is high enough (10.000 K) to fragment all molecules into their constituent atoms. Therefore, this method is not suitable for compounds and the best that is possible is a statement about the ratios of the elements (Tran et al. 2001). Some ns after the plasma generation, the ions and free electrons recombine. The different electronic states which are then occupied (highly excited atoms) generate an atomic emission spectrum which is characteristic for the respective atom. Although LIBS is a destructive method, the ablated amount is so small that it is even used in art and archaeology examinations (Anglos 2001).

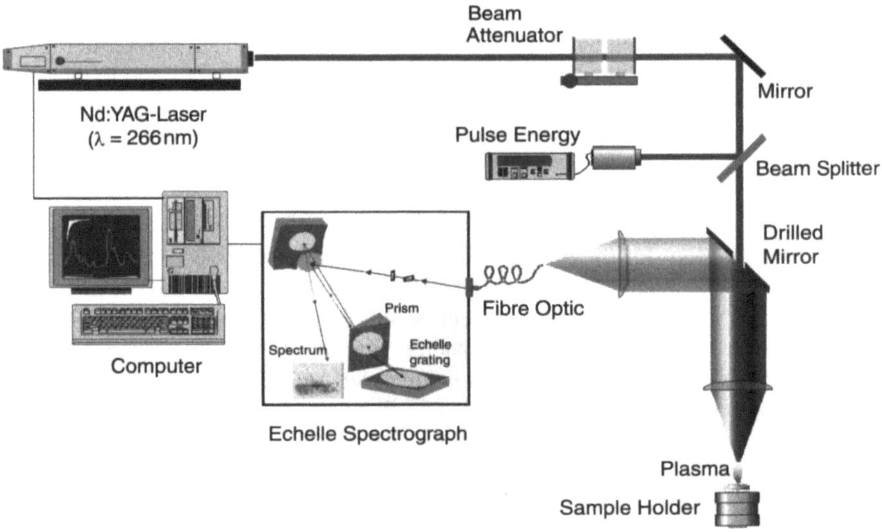

Fig. 4.5. Typical LIBS set-up (reprint with permission from Fink et al. 2001)

The common set-up of a laser-induced breakdown spectrometer consists of a pulsed laser system, optical components for beam guiding, an optical detection system consisting of a spectrometer and a time-resolved detection device (Fig. 4.5). In addition, components for the exact control of the measuring process are needed.

Basically, LIBS includes all potentials of a modern laser-based rapid in-situ analysis: With one single laser pulse and no time-consuming sample preparation (e.g. extraction), the whole sample information can be obtained. A battery powered LIBS system has been described for on-site analyses (Castle et al. 1998). But most of the optical detection units in commercial LIBS systems are based on an Echelle spectrograph (high spectral resolution: $\lambda/\Delta\lambda=10000$; spectrum range: 200–780 nm) and a gateable ICCD camera in order to reduce the background continuum emission that covers the atomic emission lines in the first μs (Bauer et al. 1998). These detection components are an essential requirement for fast multi-element measurements. In addition, the systems are suitable for remote-sensing. Thus, this technique is commonly used in environmental analysis for the detection of elements. By using their emission lines, a multi-element analysis of more than 70 elements is possible.

Although the influence of the matrix on the results is lower in comparison with LIF, matrix effects have to be taken into consideration when applying LIBS. For an essential understanding of this matter, the matrix effects were studied by shock wave propagation (Krasniker et al. 2001) as well as by multi-fibre spatial and temporal resolutions (Bulatov et al. 1998). Plasma conditions and different parameters (focal length, wavelength) were examined by spectroscopic imaging of the plasma (Bulatov et al. 1996; Stratis et al. 2001).

The LOD which is achievable by this technique lies in the range of some ppm. This is mostly sufficient for field applications. However, the absolute sensitivity of this method is high – in the ng range (Wisbrun et al. 1994; Omenetto et al. 1996). Utilizing the inductively coupled plasma technique (see LA-ICP-OES), it is possible to increase the LOD up to 3 orders of magnitude. For instance, fundamentals and applications are described in (Rusak et al. 1997).

4.3.1 Recycled Thermoplastics from Consumer Electronics Waste

In a recent study, LIBS was used for the determination of different kinds of thermoplastics which accrue by recycling consumer electronic devices (Fink et al. 2001). In this area, thermoplastics are often provided with different kinds of additives such as Cd, Pb, Cr, Sb, Ti, Sn, Zn, Ba, Si (e.g. for colouring, for flame resistance, for protection against heat and UV-radiation). With the growing number of electronic devices and the extended use of plastics, the waste is also increasing. The same statements – as mentioned in the case of wood recycling – can be made in favour of re-using these resources. A fast on-line control for the characterization and separation of the materials is necessary when working economically.

The emphasis of the investigations was laid on the applicability for an industrial environment. Therefore, the experimental set-up was designed for a maximum of flexibility. For plasma generation, a frequency quadrupled Nd:YAG laser (266 nm) was used, because of the better plasma conditions when irradiating with such wavelengths, see (Fink et al. 2001) and cited therein. For the time-resolved recording of the spectra, two different detection systems were applied. One detection unit consisted of a conventional Czerny-Turner spectrometer and an intensified CCD camera. An Echelle spectrograph was utilized for the other detection unit. Both systems have excellent spectral resolution, but the latter is also capable of recording a whole spectrum (200–780 nm) at once.

The investigated, non-prepared samples consisted of typical kinds of thermoplasts (Fig. 4.6). In order to provide a maximum of reliability, different reference and normalization techniques were applied. The LOD for the investigated heavy metals was in the range of 30 ppm (Cd) up to 750 ppm (Sb).

4.3.2 Inorganic Wood Preservatives

The LIBS technique was also successfully applied to the analysis of inorganic wood preservatives (Löbe et al. 2000) and a commercial mobile device is available (LLA Instruments GmbH, Berlin, Germany) which is also applicable to many other solids. The detection of unusual constituents in a wood sample is rather easy when using this method because the different kinds of uncontaminated wood consist of nearly the same elemental composition. Therefore, it is quite simple to detect critical substances such as heavy metals.

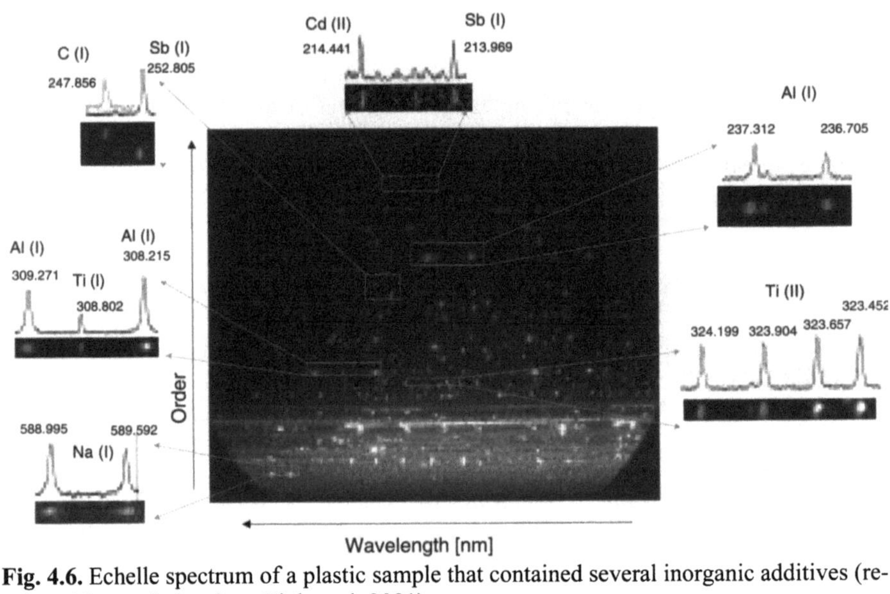

Fig. 4.6. Echelle spectrum of a plastic sample that contained several inorganic additives (reprint with permission from Fink et al. 2001)

The principle set-up of the available system is basically similar to the above described. The fundamental of a Q-switched Nd:YAG laser (1064 nm) is used for plasma generation. By using optical components and fibre optics, the atomic emission light is guided to the optical analyzer that is based on an Echelle spectrometer and an ICCD camera. The whole system is built up in a modular concept in order to guarantee easy handling under field conditions. All components are especially made to suit the conditions of an on-site analysis: protected against shock, dust, and humidity.

The emission light of the plasma is recorded in backward direction, which improves the reproducibility of the analysis. The light is coupled into a fibre and lead to the detection system. Comparing the recorded spectra with an internal reference database makes it possible to identify utilized wood preservatives.

By using LIBS, depth-resolved profiles (some 100 µm) of the contaminated wood samples are also possible and the distribution of the active component can be analysed. With each laser pulse, a small amount of the material is ablated (some µm thickness). Depending on the depth, the line intensities of a specific atomic emission line was measured. For example, the measurements proved that the wood preservatives of vacuum impregnated kinds of wood penetrate rather deeply and are evenly distributed in the wood.

For laser guiding, a sensor head based on fibre optics was also investigated. However, in order to transport the necessary high laser intensities for plasma generation via a fibre optic, a fibre diameter (> 1000 µm) is needed which limits the curvature radius of the fibre. With such a high diameter of the fibre, radii under 1000 mm are not achievable at the present time and this reduces the applicability.

For a reliable quantitative determination of the pollutants in a sample, the system has to be calibrated with known standard samples. The reabsorption process is a problem which arises when investigating rather high concentrations. Thus, the linearity and the working range are limited. The achievable limits of detection (LOD) for most elements are lower than 10 ppm. This is sufficient for detecting the threshold values of heavy metals at which the wood is classified as being unpolluted. An exception is Hg, where the detectable LOD is 4.6 ppm (threshold value: 0.3 ppm).

The applicability of the system could be demonstrated under field conditions at a recycling firm. The inorganic substances in the wood could be detected reliably. The wood samples were mainly contaminated with Cr, Cu, and B. Hg and As could only be found in a few samples. An interesting fact is the ostensibly increased concentration of the pollutants which is found when applying the LIBS method. Different volumes of the examined samples are the reason for this. As only the upper layers (some µm) are examined when applying the in-situ analysis with LIBS, 3 mm thick pieces of the wood are taken and analysed by using standard laboratory techniques. As, however, the active ingredients do not penetrate deeply into the wood and due to the greater sample volume, the standard values are lower.

These experiments demonstrated that the laser-induced breakdown spectroscopy is applicable for the fast analysis of inorganic pollutants in the matrix wood. The inorganic salts for wood protection can be detected within the framework of the usual required threshold values (2–5 ppm).

4.3.3 Transformation of Spatial in Pseudo-Temporal Resolution

The main drawback of the commercially available LIBS systems is the relatively complex and expensive set-up. The commonly used optical detector, a gateable ICCD camera for the essential temporal gating, is too expensive and has to be cooled and protected against condensation. By using several fibre optics, which are placed around the plasma in different positions and heights from the surface, the plasma propagation is measured spatially. It is possible to convert the spatial resolution into a pseudo-temporal resolution (Bulatov 2000). Thus, the demand on the experimental equipment for the analysis is reduced to a minimum: only an ordinary CCD camera and a typical imaging spectrometer are needed.

4.3.4 Multi-Element Detection of Pollutants in Soil

The heavy metal content in soils is often determined by means of atomic absorption spectroscopy (AAS) or inductively coupled plasma spectroscopy (ICP). However, these methods are not suitable for taking a great amount of measuring points on-site in order to assess the need for remediation of residual waste. Therefore, the LIBS-based systems are the obvious method for such a screening of contaminated soils.

The basics for the in-situ analysis of soil samples by using laser-based methods were tested and the influence of various measuring parameters was examined by (Hilbk-Kortenbruck et al. 2001). The laboratory set-up slightly varied from the above described, commercial LIBS devices. The fundamental wavelength of a Nd:YAG laser (1064 nm) was used in order to g the plasma. For the strived multi-element analysis, a detection system is needed which records a broad spectrum at once and which also has a high spectral resolution. For this purpose, the utilized system consisted of a special spectrometer equipped with photomultiplier tubes (PMT). The essential time resolution was realized by using a fast signal electronic.

Concerning these LIBS experiments, also reference samples were used for the calibration of the system. The limits of detection (LOD) which were achieved in the experiments with As, Cr, Cu, Ni, Pb, and Zn (some ppm) were below the threshold values of unpolluted soils. As the threshold values of Cd, Hg, and Tl are lower than 1 ppm, the LIBS technique is not sufficient to detect these elements in such low concentrations at the moment.

4.3.5 Combination Technique: LIBS-LIF

In order to achieve a remarkable improvement concerning the LOD of Cd and Tl, a combination technique of LIBS and LIF (Fig. 4.7) was tested by (Hilbk-Kortenbruck et al. 2001). In these experiments, a second tunable laser was necessary for the specific excitation of the atoms in the plasma. The wavelength of the excitation beam was tuned to an electronic transition of Tl or Cd respectively. The beam was focussed into the plasma after a short delay time of 50 µs. Thus, only those atoms were excited which should be detected. These atoms fluoresce at distinct wavelengths which were recorded by the above described LIBS detection unit for multi-element detection of pollutants in soil.

These examinations were carried out with a dye laser pumped by the 2nd harmonic of a Nd:YAG laser (532 nm). The radiation of the dye laser was also frequency-doubled and mixed again with the wavelength of 532 nm. Thereby, it was possible to attain the necessary wavelength for the excitation of Tl (228.80 nm) and Cd (276.78 nm). The LODs which could be achieved in these experiments (0.25 or 0.4 ppm respectively) are below the desired value of 1 ppm.

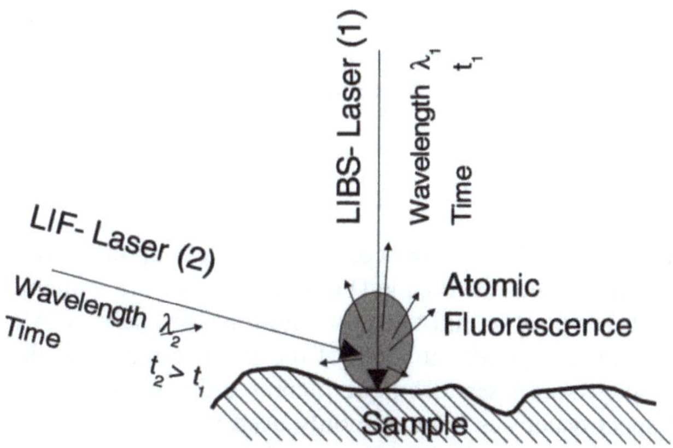

Fig. 4.7. Combination technique of laser-induced breakdown spectroscopy (LIBS) and laser induced fluorescence (LIF)

4.3.6 Conclusion

In regard to performing a rapid screening on-site, LIBS is the method of choice if a multi-element analysis (e.g. the detection of heavy metals) is sufficient and no compounds have to be detected. The apparatus available at the present time, however, are not optimized yet concerning cost-effectiveness (Echelle spectrograph, ICCD camera). Furthermore, special knowledge is necessary concerning the interpretation of the spectra, although the available data banks make this relatively easy. In contrast to LIF analysis where the samples can also be examined by using fibre optics based sensor heads (and thus increase the handling remarkably), the necessary high intensities cannot be transmitted through fibres easily. Therefore, the commercial LIBS instruments are not based on fibre optics nowadays.

4.4 Vibrational Spectroscopy

Vibrational spectroscopy is an extremely selective optical technique which is mostly used for structure analysis. Hereby, vibrational transitions in a molecule are excited. The more complex a molecule is built up, the more degrees of freedom it possesses. Additionally, the unique vibrational states of a molecule appear due to different binding forces as well as variable atomic masses. The detection of these transitions is like a fingerprint of the examined molecule. The energy which is necessary for the excitation of such a vibrational transition is in the mid infrared (MIR) range from 2.5 up to 50 μm. The excitation of the harmonics of the fundamentals is possible by irradiating the sample with light in the near infrared (NIR, 0.8–2.5 μm) range. In order to perform vibrational spectroscopy, two different methods can be used: IR spectroscopy and Raman spectroscopy (RS). Whereas the former is an absorption measurement and needs a polychromatic light source, RS is a scattering process with a very low quantum yield. Therefore, lasers with their high power densities are essentially needed for RS.

4.4.1 Infrared Spectroscopy

For some decades, the IR spectroscopy is a widely-applied analytical method which has been improved meanwhile (Fourier transform IR, tunable diode lasers) and developed to a versatile and very specific tool for analysis. Although a polychromatic light source operates in most of the IR spectrometers, the recent development in tunable diode laser technique increases the application of lasers in this field, too. IR spectroscopy is not commonly used for trace detection. Instead, it is generally applied in structure analysis and qualitative measurements. When performing IR spectroscopy for analyses of solids, the sample has to be prepared. This reduces the applicability as a rapid in-situ instrument. Nevertheless, IR spectroscopy is also used in various kinds of environmental analysis, e.g. for the determination of humic acid and flavic acid in soil as well as for oils and greases in soils and sediments. An overview is given in (Visser 1999).

4.4.2 Raman Spectroscopy

Instead of using a scanning excitation source, vibrational spectroscopy can also be performed by using a fixed wavelength: Raman scattering (RS). As the name indicates, this method is quite different to IR spectroscopy. Often, the RS is described to be the complement to IR due to the different vibrational excitation conditions. Some vibrations which are IR-active could not be detected when performing Raman scattering and vice versa. RS is a generic term for different kinds of special Raman scattering techniques: stimulated RS, surface-enhanced RS, coherent anti-Stokes RS, etc.

The Raman signal is very weak due to the inherently small cross-section. Often the weak Raman bands are covered by fluorescence. Therefore, Raman spectroscopy benefits from the high photon flux of lasers, especially when excited in the NIR in order to reduce the fluorescence yield. It has to be taken into consideration that the scattering process also depends on the wavelength of the incident radiation frequency (v^4). Thus, the choice of the laser frequency is a compromise between fluorescence reduction and intensity of the Raman signal (Schrader 1996).

Besides the usual gaseous and liquid samples, RS is also utilized for the analysis of solids. In a recent work (Lavine et al. 2001) RS was applied in combination with modern computational pattern recognition. By using genetic algorithms and the Raman bands of the scattering process, it was possible to distinguish different wood types.

For mineralogists RS is a commonly used technique. So, it was obvious to build up a RS-based device for the mineral characterization on planetary surfaces (Wang et al. 1998).

4.5 Laser Ablation, Laser Desorption, Laser Ionization

Besides the interactions of photons with the examined object and the detection of specific kinds of spectra afterwards, lasers still have other tasks when examining solids: ablation, desorption, vaporization, and ionization. As already described in the case of LIBS, these techniques are especially utilized for a certain kind of sample preparation (do not mix up with time-consuming chemical extraction methods) in order to customize the solid samples for detection. Principally, this is an ion-selective detection of vaporized material. In this connection, the laser takes advantage of its high spectral intensities and power densities (Howe et al. 2001; Bolshov and Kuritsyn 2001). In the following, a selection of these laser-based techniques shall be presented. A comprehensive overview are given in the following recent articles from (Evans et al. 2001) and (Winefordner et al. 2000); see also (Miller and Haglund 1998).

4.6 Mass Spectrometry

The measurand for mass spectrometry (MS) is the mass-to-charge ratio of an ion. A typical mass spectrometer consists of three units: an ion source, a mass-selective analyzer, and an ion detector. In this connection, the laser could serve for the vaporization process as well as for the essential ionization.

Several different techniques exist for the mass selective analyzer. Each of them has its distinct advantages. In MS the time-of-flight mass spectrometer is mostly used; sometimes as reflectron type in order to improve the resolution. Since MS is a common technique in analytic examinations, it will not be discussed here in detail; it shall be referred to the numerous literature (e.g. Linscheid 2001). The laser-based MS is mainly used in the laboratory. Basically, no sample preparation is necessary when using this method, but a vacuum is needed for the measuring process. So, this is not a typical in-situ technique and the necessary measuring apparatus is rather expensive. Therefore, only a few applications for analysing solids shall be briefly introduced and it shall be referred to detailed review articles concerning this subject (Winefordner et al. 2000; Bacon et al. 2001).

4.7 Laser Ablation Inductively Coupled Plasma Mass Spectrometry

Laser ablation inductively coupled plasma mass spectrometry (LA-ICP-MS) is a technique used for trace analysis of elements in solid samples. With this method, it is possible to determine many elements in the periodic table to high degrees of accuracy and precision. Due to the good LOD (ppb) as well as the extended working range (10^5–10^7), this technique is often used as a tool for earth and environmental science applications in the last years and it has reached a "routine" status meanwhile. Nevertheless, the systems are still limited for commercial fast in-situ measurements due to the rather costly and complex set-up. A detailed overview about the recent research projects is given in (Bacon et al. 2001).

4.8 Laser Ablation Inductively Coupled Plasma Optical Emission Spectrometry

Laser ablation inductively coupled plasma optical emission spectrometry (LA-ICP-OES) is an improvement of the above described laser-induced breakdown spectroscopy whereby an increase of the LOD and of the precision as well as a decrease of the matrix effects are feasible. When performing LA-ICP-OES, the ablation and the plasma generation process are separated: Lasers are only used for the ablation process – in contrast to LIBS, where the laser also generates the plasma that emits the atomic lines.

After the ablation, the particles are carried in a flow of an inert gas to an inductively coupled plasma, where they are ionized prior to the measurement in a spectrometer. In comparison to other techniques, which are deployed for the ablation process, the laser ablation has some essential advantages. In principle, every solid can be vaporized independent of its characteristics (shape, size, surface topography or electrical characteristics). The LA-ICP-OES is in competition with LA-ICP-MS and the above described LIBS method. Which method is appropriate depends on the special aim of the analysis. In different review articles this matter is discussed in detail (Marshall et al. 1998; Rommers and Boumanns 1996).

4.9 Resonance Enhanced Multi-Photon Ionization Time-of-Flight Mass Spectrometry

Whereas both LA-ICP methods are only capable to carry out atomic spectrometry, the resonance enhanced multi-photon time-of-flight mass spectrometry (REMPI-TOFMS) is able to analyse compounds. It represents a combination of two analysis methods: UV spectroscopy and mass spectrometry. Thus, a very selective and fast detection method is available with a low LOD (ppb) as well as with the applicability for many compounds (Weickhardt et al. 1998; Heger et al. 1999). In the case of solids, lasers serve for the desorption process as well as for the ionization process. The very selective excitation of the investigated substance can be carried out by choosing an excitation wavelength which fits exactly to an electronic transition or even to a vibrational state. From this state, further absorption of photon(s) leads to the ionized molecules that can be analyzed using mass spectrometers, whereas time-of-flight mass spectrometers are the fastest way to perform MS. Beside the capability to measure 100 mass spectra per second, the achievable mass resolution is very good.

In the course of fast laser-based soil analyses on-site, laser desorption and REMPI-TOFMS were examined (Boesl et al. 2000). In contrast to other research groups, laser desorption was combined with a pulsed gas stream that transported the desorbed molecules to the ionization laser beam. Thus, the selectivity of the resonant ionization process could be increased significantly (Fig. 4.8).

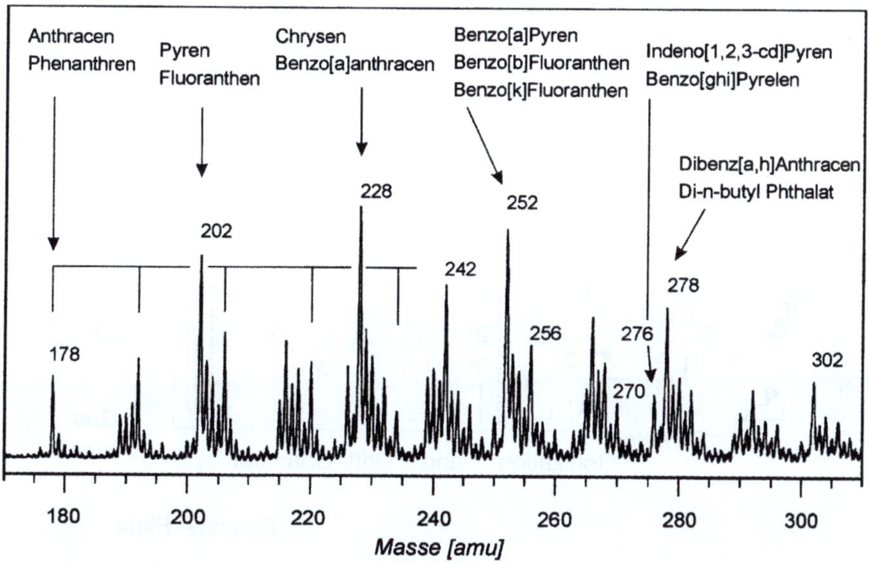

Fig. 4.8. Laser desorption/ionization mass spectrum of a soil sample with certified PAHs contamination. Measurement time: 5 minutes, LOD: better than 100 ppb (reprinted with permission from Boesl et al. 2000)

4.10 Laser-based Ion Mobility Spectrometry

Ion mobility spectrometry (IMS) is a detection method whose principle is known for more than 100 years. Nevertheless, IMS is a rather new analytical method for environmental analysis of solids (Stach 1997). The principle of the IMS is the measurement of the mobility of ions in a flow of a constant carrier gas accelerated by a constant electric field (Fig. 4.9). The various drift times and intensities allow the identification of the composition of the sample. When performing laser-based IMS, the tasks of lasers are desorption as well as ionization processes – similar to mass spectrometry.

Due to its sensitivity, IMS itself is a technique which has been used for detecting drugs, explosives and chemical warfare agents for a long time. IMS is a detection method which is mainly applied for volatile organic compounds. Less volatile compounds have to be firstly detached by a desorption process, which can be done – amongst other methods (e.g. thermal desorption) – by performing laser desorption. Compared with mass spectrometric techniques, IMS is capable of measuring at atmospheric pressure. Beyond this, small, transportable systems are available. All these properties indicate the great potential of IMS for mobile on-site analysis even of organic compounds. Only the often used radioactive NI_{63} ionization source in transportable systems limits a really user-friendly application.

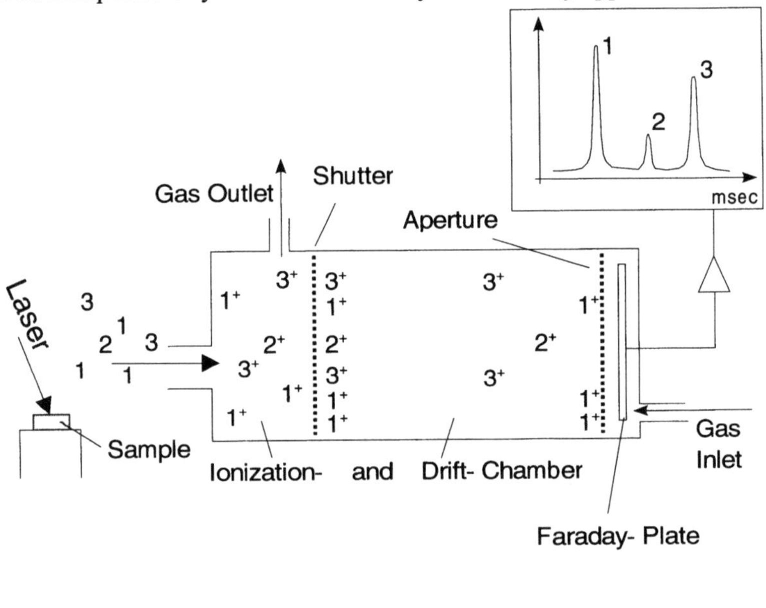

Fig. 4.9. Basic construction of a laser-based IMS

4.10.1 Organic Wood Preservatives

The detection of wood preservatives is one example where the sensitive ion mobility spectrometry was carried out in combination with laser desorption (Schumann 2000). Contrarily to LIF, it is possible to detect the often used non-fluorescent wood preservative PCP. As the wood preservatives which are of interest (especially PCP) are solid at room temperature and also have a low vapour pressure, they must be transferred into the gas phase without being decomposed before the actual analysis. For this purpose laser desorption was utilized. Often, it can be noticed that laser desorption is faster than thermodesorption, more complete, and without much fragmentation of the substances which are analysed. PCP was detected by using laser desorption in connection with a commercially available ion mobility spectrometer which was equipped with a radioactive NI_{63} ionization source.

The numerous results of laboratory and in-situ measurements proved that it is possible to detect PCP within the required threshold values (2–5 ppm). The examinations of splinter samples showed that the proof is very reliable due to three characteristic PCP peaks at relatively high polluted wood and thus the possibility of an incorrectly positive statement is very low. The analysis generally takes less than one minute, but longer rinsing can be necessary if a high sample concentration gets into the IMS.

4.10.2 Polycyclic Aromatic Hydrocarbons in Soils

The combination of laser desorption and laser ionization increases the number of detectable substances and also the selectivity of the IMS technique. As mentioned above, laser desorption can be done without fragmentation when using the "right" wavelength. This can especially be applied to the PAHs with their high molecular masses. Therefore, the capability of laser-based IMS was examined for the detection of PAHs in soils (Roch and Baumbach 1998). The results revealed that laser-based IMS is able to measure LODs in the sub-ppm range for PAHs and in the ppm range for petroleum products.

For an essential understanding of the ion-molecule interactions in the gas phase, this theoretical matter was also examined by laser-based IMS (Illenseer and Löhmannsröben 2001). These interactions were studied by analysing PAHs and petroleum products in soil, too. Desorption and ionization were carried out by the same frequency quadrupled Nd:YAG laser. Due to some limitations (saturation effects and measurement deviations), more basic research has to be done in this field. For a more reliable detection of PAHs and petroleum products in soil, a combination of LIMS and LIF spectroscopy will be presented (Illenseer et al. 2001).

4.11 Photo- and Optoacoustic Spectroscopy

Photoacoustic spectroscopy (PAS) and optoacoustic spectroscopy (OAS) belong to the photothermal spectroscopic methods. The basic principle of both methods is the photoacoustic effect. Although this weak effect is known for more than 100 years, an increase of applications, however, was only possible due to the improvement of detection methods as well as the use of lasers. Thereby the combination of tunable narrowband laser sources and the highly sensitive photoacoustic effect offers the potential for a sensitive and selective multi-component analysis.

This non-invasive technique is applicable to samples of any state of aggregation and the technical expenditure is quite low. These are best conditions for a modern fast analysis. The principle of PAS and OAS is very simple. The absorption of laser light causes a slight heating of the material which is connected with an expansion at the same time. The pressure transients, which are generated due to the short pulses, can be detected by pressure sensors. The measured signals are directly correlated with the absorption coefficient of the examined object.

The difference between both techniques is not quite easy to see, the more so as no conformity considering this matter can be noticed in the literature. In the following, PAS is described as a technique which utilizes a modulated laser beam in order to induce a periodic pressure alteration in a resonant gas cell, whereas, when using OAS, the acoustic waves which are generated in the sample are detected directly. Two examples shall demonstrate each technique. A detailed overview is given in (Bolshov and Kuritsyn 2001).

4.11.1 PCP Detection on Wood

The in-situ proof of the fungicide PCP on wood was examined by laser based PAS (Beck et al. 2000). For this purpose, the contaminated wood sample was heated by a halogen lamp in order to carry out thermodesorption. The desorbed material was transferred to a cell where it was excited by an amplitude-modulated laser beam. The tunable diode laser (1.41–1.48 μm) which was used for that purpose was adjusted to the absorption maximum of the gaseous PCP (1.44 μm). The modulation of the laser beam was tuned to the resonance frequency of the photoacoustic cell. The performed non-destructive measurements showed that it was possible to detect PCP even in a concentration of about 10 $\mu g/cm^2$ (43 ppm).

4.11.2 Determination of the Optical Properties of Human Skin

Lasers are ideal instruments for investigating interactions between UV radiation and human skin in vivo because of their high spectral, spatial and time resolution (Anders et al. 95). The optical properties of human skin as well as the interactions of human skin with UV light are not completely understood yet. From a topical point of view (alteration of the ozone layer associated with an increased amount of UV radiation at the earth's surface; changes in leisure time behaviour: vacation, solarium) detailed information is necessary in order to estimate the impact of UV on the skin.

In a current research project (Krebs et al. 2001) a laser system consisting of a frequency-doubled optical parametric oscillator (OPO) was used for the determination of the absorption coefficients of different skin areas and layers in vivo by using optoacoustic tissue differentiation. The pulsed laser light (280 – 330 nm) was led through an optical fibre and thus directly on the skin. The pressure transients which were measured by a piezoelectric transducer, provided information about the optical parameters of human skin., e.g. about the penetration depths of UV light. The influence of sunscreens on the skin was also studied.

4.12 Other Techniques and Outlook

The presented laser-based methods for the analysis of solids are only a specific selection with regard to a modern rapid analysis (see introduction). Several other techniques, which benefit from using a laser, are available for the examination of solids. However, these techniques often need a sample preparation to a greater or lesser extent (e.g. extraction) which reduces the advantages when applying laser technology.

Some other techniques where lasers are utilized shall briefly mentioned: the widely applied atomic absorption spectroscopy (AAS) is usually not suitable as a mobile measuring instrument for on-site analysis of solids and normally uses (like IR spectroscopy) a polychromatic light source for this purpose. In (Bolshov and Kuritsyn 2001) the feasibility and the advantages of diode lasers in AAS are discussed. Furthermore, the properties of thermal lens spectroscopy (TLS) are described. Cavity ring-down spectroscopy (CRDS) is an interesting absorption method, too; see (Wheeler et al.1998). But for the investigation of solids, a sample preparation is normally necessary.

A review about the different kinds of atomic spectroscopy in the environment is given in (Cave et al. 2000). The articles of (Winefordner et al. 2000) and (Omenetto 1998) also give a detailed overview of different laser-based techniques in atomic spectroscopy. Laser-based spectroscopy of organic molecules is reviewed in (Imasaka 1996). A combined review about laser-based techniques in environmental analysis including atomic and molecular spectroscopy is given in (Panne and Nießner 1997).

When investigating solids by MS, the substance has to introduce as a gas into the mass spectrometer. But macromolecules (DNA, proteins, polymers) are often decomposed by the "normal" treatments for desorption (thermal) and for ionization (electron impact or chemical methods). Matrix assisted laser desorption and ionization (MALDI) can be a solution for this problem (Carroll and Beavis 1998).

Keeping the key words of a modern rapid analysis in mind, the first question which has to be answered when utilizing lasers for the analysis of solids is: What shall be examined and what is the purpose? So, for instance, it is not always necessary to detect substances in the ppb trace range. Often, rather a rapid analysis of many objects is demanded, e.g. for an area-wide screening or for process controls. In this sense, the laser analysis has already found its way to be used "outside". Commercial devices are available which are based on the relatively "simple" laser-based techniques LIBS and LIF. Both are successfully employed for their distinct tasks (e.g. PAHs detection in soil) despite of their restrictions.

Further promising techniques – e.g. combination techniques – are almost ready to be made accessible for further analysis areas. But these systems should be mainly based on cost-effective and plug and play techniques as well as turnkey simplicity in order to be used in the mobile in-situ analysis. This applies to the lasers as well as to the detection systems. For economic working companies, a further attribute is the reliable and maintenance free utilization of the laser system. The need for cost-effective, easy to handle, and tunable lasers could e.g. be fulfilled by the further developments of diode lasers in the blue or the UV range respectively (Bolshov and Kuritsyn 2001).

References

Anders A, Altheide HJ, Knälmann M, Tronnier H (1995), Action spectrum for erythema in humans investigated with dye lasers. Photochem Photobiol 61: 200–205

Anglos D (2001) Laser-Induced Breakdown Spectroscopy in Art and Archaelogy, Applied Spectroscopy 55 (6):186A–205A

Bacon JR, Crain JS, van Vaeck L, Williams JG (2001) Atomic Spectrometry Update. Atomic mass spectrometry. J Anal At Spectrom 16:879–915

Bauer HE, Leis F, Niemax K (1998) Laser induced breakdown spectroscopy with an échelle spectrometer and intensified charge coupled device detection. Spectrochimica Acta Part B 53: 1815–1825

Beck AH, Bozóki Z, Niessner R (2000) Screening of Pentachlorophenol-Contaminated Wood by Thermodesorption Sampling and Photoacoustic Detection. Anal Chem 72:2171–2176

Bloch J, Johnson B, Newbury N, Germaine J, Hemond H, Sinfield J (1998) Field Test of a Novel Microlaser-Based Probe for *in Situ* Fluorescence Sensing of Soil Contamination. Applied Spectroscopy 52(10):1299–1304

Boesl U, Rink J, Zimmermann R, Püffel P (2000) Resonante Lasermassenspektrometrie: Neue Möglichkeiten für schnelle und mobile chemische Analytik in Industrie und Umwelt. VDI Berichte Nr. 1551: 85-90

Bolshov MA, Kuritsyn YA (2001) Laser Analytical Spectroscopy, In: Handbook of Analytical Techniques, Günzler H, Williams A (eds.), Wiley-VCH, Weinheim, pp. 727–752

Bujewski G, Rutherford B (1997) The Site Characterization and Analysis Penetrometer System (SCAPS) Laser-Induced Fluorescence (LIF) Sensor and Support System. In: EPA Rep. EPA/600/R-97/520, Environmental Protection Agency, Washington DC

Bujewski G, Rutherford B (1997) The Rapid Optical Screening Tool (ROST) Laser-Induced Fluorescence (LIF) System for Screening of Petroleum Hydrocarbons in Subsurface Soils. In: EPA Rep. EPA/600/R-97/020, Environmental Protection Agency, Washington DC

Bulatov V, Krasniker R, Schechter I (2000) Converting Spatial to Pseudo-Temporal Resolution in Laser Plasma Analysis by Simultaneous Multifiber Spectroscopy. Anal Chem 72: 2989–2994

Bulatov V, Krasniker R, Schechter I (1998) Study of Matrix Effects in Laser Plasma Spectroscopy by Combined Multifiber Spatial and Temporal Resolution. Anal Chem 70: 5302–5311

Bulatov V, Xu L, Schechter I (1996) Spectroscopic Imaging of Laser-Induced Plasma. Anal Chem 68:2966–2973

Bünting U, Lewitzka F, Karlitschek P (1999) Mathematical Model of a Laser-Induced Fluorescence Fiber-Optic Sensor Head for Trace Detection of Pollutants in Soil. Applied Spectroscopy 53 (1): 49–56

Carroll JA, Beavis RC (1998) Matrix-Assisted Laser Desorption and Ionization. In: Miller JC, Haglund RF (eds.), Laser Desorption and Ionization, Experimental Methods in the Physical Sciences Vol 30, Academic Press, San Diego

Castle BC, Knight AK, Visser K, Smith BW, Winefordner JD (1998) Battery powered laser-induced plasma spectrometer for elemental determinations. J Anal At Spectrom 13: 589–595

Cave MR, Butler O, Cook JM, Cresser MS, Garden LM, Miles DL (2000) Environmental analysis. J Anal At Spectrom 15:181–235

Evans EH, Dawson JB, Fisher A, Hill SJ, Price WJ, Smith CMM, Sutton KL, Tyson JF (2001) Atomic Spectrometry Update. Advances in atomic emission, absorption, and fluorescence spectrometry, and related techniques. J Anal At Spectrom 16:672–711

Fink H, Panne U, Niessner R (2001) Analysis of recycled thermoplasts from consumer electronics by laser-induced plasma spectroscopy. Analytica Chimica Acta 440:17–25

Handbook of Analytical Chemistry (2001), Günzler H, Williams A (eds.), Wiley-VCH, Weinheim

Linscheid M (2001) Mass Spectrometry. In: Handbook of Instrumental Techniques for Analytical Chemistry (1997), Settle F (ed.), Prentice Hall, Upper Saddle River

Hart SJ, Chen YM, Kenny JE, Lien BK, Best TW (1997) Field Demonstration of a Multichannel Fiber-Optic Laser-Induced Fluorescence System in a Cone Penetrometer Vehicle. Field Analytical Chemistry and Technology 1(6):343–355

Harvey RG (1991) Polycyclic aromatic hydrocarbons: chemistry and carcinogenicity. Cambridge University Press, Cambridge, England, 396pp

Heger HJ, Boesl U, Zimmermann R, Dorfner R, Kettrup A (1999) On-line resonance-enhanced multiphoton ionization time-of-flight laser mass spectrometry for combined multi-component-pattern analysis and target-compound monitoring: non-chlorinated aromatics and chlorobenzene in flue gases of combustion processes. Eur Mass Spectrom 5: 51–57

Hennig K, de Vries T, Paetzold R, Jantos K, Voss E, Anders A (2001) Multi sensor system for fast analyses in environmental monitoring with an application in waste water treatment. In: A decade of trans-european remote sensing, Proceedings of the 20th EARSeL Symposium, Buchroithner MF (ed.), Balkema

Hilbk-Kortenbruck F, Noll R, Wintjens P, Falk H, Becker C (2001) Analysis of heavy metals in soils using laser-induced breakdown spectrometry combined with laser-induced fluorescence. Spectrochimica Acta B 56:933–945

Howe T, Shkolnik J, Thomas R (2001) A solid sampling tool finally reaches maturity. Spectroscopy 16(2):194–235

Illenseer C, Löhmannsröben HG (2001) Investigation of ion-molecule collisions with laser-based ion mobility spectrometry. Phys Chem Chem Phys 3:2388–2393

Illenseer C, Löhmannsröben HG, Roch T, Zimmermann U (2001) Combination of Laser Induced Fluorescence (LIF) Spectroscopy and Laser-Based Ion Mobility Spectrometry (LIMS) for the Detection of Polycyclic Aromatic Compounds (PAC) and Petroleum Products in Soils. J Environ Monit, submitted

Imasaka I (1996), Laser spectroscopy on organic molecules. Fresenius J Anal Chem 355: 216–221

Jantos K, Paetzold R, Anders A (2000) Nachweis von Pestiziden auf Blattoberflächen durch laserinduzierte Fluoreszenzspektroskopie und Reflexionsmessungen. To be published

Karlitschek P, Lewitzka F, Bünting U, Niederkrüger M, Marowsky G (1998) Detection of aromatic pollutants in the environment by using UV-laser-induced fluorescence. Appl. Phys. B 67:497–504

Krasniker R, Bulatov V, Schechter I (2001) Study of matrix effects in laser plasma spectroscopy by shock wave propagation. Spectrochimica Acta Part B 56: 609–618
Krebs R, Meinhardt M, Lubatschowski H, Anders A (2001) Investigation of optical properties of human skin in the UV. Proceedings of SPIE 4618, to be published
Lakowicz JR (1999) Principles of Fluorescence Spectroscopy. Kluwer Academic / Plenum Publishers, New York
Lavine BK, Davidson CE, Moores AJ, Griffiths PR (2001), Raman Spectroscopy and Genetic Algorithms for the Classification of Wood Types. Applied Spectroscopy 55(8):960–967
Löbe K, Lucht H, Kreuchwig L, Uhl A (2000) Schnellanalyse von Schadstoffen in Althölzern mittels Laser-Plasma-AES. In: Bestimmung von Holschutzmitteln in Gebrauchtholz, WKI-Bericht Nr. 36
Löhmannsröben HG, Roch T (1997) Laserfluoreszenzspektroskopie als extraktionsfreies Nachweisverfahren für PAK und Mineralöle. In: Analytiker Taschenbuch 15, Springer Verlag, Berlin, pp 217–253
Löhmannsröben HG, Schober L (1999) Combination of laser-induced fluorescence and diffuse-reflectance spectroscopy for the *in situ* analysis of Diesel-fuel-contaminated soils. Applied Optics 38(9):1404–1410
Löhmannsröben HG, Roch T (2000) *In situ* laser-induced fluorescence (LIF) analysis of petroleum product contaminated soil sample. J Environ Monit 2:17–22
Löhmannsröben HG, Roch T, Schultze RH (1999) Laser-induced fluorescence (LIF) spectroscopy for *in situ* analysis of petroleum products and biological oils in soils. Polycyclic Aromatic Compounds 13:165–174
Marshall J, Chenery S, Evans EH, Fisher A (1998) Atomic spectroscopy update – atomic emission spectrometry. Analytical Atomic Spectrometry 13:107R–130R
Miller JC, Haglund RF (1998) Laser Ablation and Desorption, Experimental Methods in the Physical Sciences Vol. 30, Academic Press, San Diego
Omenetto N (1998) Role of lasers in analytical atomic spectroscopy: where, when and why. J Anal At Spectrom 13:385–399
Omenetto N, Petrucci GA, Cavalli P, Winefordner JD (1996) Absolute and/or relative detection limits in laser-based analysis: the end justifies the means. Fresenius J Anal Chem 355:878–882
Panne U, Nießner R (1997) Laserverfahren in der Umweltanalytik. In: Analytiker Taschenbuch Bd. 16, Springer Verlag, Berlin, pp. 155–270
Paetzold R, Anders A (2001) Detection of DDT and Hylotox 59 on wood by means of laser induced fluorecence. To be published
Roch T, Baumbach JI (1998) Laser-based ion mobility spectrometry as an analytical tool for soil analysis. International Society for Ion Mobility Spectrometry 1, 43–47
Rommers P, Boumans P (1996) ICP-AES versus (LA-)ICP-MS: Competition or a happy marriage? – A view supported by current data. Fresenius J Anal Chem 355:763–770
Rusak DA, Castle BC, Smith BW, Winefordner JD (1997) Fundamentals and Applications of Laser-Induced Breakdown Spectroscopy. Crit Rev Anal Chem 27(4):257–290
Schrader B (1996) Raman spectroscopy in the near infrared – a most capable method of vibrational spectroscopy. Fresenius J Anal Chem 355:233–239
Schechter I (1997) Laser Induced Plasma Spectroscopy. A Review of Recent Advances. Reviews in Analytical Chemistry 16:173–298
Schumann A (2000) Schnelle Bestimmung von organischen Holzschutzmitteln in Gebrauchtholz mittels Ionenmobilitätsspektrometrie. In: Bestimmung von Holschutzmitteln in Gebrauchtholz, WKI-Bericht Nr. 36
Stach J (1997) Ionenmobilitätsspektrometrie – Grundlagen und Applikationen. In: Analytiker Taschenbuch Bd. 16, Springer Verlag, Berlin, pp. 119–154

Stratis DN, Eland KL, Carter JC, Tomlinson SJ, Angel SM (2001) Comparison of Acousto-optic and Liquid Crystal Tunable Filters for Laser-Induced Breakdown Spectroscopy. Applied Spectroscopy 55(8):999–1004

Tran M, Sun Q, Smith BW, Winefordner JD (2001) Determination of C:H:O:N ratios in solid organic compounds by laser-induced plasma spectroscopy. J Anal At Spectrom 16: 628–632

Visser T (2000) Infrared Spectroscopy in Environmental Analysis. In: Encyclopedia of Analytical Chemistry, Meyers RA (ed), Wiley & Sons, Chichester, pp 1–21

Wang A, Haskin LA, Cortez E (1998) Prototype Raman Spectroscopic Sensor for *in Situ* Mineral Characterization on Planetary Surfaces. Applied Spectorscopy 52 (4): 477–487

Weickhardt C, Grun C, Grotemeyer J (1998) Fundamentals and features of analytical laser mass spectrometry with ultrashort laser pulses. Eur Mass Spectrom. 4: 239–244

Wheeler MD, Newman SM, Orrewing AJ, Ashfold MNR (1998) Cavity Ring-Down Spectroscopy. J Chem Soc, Faraday Trans, 94: 337–351

Winefordner JD, Gornushkin IB, Pappas D, Matveev OI, Smith BW (2000) Novel uses of lasers in atomic spectroscopy. J Anal At Spectrom 15:1161–1189

Wisbrun R, Schechter I, Niessner R, Schröder H, Kompa KL (1994) Detector for Trace Elemental Analysis of Solid Environmental Sample by Laser Plasma Spectroscopy. Anal Chem 66:2964–2975

5 Laser-Induced Fluorescence (LIF) Spectroscopy for the In Situ Analysis of Petroleum Product-Contaminated Soils

R.H. Schultze, M. Lemke and H.-G. Löhmannsröben

5.1 Introduction

Laser-induced fluorescence (LIF) spectroscopy is of great importance for environmental monitoring. Besides outstanding sensitivity and good selectivity, particular advantages of the LIF technique include the capabilities for in situ analysis and remote sensing. The major advantage of in situ LIF measurements is the lack of sampling and clean-up procedures preceding the analysis. Such procedures are error-prone, time consuming and expensive. The contamination of water or soil with petroleum products (oils) represents a major environmental risk. Since most petroleum products exhibit distinct native fluorescence it is promising to apply LIF analysis to the detection and characterization of oils in environmental compartments. The LIF investigation of oil-polluted waters with LIDAR and fiber optical sensing techniques is well advanced and appropriate instrumentation is commercially available. The employment of fluorescence techniques for in situ analysis of soil contaminations has received considerable attention only during the last decade. The combination of LIF instrumentation with geotechnical drilling equipment for real-time subsurface detection of oil pollutions, as pioneered by Lieberman et al. (Lieberman 1990), and field demonstrations have recently been described in detail in technology reviews and a monograph (Hart 1997, Lieberman 1998, Balshaw-Biddle 2000). Importantly, LIF-based techniques find increasing regulatory acceptance and have been verified as field screening methods for petroleum products by the US Environmental Protection Agency (US-EPA). For assessment and control of the environmental damage imposed in Kuwait during the Gulf War LIF investigations play an important role in what may become the "biggest environmental remediation project ever attempted" (Shouse 2001, Quinn 1995). In Germany, various aspects of LIF analysis of soils have been addressed in works of, among others, Niessner et al., Schade et al., Marowsky et al., Zimmermann and Lucht (Baumann 2000, Schade 1996, Marowsky 2001, Zimmermann 1997, and references therein), as well as by our group (Löhmannsröben 1996, 1997, 1999, 2000).

As for most other analytical techniques, calibration is essential for quantitative LIF analysis. It is therefore important to consider the signal/concentration-relationship. Under appropriate experimental conditions the measured LIF signal intensities $I_F(\lambda_{em})$ at the emission wavelengths λ_{em} can approximately be described by the following general equation:

$$I_F(\lambda_{em}) = \text{const.}\ I_0(\lambda_{ex})\ \varepsilon^*(\lambda_{ex})\ \eta_F(\lambda_{ex})\ f(M)\ c \tag{5.1}$$

Here, the term const. includes constant apparative and experimental parameters such as detection geometry, detector sensitivity, etc., $I_0(\lambda_{ex})$ is the excitation intensity at the excitation wavelength λ_{ex}, $\varepsilon^*(\lambda_{ex})$ and $\eta_F(\lambda_{ex})$ are effective extinction coefficients and fluorescence efficiencies, respectively, and c is the concentration of the analyte, i. e. the oil. Since petroleum products are complex multicomponent mixtures both $\varepsilon^*(\lambda_{ex})$ and $\eta_F(\lambda_{ex})$ are excitation wavelength-dependent and cannot directly be related to molecular parameters. The influence of various matrix parameters (M) on the LIF signal is symbolized by heuristic introduction of the function $f(M)$. Without this term, Equation 5.1 is derived from the Lambert-Beer law under the condition of small overall absorbance $A(\lambda_{ex})$ over the optical interaction length d:

$$A(\lambda_{ex}) = \varepsilon^*(\lambda_{ex})\ c\ d << 1 \tag{5.2}$$

For one given analyte, with the excitation conditions kept constant, Equation 5.1 can be simplified to a calibration function which expresses the linear relationship between LIF signal intensity and analyte concentration:

$$I_F(\lambda_{em}) = m(M)\ c \tag{5.3}$$

The slope of the calibration function $m(M)$ is a direct measure of the detection sensitivity and thus an important analytical benchmark. Comparison of Equations 5.1 and 5.3 shows that $m(M)$ depends both on matrix parameters and on analyte properties, namely on the product $[\varepsilon^*(\lambda_{ex}) \times \eta_F(\lambda_{ex})]$. As to matrix parameters, we have previously shown that for different Diesel fuel-spiked soils with varying optical properties it is important to take into account the soils' diffuse reflectances (R) in the spectral range of oil emission (Löhmannsröben 1999). The current report will focus on photophysical properties of the oils, i. e. on the expected direct proportionality between $m(M)$ and $[\varepsilon^*(\lambda_{ex}) \times \eta_F(\lambda_{ex})]$. It is our aim to derive an understanding of the LIF calibration behavior by connecting oil properties with LIF measurements of oil-spiked soils as laboratory reference (LR) materials. This work is thus part of our attempt to employ laser-based techniques for qualitative and quantitative in situ analysis of oil-contaminated soils. Based on the knowledge of the photophysical properties of oils as analytes and of the optical properties of soils as matrices, it is our strategy to use LR and certified reference (CR) materials to make way from the investigation of spiked and real-world samples in the laboratory towards on site field applications.

5.2 Experimental Techniques

5.2.1 The LIF demonstrator unit

In the course of a research project, a mobile LIF spectrometer was developed (Fig. 5.1). As a first step, a demonstrator unit was set up at the University of Erlangen-Nürnberg. The unit used a N_2-Laser (MSG 500, LTB Berlin, pulse energy ca. 400 μJ at 337 nm, pulse duration ca. 500 ps) as excitation light source. For detection, a photomultiplier (Hamamatsu R 2496) in combination with a monochromator (Acton Research, Spectra-Pro 275) was employed. A cut-off filter (GG 368, Schott) was placed in front of the monochromator. Data acquisition was performed by a 14 bit AD/DA-conversion card (Decision Computer Inc.). The card was later replaced by a Keithley DAS 1601 Data Aquisition Board. LIF measurements from soil surfaces were performed using a simple sensor head with bifurcated fiber bundles (quartz fibers with ca. 100 μm diameter for excitation and detection) connected to a plain circular metal plate. This resulted in an anti-parallel orientation of excitation and emission optical paths (180° geometry). Towards the soil surfaces, the fiber was protected by a 3 mm quartz glass window. At the soil remediation plant additional subsurface measurements were performed in soil heaps using a metal lancet equipped with bifurcated quartz fiber bundles and a sapphire window.

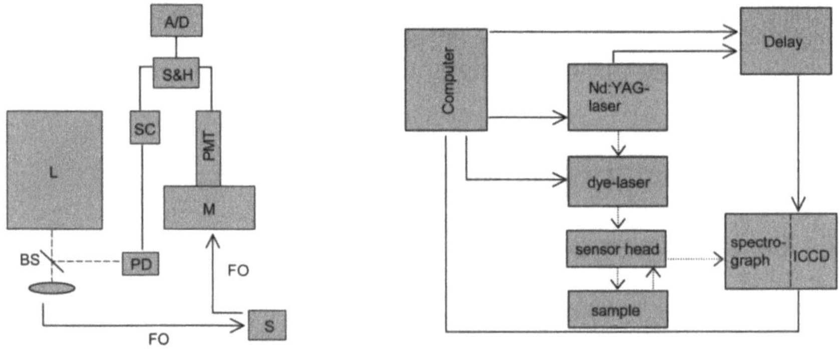

Fig. 5.1. Schematic representation of the LIF spectrometers. Left: LIF demonstrator unit (L: N_2-laser, BS: beam splitter, FO: fiber optics, S: sample, M: monochromator, PMT: photomultiplier, PD: photodiode, SC: signal converter, S&H: sample and hold amplifier, A/D: analog/digital converter), right: mobile LIF spectrometer OPTIMOS.

5.2.2 The mobile LIF spectrometer OPTIMOS

Based on the experiences with the demonstrator unit, a mobile LIF system was designed and manufactured at Optimare GmbH. This device, called OPTIMOS (oil pollution transportable independent monitoring system), includes a diode laser-pumped Nd:YAG-laser (B.M.Industries, pulse energies ca. 5 mJ at 355 nm and 700 μJ at 266 nm, pulse duration ca. 10 ns) as excitation source (Fig. 5.1). The OPTIMOS can additionally be equipped with a custom designed dye laser. An intensified CCD camera (Andor, ICCD DH501-18, 1024×128 pixels, minimum gate width less than 5 ns) in combination with a spectrograph (LOT-Oriel, MS125, F/3.7) was used as detection system. For the measurements of the real-world soil samples in the laboratory, the OPTIMOS was operated at an excitation wavelength of 355 nm with a cut-off filter GG 385 (Schott) to reduce backscattered excitation light. For the experiments at the gravel pit, a flash lamp-pumped Nd:YAG laser (Minilite 2, Continuum) with 266 nm excitation (cut-off filter GG 368) was chosen. To demonstrate the capability of real-time subsurface measurements, the OPTIMOS was also combined with a truck-based cone penetrometer and direct-push measurements were performed at the pit site using a specifically designed penetrometer tip.

5.2.3 Investigated petroleum products and soil samples

As petroleum products, 10 different oils, provided by the Wilhelmshavener Raffineriegesellschaft, were investigated. The samples included various petroleum products (distillates and reformates) as well as crude oils of different provenience (see Table 5.1). For absorption measurements (Perkin Elmer, Lambda 2, slit widths 2 nm) of the oils diluted with cyclohexane, the concentrations were ca. 1, 5, 100 g / L for the crude oils, distillates and reformates, respectively. Emission properties of dilute and neat oils were investigated with the LIF systems (180° geometry) and a conventional fluorescence spectrometer (Perkin Elmer, LS50 with 1P28 photomultiplier, slit widths 2.5 nm, 90° geometry).

Table 5.1 Photophysical and calibration data of the petroleum products

	Absorption		**Fluorescence**		**Calibration**			
					Quartz Sand		A_h Soil	
Oil (a)	ε^* (b)	U (c)	η_F (d) (dil.)	η_F (d) (neat)	m (e)	LOD (f)	m (e)	LOD (f)
	$L\,g^{-1}\,cm^{-1}$	cm^{-1}				ppm		ppm
Df 1	0.039	3.2	0.50	1.0*	1.4	600	0.4	>2,000
Df 2	0.012	2.4	0.58	0.65	n.d.(g)	n.d.	n.d.	n.d.
Fo	0.076	3.8	0.74	0.91	16	50	4	530
Gas 1	0.005	5.3	0.67	0.67	n.d.(g)	n.d.	n.d.	n.d.
Gas 2	0.013	3.5	0,70	0.78	7.8	110	0.6	>2,000
Gcr	3.5	10.8	0.71	0.07	36	≈23	6	340
Icr	3.2	10.5	0.52	0.06	62	≈13	12	190
Lcr	1.8	11.8	0.58	0.15	n.d.(g)	n.d.	n.d.	n.d.
Acr	1.5	8.3	1.0*	0.54	220	≈4	23	100
NScr	0.9	9.0	0.79	0.27	110	≈8	34	60

[a] Df, Diesel fuel; Fo, fuel oil; Gas, gasoline; Gcr, German crude; Icr, Iranian crude; Lcr, Libyan crude; Acr, Algerian crude; NScr, North Sea crude, [b] effective extinction coefficients measured for dilute oils in cyclohexane at $\lambda_{ex} = 337$ nm, [c] Urbach decay width obtained from a linear fit of logarithmic dilute oil absorbances versus the photon energy (cf. Equation 5.3) in 10^3 cm^{-1}, [d] fluorescence efficiencies were measured with $\lambda_{ex} = 337$ nm (180° geometry) and are given relative to the oils (marked by an asterisk) showing the highest intensities, [e] slopes of the calibration functions (cf. Equation 5.2, in 10^3 rel. units/ppm, for crude oils on quartz sand determined in low concentration ranges) as measured with the OPTIMOS unit ($\lambda_{ex} = 355$ nm), [f] detection limits were calculated from the ratios of the threefold standard deviations of blanks (uncontaminated soils, 10 replicate measurements) and m, [g] n.d.: not determined.

The emission spectra obtained under 180°-geometry were superimposed by a strong scattering signal from reflections on the cuvettes surfaces which was eliminated by subtraction of a cyclohexane reference spectrum. Emission measurements were not quantum-corrected. Standard 1 cm ×1 cm cuvettes (Helma) were used for the absorption and emission experiments. For the investigation of oil-spiked LR soils, two different materials were used, namely a commercially available quartz sand (Lugato Chemie Hamburg) and a soil taken from an A_h-horizon. The A_h-soil, a pseudogley podsol, was provided and characterized by the Institute of Geology and Mineralogy of the University Erlangen-Nürnberg. For laboratory investigations, 20 g of soil with oil contents of 250 – 5,000 ppm, as determined gravimetrically, were used. The samples were mixed by stirring with a spatula for 5 min without addition of any solvent.

In addition to the LR materials, real-world contaminated soil and waste samples were investigated. These samples were obtained from Umweltschutz Nord (Ganderkesee) and the University of Applied Sciences in Wilhelmshaven. The soil samples, which were taken at five different locations, were subsequently dried and ground in the laboratory. The waste samples were also taken from various locations and consisted of completely different materials, e. g., natural soils, slurry, rubble, industrial deposits etc. Umweltschutz Nord analyzed the soils and wastes for their total petroleum hydrocarbon (TPH) contents with IR-spectroscopy as reference analysis according to the German DIN 38409-H18 (corresponding to US-EPA method 418.1). TPH contents of the real-world samples ranged from < 100 ppm up to ca. 20,000 ppm.

5.3 Results and Discussion

5.3.1 Photophysical properties of the petroleum products

Absorption spectra of some of the oils diluted in cyclohexane are shown in Figure 5.2, and extinction coefficients at one laser excitation wavelength (λ_{ex} =337 nm) are given in Table 5.1.

As is characteristic for such complex multi-component mixtures, the petroleum products show a structureless absorption behavior with a strong increase from low absorbances in the NIR to strong absorbances in the UV spectral region. Only in the spectra of the gasolines, absorption bands at 360 nm and 380 nm can be observed (inset in the left part of Figure 5.2). Considering the oil classes, it is obvious that the colored crude oils exhibit the strongest absorbances. At 337 and 420 nm (data not given), the effective extinction coefficients $\varepsilon^*(\lambda_{ex})$ of the oils vary by about 2 and 3.5 orders of magnitude, respectively. The absolute values of $\varepsilon^*(\lambda_{ex})$ can be regarded as typical for the oils and are consistent with earlier analytical determinations (Löhmannsröben 1996, 1997).

Extended long-wavelength absorption tails are often described in terms of the so-called Urbach phenomenology. Originally, this concept was developed for the analysis of alkali halide absorption, and later successfully extended to evaluate absorption behavior of such diverse materials as liquid water, glasses, organic dyes, amorphous semiconductors and even solvated electrons.

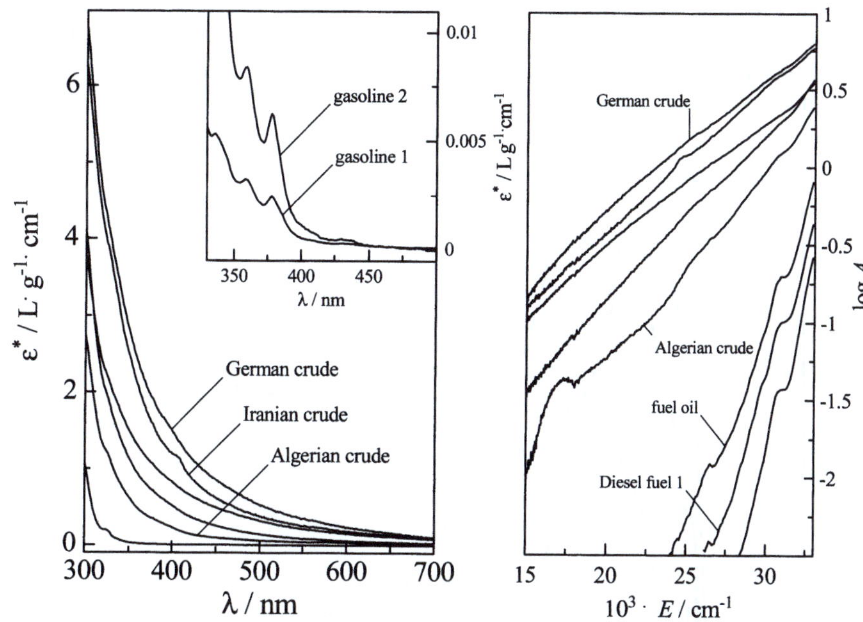

Fig. 5.2. Absorption spectra (left) and Urbach representation (right) of selected dilute oils. Left: Figure of the extinction coefficient plotted against the wavelength. The absorbance A refers to oil concentrations of 0.1 $g{\cdot}L^{-1}{\cdot}cm^{-1}$ and 1 cm cuvettes.

It is based on the idea that a thermally or structurally disordered system with a distribution of absorption sites can produce an absorption profile which depends exponentially on the photon energy E (the Urbach rule):

$$A(E) \propto \exp(E/U) \qquad (5.4)$$

Here, the so-called Urbach decay width U is an energy term consisting of thermal and structural elements. We have interpreted the absorption spectra of humic substances, which are important constituents of dissolved organic materials and ubiquitous in natural waters and soils, with the simple Urbach Ansatz (Illenseer 1999). Mullins et al. were first to discuss the absorption spectra of different crude oils and asphaltenes in the framework of the Urbach theory (Mullins, Zhu 1992). The dependence of log A vs. E for some of our oils is shown in the right part of Figure 5.2. Despite some residual spectral features overlaid, the expected linearity is found, and it can be concluded that the long-wavelength absorption tails are consistent with the Urbach rule. The other oils exhibited similar behavior. The Urbach decay widths calculated from these spectra are in the range $U = 2{,}400 - 12{,}800$ cm^{-1}, which is in complete agreement with the results reported by Mullins et al. for different crude oils (Mullins et al. 1992). It is notable that these values of U are much larger than the thermal energy at room temperature ($kT = 207\ cm^{-1}$). This is indicative of components with high structural disorder. It is reasonable that such constituents are predominantly present in the crude oils which are expected to contain biogenic macromolecules and polymers. Further investigations are necessary to elucidate the potential of the parameter U to characterize and understand petroleum product properties.

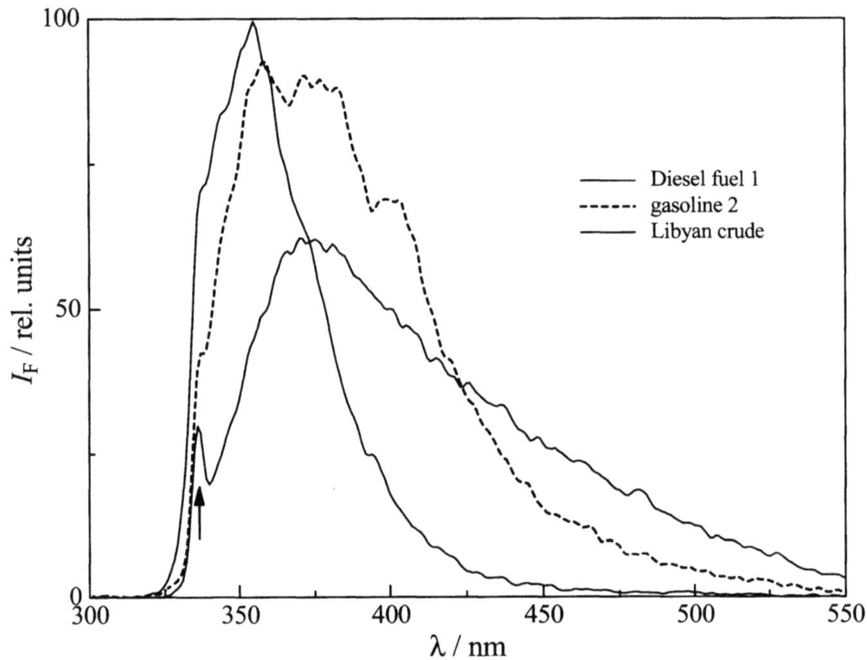

Fig. 5.3. Emission spectra of three dilute oils. The measurements were performed in 1 cm cuvettes with the conventional fluorescence spectrometer (Perkin Elmer LS50, $\lambda_{ex} = 337$ nm, absorbance ca. 0.1). The arrow marks residual stray light.

The fluorescence spectra of dilute and neat oils were recorded with an excitation wavelength of $\lambda_{ex} = 337$ nm. Typical spectra of dilute oils are shown in Figure 5.3. The distillates and the crude oils show broad structureless emission spectra with maxima at $\lambda_{em} \approx 355$ nm (distillates), resp. $\lambda_{em} \approx 370$-380 nm (crude oils). Only for the reformates more structurized emission spectra with maxima at $\lambda_{em} \approx 360$, 380 and 410 nm were observed. In Figure 5.4, the fluorescence spectra of dilute and neat Algerian crude oil and fuel oil are compared as representative examples. For all neat oils a red spectral shift of the emission with respect to the dilute oils was observed. In comparison to the dilute oils, the fluorescence signal intensities $I_F(\lambda_{em})$ were lower for the neat crude oils, but larger for the distillates and reformates. In order to obtain relative fluorescence efficiencies $\eta_F(\lambda_{ex})$ of the dilute oils, samples were prepared with approximately the same absorbance at the excitation wavelength ($A \approx 0.1$ at 337 nm).

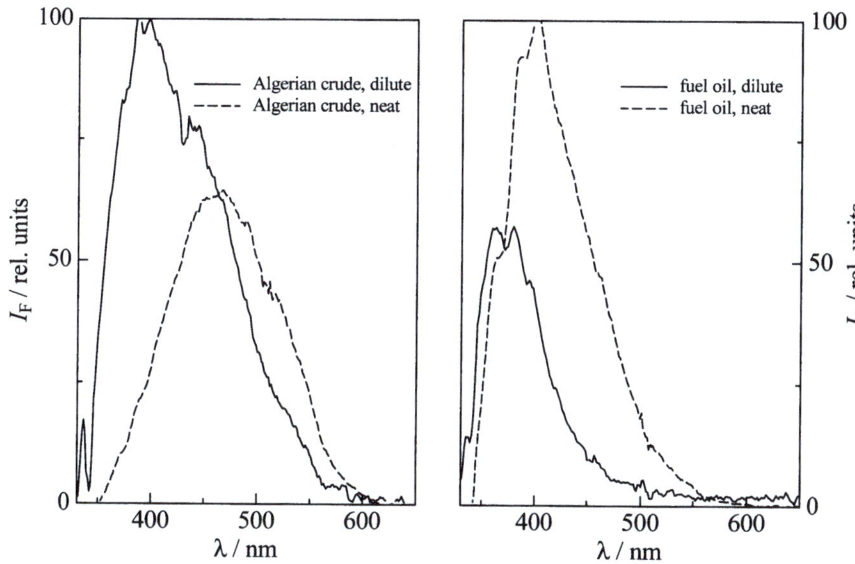

Fig. 5.4. Comparison of dilute and neat oil fluorescence spectra for Algerian crude (left) and fuel oil (right).

For the 180° excitation geometry employed we also assume same excitation conditions among all the neat oils, where the high absorbances garantee complete absorption of the excitation light within a very small volume. Therefore, relative fluorescence efficiencies were simply determined by integration of the emission spectra of dilute and neat oils. For the dilute oils under consideration the $\eta_F(\lambda_{ex})$ values differ not more than by a factor of 2, and no systematic variation among the oil classes is evident. This is in agreement with our earlier investigations, in which absolute values of $\eta_F(\lambda_{ex})$ ranging from 0.35 for a Diesel fuel to 0.52 for Brent crude oil were determined (diluted in cyclohexane, under N_2, λ_{ex} 337 nm) (Löhmannsröben et al. 1997). In contrast, the fluorescence efficiencies of the neat crude oils are approximately one order of magnitude lower than those of the neat reformates and distillates. These results indicate the clear difference between fluorescence properties of dilute and neat oils, i. e a distinct concentration dependence of fluorescence spectral properties and efficiencies. Such a behavior can be caused by fluorescence quenching processes between different oil components, as shown, e.g., for benzene and naphthalene (Lieberman 1998), or in time-resolved fluorescence measurements of crude oils in different concentrations (Wang 1994). Also, reabsorption of fluorescence photons by the long-tailing absorption bands, the so-called inner filter effect, has to be considered. While it is plausible to consider neat oil characteristics as "native" oil properties, it is also clear that particularly reabsorption will be much lower in dilute oils. It is therefore instructive to normalize neat oil- with respect to dilute oil- fluorescence efficiencies and to relate this ratio with the oils' effective extinction coefficients. Figure 5.5 shows an approximately logarithmic correlation between [η_F(neat)/η_F(dilute)] and ε^*, which suggests that the reduced fluorescence capabilities of crude oils are mainly due to the inner filter effect.

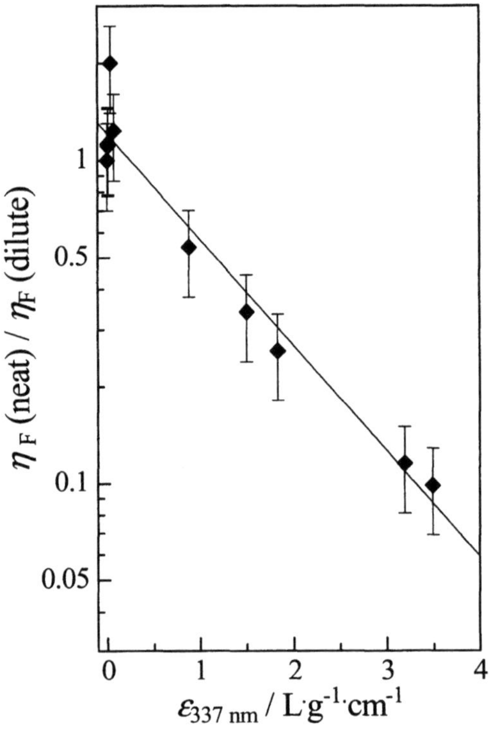

Fig. 5.5. Semi-logarithmic representation of the ratio of the relative fluorescence efficiencies of neat and dilute oils vs. effective extinction coefficients at 337 nm.

5.3.2 LIF spectroscopic investigations of oil-spiked samples

LIF calibration measurements were performed with oil-spiked quartz sand and A_h soil as LR materials. For practical purposes it is convenient to employ the back-scattered excitation light, detected in the second diffraction order (710 nm), for normalization of the LIF signals. This can be used to correct for different experimental settings, as e. g. detection sensitivities (Fig. 5.6), or for strongly varying soil properties (see below). If not mentioned otherwise all LIF measurements of LR and real-world soil presented below were corrected in such a way. LIF spectra measured for various concentrations of Algerian crude oil on quartz sand, and the concentration dependence of the integrated LIF signals together with the calibration function I_F vs. c, are shown in Figure 5.6 (middle and right part).

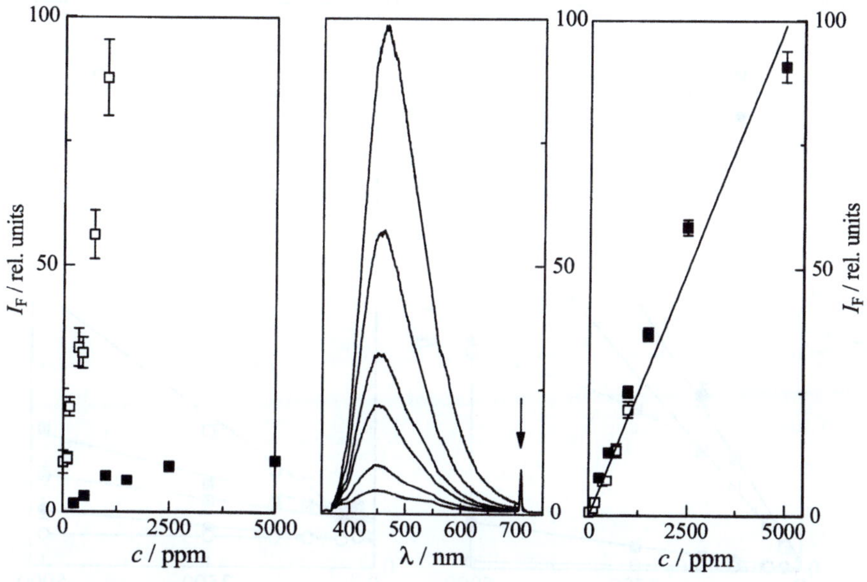

Fig. 5.6. LIF intensities (left) and spectra (middle) of Algerian crude oil in a concentration range of 250-5,000 ppm on quartz sand as measured with the OPTIMOS (λ_{ex} = 355 nm). In the left, results obtained with two different gains of the ICCD-camera are displayed. The spectra in the middle were normalized to the intensity of residual laser stray light (arrow, detected in second diffraction order at 710 nm). Right: Calibration function obtained with the normalized fluorescence intensities.

More examples of the calibration functions obtained are shown in Figure 5.7. From these results the following qualitative observations can be made: (i) Most of the calibration measurements yielded the expected linear relationship between the integrated LIF signal intensities I_F and oil concentrations c (cf. Equation 5.1). With quartz sand as matrix for the crude oils a nonlinear behavior at higher concentrations was observed. This is due to changes of the quartz sand optic properties (coloration) with high oil dosages. In these cases, the slopes of the calibration functions were determined in low concentrations ranges ($c < 2{,}000$ ppm). (ii) For the A_h soil as matrix, the calibration functions are significantly flatter than for quartz sand. This indicates the direct influence of the lower reflectance R of the darker soil. It is notable that the non-spiked A_h soil produced a significant background signal, which presumably stems from fluorescence of soil organic matter. The analytical parameters obtained, namely the slopes of the calibration function m (sensitivities) and the detection limits (LOD), are also included in Table 5.1. It can be seen that for the different oil types the sensitivities differ strongly, varying, e. g., on quartz sand from 1.4 to 220 rel. units/ppm for Diesel fuel 1 and North Sea crude oil, respectively. With two exceptions (see below), the detection limits determined range from LOD $\approx 4 - 23$ ppm for the crude oils on quartz sand to ca. 500 - 600 ppm for Diesel fuel 1 on quartz sand and fuel oil on the A_h soil.

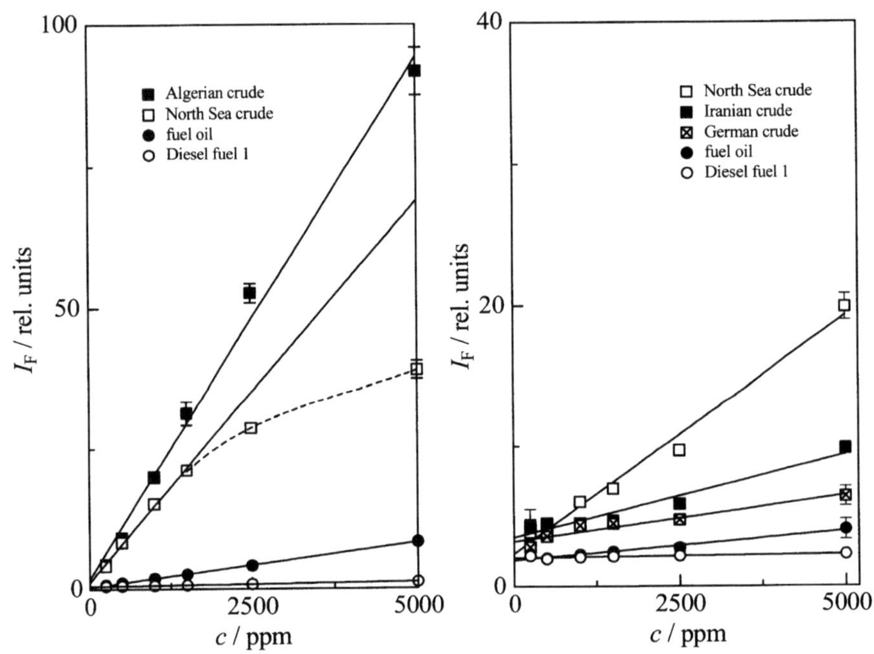

Fig. 5.7. Calibration functions of selected oils on quartz sand (left) and the A_h soil (right) as measured with the OPTIMOS ($\lambda_{ex} = 355$ nm). Also shown in the left part is an example of the non-linear response (North Sea crude oil) and the extrapolated line from the linear initial part at low concentrations.

These findings are in accordance with earlier results from the investigation of LR materials (Löhmannsröben 1999) and from subsurface field screening measurements (Liebermann 1998). Because of their low absorbances with respect to a dark matrix, Diesel fuel 1 and gasoline 2 on the A_h soil represent examples for which the sensitivities are so low ($m < 0.6$ rel. units/ppm) that only lower boundaries of the LOD values can given. Generally, for the LIF analysis of reformates and distillates the usage of the frequency-quadrupled emission of Nd:YAG lasers (λ_{ex} = 266 nm) yields much better results.

The highly varying calibration behavior of the oils under investigation can be traced back to their absorption and emission properties, viz. to the product of effective extinction coefficients ε* and relative fluorescence efficiencies η_F (cf. Equation 5.1). Since we have determined η_F with an excitation at λ_{ex} = 337 nm, also ε*(337 nm) is employed in the following. For neat and dilute oils, plots of the products (ε* x η_F) vs. *m* determined on quartz sand are shown in Figure 5.8 (left and right parts). Obviously, introduction of the dilute oil properties does not yield a picture consistent with Equation 5.1. On the contrary, if the neat oil properties are employed the data suggest the expected direct proportionality between (ε* x η_F) and *m* (left part of Figure 5.8). The situation is completely analogous for the sensitivities determined on the A_h soil. The results presented in Figure 5.8 therefore provide a convincing illustration that the neat oil properties are the analyte parameters relevant for the calibration behavior of the oils.

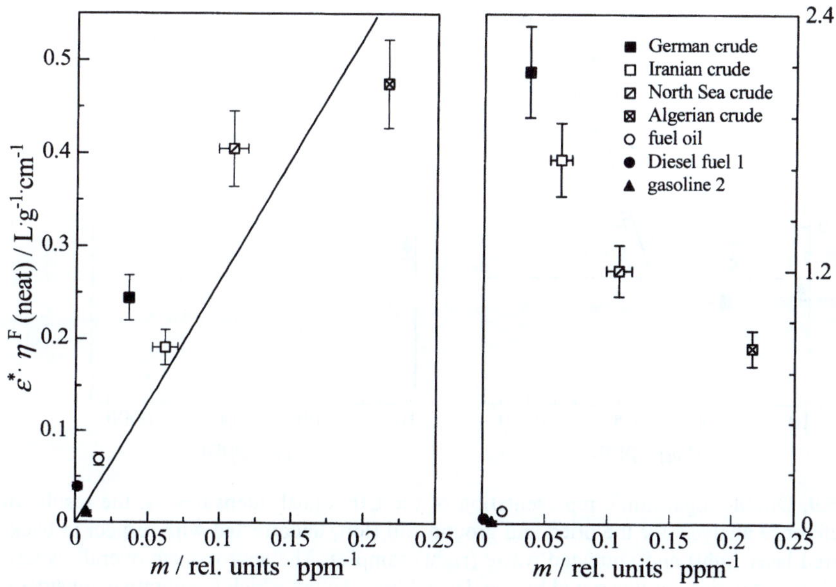

Fig. 5.8. Correlation of photophysical and calibration properties. Plots of the products of effective extinction coefficients and relative fluorescence efficiencies (both determined with the LIF demonstrator unit at λ_{ex}=337 nm) for neat (left) and dilute (right) oils vs. the slopes of the quartz sand calibration functions measured with the OPTIMOS (λ_{ex}=355 nm).

5.3.3 LIF spectroscopic investigations of real-world soils

In order to evaluate the potential of LIF spectroscopy in practice, two sets of real-world soils were investigated. First, the treated (dried and ground) soil samples were analyzed with the LIF demonstrator unit ($\lambda_{ex} = 337$ nm). Secondly, the untreated waste materials were investigated with the mobile LIF spectrometer OPTIMOS. All samples exhibited highly varying matrix properties, as was already evident from visual perception. In most cases the LIF spectra obtained from the real-world samples were broad and structureless with emission maxima around 450 nm. Figure 5.9 shows comparisons of the results obtained by in situ LIF and ex situ IR analyses in double-logarithmic representations.

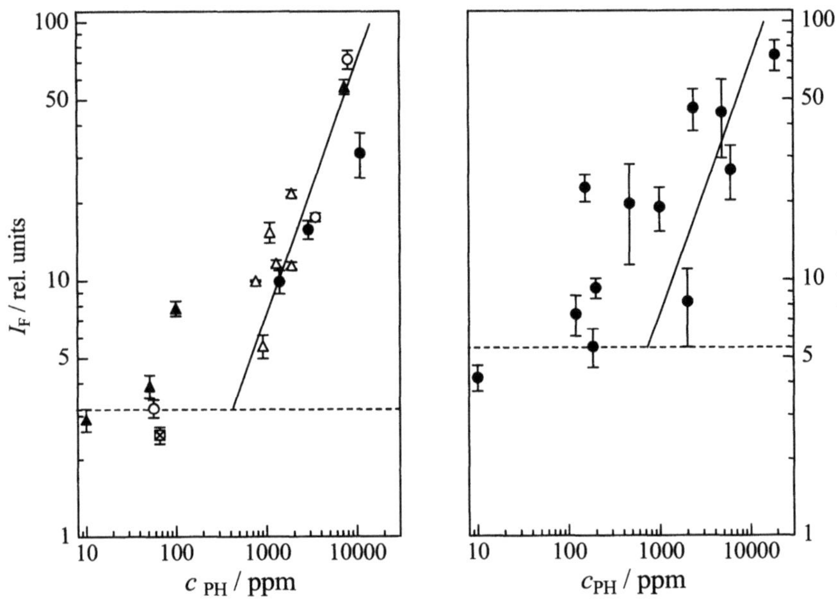

Fig. 5.9. Double-logarithmic representation of the LIF signal intensities vs. the results of the reference analysis for the dried and ground soil (left, uncorrected with respect to back-scattered laser light) and untreated waste (right) samples. The approximate overall experimental uncertainties are illustrated by the dashed lines below which quantitative considerations are not possible. For the logarithms of the I_F and c values, the solid lines have slope unity.

For the treated soils (left part) with TPH contents higher than 500 ppm a good correlation between I_F and TPH content is observed. The superimposed straight line with slope unity illustrates the concentration dependence of the LIF signal as expected from Equation 5.1. The data points scatter regularly around this line, which suggests that no systematic deviations occurred. The overall experimental uncertainties, mainly related to variations of background signals and reproducibility problems, preclude a quantitative interpretation of LIF signals below the limits shown as dashed lines in Figure 5.9. This is in accordance with an estimated detection limit of ca. 100 ppm for our LIF analysis of soils (Löhmannsröben 2000). The analogous plot of I_F vs. TPH content for the waste materials (right part of Figure 5.9) clearly depicts enhanced deviations of the data points from the line with slope unity. However, the clear correlation between LIF signal intensity and degree of contamination is evident. It is emphasized that the waste materials were of very heterogenous nature, and would thus be difficult to characterize by any analytical technique. For us it was an important benchmark test that even under these difficult conditions for samples with TPH contents over ca. 1,000 ppm a reliable in situ LIF analysis is feasible. We have previously shown that for different Diesel fuel-spiked LR soils with varying optical properties the LIF calibration behavior was improved by normalization with averaged values of the soils' diffuse reflectances (R) in the spectral range of oil emission (Löhmannsröben 1999). This simple approach did not work for the real-world soils, i. e. did not improve the correlations shown in Figure 5.9. This is probably due to the fact that averaged R values do not sufficiently describe the optical properties of these samples. Further investigations are under way to improve our understanding how optical properties of untreated real-world soils can be introduced into LIF calibration procedures.

5.3.4 Field investigations

We have performed field experiments at two different sites in southern Germany. All LIF data obtained were only interpreted qualitatively, i. e. no calibration was attempted. First, measurements were performed at a soil remediation plant in Roth (near Nürnberg) employing the LIF demonstrator unit. Soil remediation plants have a continuous demand for analytical investigations in order to control and optimize the remediation procedure. At the plant site various soil batches were examined, either by employing the fiber optical sensor head to investigate soils in buckets or by inserting the lancet into soil heaps. As an example, LIF spectra obtained from two soil charges in buckets, with measurements at three different surface positions each, are displayed in the left part of Figure 5.10. The upper traces show the LIF signals obtained directly from a contaminated soil at delivery, whereas the lower traces were from material after remediation. The rather slight variations of I_F at the three measurement positions indicate reasonable sample homogeneities. Subsequent laboratory analysis performed revealed that the TPH content of the soil at delivery was 2,100 ppm, whereas in the remediated material it was below 400 ppm.

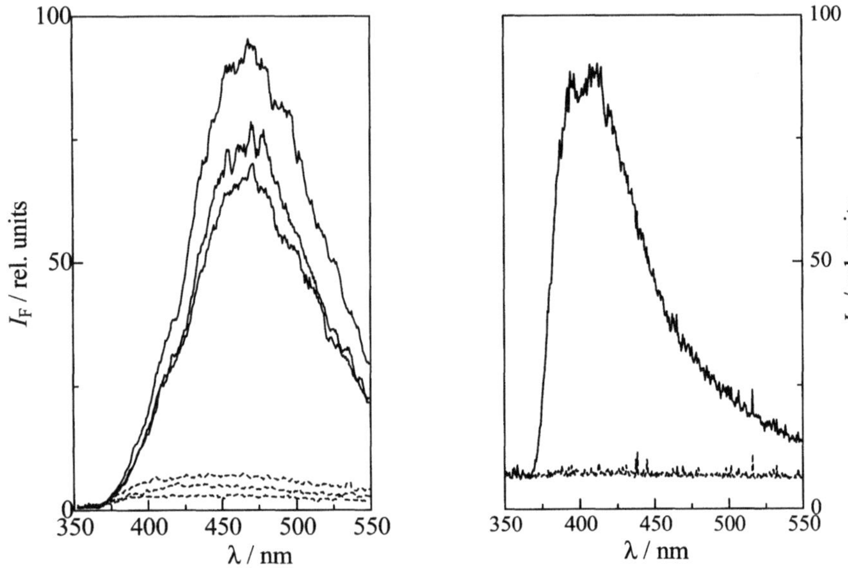

Fig. 5.10. LIF spectra obtained in field investigations at a soil remediation plant (left, LIF demonstrator unit, λ_{ex}=337 nm) and a gravel pit (right, OPTIMOS, λ_{ex}=266 nm).

In a second campaign measurements with the OPTIMOS spectrometer were performed at a gravel pit near Bamberg. It was suspected that in this pit slurries from paper recycling processes had been discharged and buried. For LIF measurements of subsurface profiles soil cores were brought up with a core barrel sampler and scanned along the rod with the fiber optical sensor head (Fig. 5.11). Shown in the right part of Figure 5.10 are LIF signals from two positions of such a core, namely from the surface (lower trace) and from ca. 50 cm depth (upper trace).

These results suggest an uncontaminated surface soil layer and the presence of fluorescent contamination underneath. We were thus able to predict a depth range of suspected high contamination. Based on these investigations the cores were sampled and high levels of contamination (up to TPH contents of 30,000 ppm) were later verified by laboratory analysis.

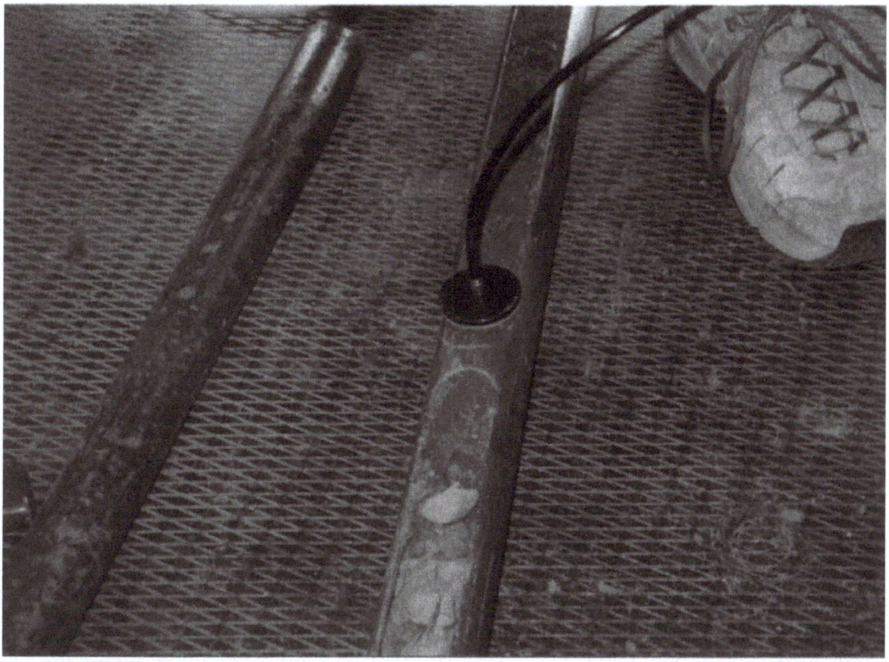

Fig. 5.11. Photograph of the core barrel sampler and the LIF sensor head used for scanning the soil core.

These examples demonstrate that in situ LIF analysis is a powerful method for on site screening investigations of contaminated soils. In soil remediation plants LIF measurements have promising potential for entrance and quality control. In combination with drilling techniques, in situ LIF is without competition for chemical analysis of organic pollutants in real-time subsurface characterization of contaminated sites, landfills etc. Undoubtedly, modern MWD ("measuring while drilling") concepts will increasingly make use of laser-based techniques.

5.4 Conclusions

Based on the photophysical characterization of the oils, the investigations of oil-spiked LR materials provided information on the LIF calibration behavior. It was found that absorption and emission properties of the neat oils were the parameters relevant for detection sensitivity. The direct proportionality between the product (ε^* x η_F) and the slopes m of the calibration functions, as expected from fundamental considerations and as suggested by the results presented, e. g., in the left part of Figure 5.8, could have great practical advantages. Therefore the analysis of real-world contaminated soils is facilitated by solution-phase measurements of oils with appropriate photophysical properties. As an ultimate aim it can be envisaged that the analytical parameters for in situ LIF analysis of various oils can thus be predicted and the number of necessary calibrations would be significantly reduced. With two exceptions, the oil detection limits determined ranged from 4–23 ppm for the crude oils on quartz sand to 500 – 600 ppm for Diesel fuel 1 on quartz sand and fuel oil on the A_h soil. Higher values were mainly due to excitation of reformates or distillates in spectral regions of too low absorbance. We feel confident that with optimized excitation conditions, detection limits of ca. 100 ppm can be achieved for most oils on many soils. Of special interest are concentrations with regulatory relevance, as, e. g. contained in the so-called Dutch List for soil contaminants (Rosenkranz 1995). Herein, two values are given for petroleum products in soils: a reference concentration (R value) of 1,000 ppm, above which additional investigations of the soils are recommended, and an intervention concentration (I value) of 5,000 ppm, above which soil remediation is proposed. Evidently, the sensitivity achieved in the in situ LIF analysis of LR materials is fully sufficient to monitor these regulatory concentrations. For the real-world soil and waste samples, only qualitative LIF signal intensities I_F were recorded, i. e. no calibration attempt was made here. The clear correlation between I_F from in situ LIF measurements and the results of the ex situ reference analysis seems rewarding. Soil extraction and the quantification of petroleum products with the reference IR analysis is tedious and difficult. Round robin tests have shown that TPH contents reported by different laboratories have to be interpreted with caution (Fischer 1994, Heininger 1998). It is unclear to what extent the significant scatter of the data points in Figure 5.9 may be related to experimental uncertainties of the reference analysis. With in situ LIF investigation, disadvantages of sample pretreatment procedures can be avoided, and real-time, rapid and reliable soil analysis with high spatial resolution, even in subsurface measurements, can be performed. It is therefore expected that LIF measurements will acquire increasing importance for laboratory and on site soil analysis.

Acknowledgment

This work was financially supported by the Deutsche Bundesstiftung Umwelt (DBU, research programme OPTIMOS), the Bundesministerium für Bildung und Forschung (BMBF, research programme FLUTRAS), Bayern Innovativ and the University of Potsdam (Innovationsfond). For fruitful cooperation and support, thanks are due to Geosond GmbH, Bilfinger&Berger AG, and Th. Hengstermann and F. Storck (both Optimare GmbH).

References

Apitz SE, Borbridge LM, Theriault GA, Lieberman SH (1992) Remote in situ determination of fuel products in soils: field results and laboratory investigations. Analusis 20: 461-474

Balshaw-Biddle, K (ed.) (2000) Subsurface monitoring using laser fluorescence. Lewis Publishers, Boca Raton

Baumann T, Haaszio S, Niessner R (2000) Applications of a laser-induced fluorescence spectroscopy sensor in aquatic systems. Water Res. 34: 1318-1326

Fischer H, Kretschmar H-J, Christoph G, Neyen V (1994) Bestimmung von Mineralölkohlenwasserstoffen im Boden - Bodenstandards und Ringanalysen. Z. Umweltchem. Ökotox. 6: 189-195

Hart SJ, Chen Y-M, Kenny JE, Lien BK, Best TW (1997) Field demonstration of a multichannel fiber-optic laser-induced fluorescence system in a cone penetrometer vehicle. Field Anal. Chem. Technol. 1: 343-357

Heininger P, Pelzer J, Henrion R, Henrion G (1998) Results of a complex round robin test with four river sediments. Fresenius J. Anal. Chem. 360: 344-347

Illenseer C, Löhmannsröben H-G, Skrivanek Th, Zimmerman U (1999) Laser spectroscopy of humic substances. In: Understanding humic substances: advanced methods, properties and applications, Ghabbour EA, Davies G (eds.), Royal Society of Chemistry, London, 129-145

Lieberman SH, Inman SM, Theriault GA, Cooper SS, Malone PG, Lurk PW (1990) Fiber optic-based chemical sensors for in situ measurements of metals and aromatic organic compounds in seawater and soil systems. Proc. SPIE 1269: 175-184

Lieberman SH (1998) Direct-push, fluorescence-based sensor system for in-situ measurement of petroleum hydrocarbons in soils. Field Anal. Chem. Technol. 2: 63-73

Löhmannsröben H-G, Roch Th (1996) In-situ LIF analysis of polynuclear aromatic compounds (PAC) and mineral oils in soils. Proc. SPIE 2835: 128-134

Löhmannsröben H-G, Kauffmann C, Roch Th (1997) Spectroscopic properties of petroleum products in solution for in-situ analysis of oil contaminations. Proc. SPIE 3107: 305-314

Löhmannsröben H-G, Roch Th (1997) Laserfluoreszenzspektroskopie als extraktionsfreies Nachweisverfahren für PAK und Mineralöle in Bodenproben. In: Analytiker Taschenbuch, Springer-Verlag, Heidelberg, vol. 15, 217-253

Löhmannsröben H-G, Schober L (1999) Combination of laser-induced fluorescence and diffuse reflectance spectroscopy for the in situ analysis of Diesel fuel-contaminated soils. Applied Optics 38: 1404-1410

Löhmannsröben H.-G, Roch Th (2000) In situ laser-induced fluorescence (LIF) analysis of petroleum product-contaminated soil samples. J. Environ. Monit. 2: 17-22

Marowsky G, Lewitzka F, Bünting U, Niederkrüger M (2001) Quantitative analysis of aromatic compounds by laser induced fluorescence spectroscopy. Proc. SPIE 4205: 218-223

Mullins OC, Zhu Y (1992) First observation of the Urbach tail in a multicomponent organic system. Appl. Spec. 46: 354-356

Mullins OC, Mitra-Kirtley S, Zhu Y (1992) The electronic absorption edge of petroleum. Appl. Spec 46: 1405-1411

Quinn MF, Al-Otaibi AS, Abdullah A, Sethi PS, Al-Bahrani F, Alameddine O (1995) Determination of intrinsic fluorescence lifetime parameters of crude oils using a laser-fluorosensor with a streak camera detection system. Instrum. Sci. Technol. 23: 201-215

Ralston CY, Wu X, Mullins OC (1996) Quantum yields of crude oils. Appl. Spec50: 1563-1568

Rosenkranz D, Einsele G, Harreß H-M (1995) Bodenschutz. E. Schmidt Verlag, Berlin, vol. 2, 1-4

Schade W, Bublitz J (1996) On-site laser probe for the detection of petroleum products in water and soil. Environ. Sci. Technol. 30: 1451-1458

Shouse B (2001) Kuwait unveils plan to treat festering desert wound. Science 293: 1410

Wang X, Mullins OC (1994) Fluorescence lifetime studies of crude oils. Appl. Spec. 48: 977-984

Zimmermann B, Lucht H (1997) Field measuring devices for in-situ analysis of fluorescence contaminants in water and soil. In: Field Screening Europe, Gottlieb J. (ed.), Kluwer Academic Publishers, Dordrecht, 381-384

6 Laser Induced Breakdown Spectroscopy (LIBS) in Environmental and Process Analysis

Ulrich Panne

6.1 Introduction

Laser induced breakdown spectroscopy, LIBS, is a method utilizing laser ablation and the subsequent atomic emission from the plasma for elemental analysis. Besides the acronym LIBS, today other designations such as LIPS (laser-induced plasma spectroscopy), LA-OES (laser ablation optical emission spectroscopy), or LSS (laser spark spectroscopy) can be found in the literature. Laser ablation is at present the only analytical method that offers direct sampling from any kind of material without sample preparation. So LIBS allows a multielement analysis of virtually all type of materials (gas, solids, liquids) through atomic emission spectroscopy. Today's availability of reliable and less costly laser sources and improved detectors permits a rapid, on-line, and in-situ analysis with LIBS. This makes LIBS especially attractive for all kind of process analysis and environmental screening and monitoring.

The first laser ablation experiment was reported 1962 by Brech et al. (Brech and Cross 1962) and describes already an analytical application, i.e. the microanalysis of solid samples. The plasma was generated with a ruby laser and additionally excited with a electrical spark. Until 1970, it followed a period of lively research in laser plasmas to develop a laser microprobe, inspired by the popular electron microprobe. Although the laser ablation seemed to be conceptual and experimental simple on a first glance, the early results on the laser-matter interaction, the plasma dynamics, and the associated shock waves revealed a much more complex picture of the laser ablation (Ready 1971). The early analytical efforts are summarized in the monograph by Moenke-Blankenburg (Moenke-Blankenburg 1989) and were mainly devoted to analysis of metals. Although three commercial LIBS systems (Moenke-Blankenburg 1989) were available by 1975, quantitative analysis by LIBS was still in an early stage due to problems such as self-absorption, self-reversal, non-stoichiometric ablation, and variable laser-matter interaction (L. J. Radziemski and Cremers 1989). In contrast, laser ablation for atomization, and ionization in mass spectrometry (LAMMA, laser microprobe mass analyzer) or purely sample introduction e.g. for inductively coupled plasma mass spectrometry (ICP-MS), became quite popular in the early eighties. After a period of stagnation in LIBS research, the advances in laser technology and optoelectronics in the late eighties triggered a renaissance in LIBS with a special focus on monitoring and screening instrumentation in environmental and process analysis (Y. I. Lee and Sneddon 1999;L.J. Radziemski 1994;L. J. Radziemski and Cremers 1989;Rusak et al. 1998;Song et al. 1997).

6.2 Plasma Generation

During laser-induced desorption or ablation the energy of excited vibronic and electronic states is converted to kinetic energy, which allows to remove atoms, ions, molecules, and clusters from the surface respective the probed interface. Typically, laser desorption does not involve a mesoscopic modification of the surface and the species yield is a linear function of the number of excited states. In contrast, laser ablation (LA) of a solid is accompanied by the formation of a weakly ionized expanding plasma. The observed modification of the surface is beyond a monolayer and non-linearly related to the number of excited states. Further, an irradiance threshold is observed for laser ablation (Miller 1994).

While thermal laser-matter interactions were favored for laser ablation until the eighties, the advent of fs-lasers revealed the importance of alternative electronic mechanisms. As for LIBS mainly ns-laser pulses are utilized, the discussion will be limited to this time scale (see ref (Amoruso et al. 1999;Griem 1997;Miller 1994;Miller and Haglund 1998;Russo 1995) for more details on plasmas and laser ablation). Due to its mesoscopic character, the macroscopic characteristics of the sample as well as the gas atmosphere can influence the plasma. The plasma generation and dynamics itself additionally influences the ablation (crater morphology, amount of ablated mass etc.) in contrast to other beam techniques utilizing electrons or ions. For ns-pulses (pulse width between 1–30 ns), the laser-matter interaction can be treated separately from the adiabatic expansion of the plasma into a buffer gas or vacuum.

By definition a plasma describes an ensemble of atoms, molecules, ions, and electrons, which as a whole is electrical neutral. However, due to the large number of charged species in the plasma, its characteristics are determined by internal or external electromagnetic interactions. For analytical laser plasmas, which are often considered to be in a local thermodynamic equilibrium (LTE, see below), the electron density n_e and electron temperature T_e are the most important parameters.

The chronology of a laser ablation can be roughly divided in three phases (see Fig. 6.1): (i) primary laser-matter interaction, i.e. heating and evaporation, (ii) plasma generation and laser-plasma interaction (in case of ns laser pulse width), i.e. multiphoton ionization (MPI), invers bremsstrahlung (IB), and photoionization of excited states (PI), (iii) adiabatic expansion of the plasma. Besides the linear scheme in Fig. 6.1, the phenomena are temporally overlapping during laser ablation.

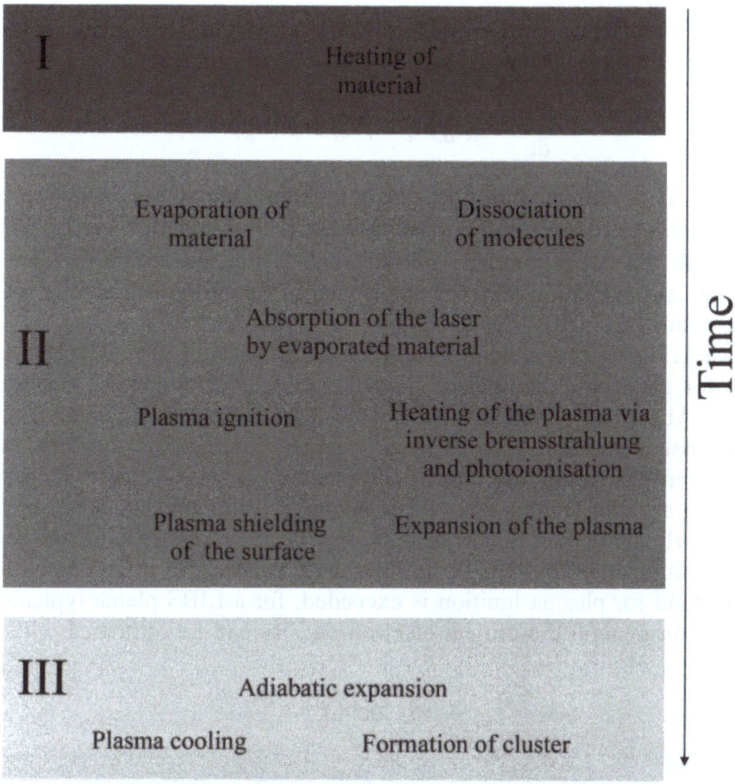

Fig. 6.1. Chronology of a laser ablation

The following elementary processes are of fundamental importance to all three phases: Multiphoton(MPI) or photonionization (PI),

$$A^{+} + e \leftrightarrow A + m\,h\upsilon \tag{6.1}$$

three-body recombination (TBR) resp. electron impact ionization,

$$A^{+} + e + e \leftrightarrow A + e \tag{6.2}$$

collisional excitation and deexcitation, and

$$A + e \leftrightarrow A^{*} + e \tag{6.3}$$

absorption and spontaneous and induced emission.

$$A + h\upsilon \leftrightarrow A^{*} \tag{6.4}$$

The first step is a linear or non-linear absorption of the laser beam via electronic and vibronic states of the investigated material. For plasma ignition, a critical electron density n_e is necessary, which is generated either through a cascade of electron impact ionization (Eq. 6.2) and/or multiphoton ionization (Eq. 6.1). The development of the electron density is then given by,

$$\frac{dn_e}{dt} = \eta(I_B)n_e + \beta I_B^m - R(n_e) \tag{6.5}$$

where the term $\eta(I_B)n_e$ describes the ionization with the avalanche coefficient $\eta(I_B)$, which is a funtion of the irradiance I_B. The second term give the photoionization with the corresponding coefficient β, while the last term represents the losses via recombination (Eq. 6.3 and 4) and diffusion (which is only relevant for a pulse width > 30 ns).

Considering laser wavelength between 200 nm and 1 μm, the plasma ignition differs for absorbing samples, e.g. metals and semiconductors and transparent samples. Absorbing samples dissipate the laser energy through thermal conduction, so that in the case of a ns-pulse a larger volume than the focal volume is heated. Transparent samples absorb photons via non-linear effects such as MPI or absorption at color centers and other defects. All this processes can provide free electrons which are than further heated via inverse bremsstrahlung and result in an avalanche ionization through electron impact. At a certain critical electron density the threshold for plasma ignition is exceeded, for a LIBS plama typically around 10^{18} cm^{-3}. In general the critical electron density can be estimated with the laser frequency ν_L via

$$n_e = \frac{m_e \, 4\pi^2 \, \upsilon_L^2}{\varepsilon_0 \, e^2} \, . \tag{6.6}$$

Eq. 6.6 defines also the plasma frequency ν_p, which limits the absorption via inverse bremsstrahlung. This also explains the differences in the ablation with NIR wavelengths and UV wavelengths: The plasma is transparent for UV wavelengths so that the laser can interact with the sample during the whole pulse, while for NIR wavelengths (e.g. fundamental wavelength of a Nd:YAG laser) the plasma absorbs most the laser pulse. The ablation with UV wavelength permits higher mass ablation and significant higher electron densities, while ablation with NIR wavelengths result in higher plasma temperatures due to IB. For nanosecond laser pulse widths, the threshold for plasma ignition can be defined only by a 50% probability for plasma generation. This stochastic nature of the cascade ionization, is due to the low number of initial electrons in the focal volume (typically about one). The general order for the breakdown threshold of different materials is: solids < liquids < gas (Bettis 1992).

The heated volume of the sample and the temperature of the surface depends for a ns-pulse on the heat diffusion, i.e. thermal effects dominate the mass ablation. Not surprisingly, this can result in non-stoichiometric ablation, which is often observed for ablation with NIR wavelength, but is less important for UV ablation. The ablation itself is a hierarchical process so that thermal, electronic, and structural modifications in the early phase can influence the ablation process as a whole. For example surface structures can lead to enhanced absorption ('hot val-

leys') and subsequent ablation at structural defects. This can result in a dynamic structuring of the surface during the ablation as function of the irradiance and sample characteristics.
After the plasma ignition, the laser interacts with the plasma via inverse bremsstrahlung and photoionization. For UV wavelengths, the electron density is significantly influenced through PI of excited states or MPI of ground states. For an irradiance in the order of 10^8-10^9 W cm^{-2} often a saturation of T_e and n_e is observed due to a self-regulating regime. Through a dynamic equilibrium between the absorption of energy in the plasma and transfer of thermal in kinetic energy, an isothermic expansion of the plasma occurs, while the mass input into the plasma is reduced. With the reduction of charged species in the plasma, the absorption of laser radiation by the plasma is reduced so that again more sample can be ablated. In this way, the size and expansion of the plasma respective the electron density and electron temperature is self regulated during the laser pulse, so that a 'roll-off' of the mass ablation in dependence from the irradiance is observed (Russo 1995).

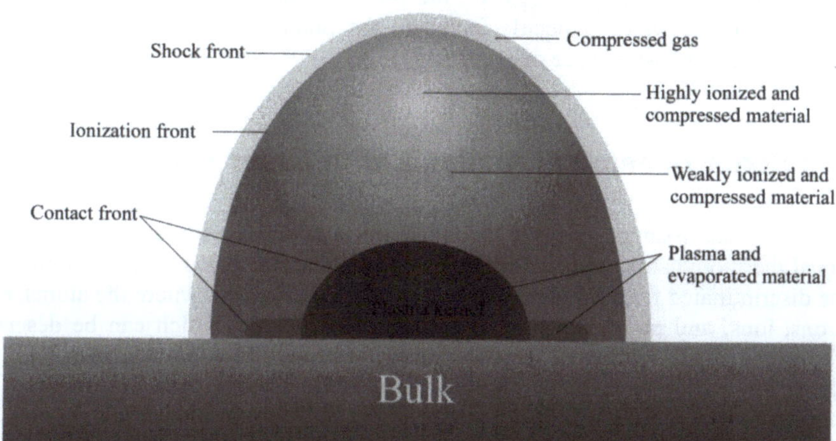

Fig. 6.2. Plasma expansion and formation of a shockwave after plasma ignition

During the phase III in Fig. 6.1, the plasma expands adiabatic, for pressures above 100 Pa a separation of a shock wave front form the luminescent part of the plasma can be observed (see Fig. 6.2). A buffer gas atmosphere regulates the dynamic of the plasma expansion according to the density of the gas and mass input in the plasma. The initial planar shock wave changes with increasing distance to the sample into a spherical wave front following a point source explosion model. Behind the shock wave front the luminescent zone of the plasma is initially further accelerated by IB and PI. The buffer gas also controls the luminescence of the plasma (Ar > Ne, air > He). Generally, Ar as buffer gas results in the highest electron temperatures and electron densities. For He, which is less ionized as Ar, a higher mass ablation is observed due to the reduced shielding of the sample. Polyatomic gases such as oxygen and nitrogen reduce the intensity due to their cross-sections for collisional deexcitation. The plasma ends with the formation of clusters through condensation.

While the plasma generation in gases is very similar to the plasma formation on a solid, the plasma generation in liquids follows a different mechanism, the 'moving-breakdown' model (Sacchi 1991;VogelBusch et al. 1996;VogelNahen et al. 1996). This approach assumes that multiple plasmas are ignited independently along the optical axis, typically the Rayleigh distance of the focal volume, when the critical electron density is reached. With increasing laser pulse energy during the pulse, the breakdown moves in direction of the laser and so that a fast expanding plume (expansion velocity 10^4–10^6 ms^{-1}) is observed, although the shockwave in water is neglectable. This models assume not only that each plasma is not influenced by others, but also that water does not reflect or absorbs the laser pulse in comparison to absorption by inverse bremsstrahlung. The initial electron is generated in ultrapure water through MPI, while the plasma is ignited with ns-pulse width via a cascading impact ionization. In general, the plasma lifetimes in water are reduced by the higher pressure the plasma experiences. This and the dipole moment of the water molecule increases the Stark-broadening of the lines and leads to a fast non-radiative deexcitation.

6.3 Spectrochemical Analysis with Laser Plasmas

The essential assumption for a spectrochemical analysis with laser plasmas is the local thermodynamic equilibrium (LTE) of the plasma. A local equilibrium has to be discriminated from an ideal thermodynamic equilibrium, where the atoms, electrons, ions, and photons are in a reversible equilibrium which can be described through a small number of thermodynamic state variables. In case of a LTE, the mean free path length of the photons exceeds the characteristic dimension of the plasma so that radiation field departs from the equilibrium Planck law. For a sufficient electron density, the electrons as well as the other plasma species still follow a Maxwell distribution with the same temperature T_e. With the electron temperature, the ionization and the distribution of excited states is fixed through the Saha equation

$$\frac{n_e n_i}{n_a} = \frac{2Z_i(T_e)}{Z_a(T_e)} \left(\frac{2\pi m_e k T_e}{h^2} \right)^{3/2} \exp\left(-\frac{\chi_i - \Delta\chi_i}{kT_e} \right) \tag{6.7}$$

and the corresponding

$$\frac{n_m}{n} = \frac{g_m}{Z(T_e)} \exp\left(-\frac{E_m}{kT_e} \right) \tag{6.8}$$

Boltzmann distribution. The indices i resp. a denote atomic and ionic species with their respective partition functions $Z(T_e)$. The ionization potential χ_i is lowered by a reduction $\Delta\chi_i$ due to the plasma. Usually, it is assumed that under the typical experimental LIBS conditions only atoms and singly ionized atoms can be found in the plasma.

Due to their finite size, macroscopic gradients of T_e and n_e are expected in LIBS plasmas, so that an equilibrium can be expected only locally. T_e and n_e are adjusted instantaneously and are unequivocally defined for every moment. As the plasma is assumed to be optically thin, the radiative deexcitation is not compensated through absorption. For an local equilibrium, the excited states have to be populated through collisions. The rate constants for collisional excitation in Eq. 6.3 exceed all other rate constants, so that only the radiation field is decoupled from the equilibrium. Already a small degree of ionization (0.01-1%) is sufficient for the existence of an electron temperature according to the Boltzmann distribution. Griem defined a threshold level of n_e for LTE via

$$n_e \geq 1 \times 10^{12} \sqrt{T_e} \left(E_k - E_i\right)^3 \tag{6.9}$$

with n_e in cm^{-3} and T_e in K. The difference E_k-E_i (in eV) denotes the largest energy difference in the term system of all plasma species. As for all laser plasmas

$$\left(E_k - E_i\right)_{max} < 0.8\chi_i \tag{6.10}$$

holds, a lower electron density can be defined with the largest ionization potential $\chi_{i,max}$ of all plasma species

$$n_e \geq 1 \times 10^{11} \sqrt{T_e}\, \chi^3_{i,max}\,. \tag{6.11}$$

For lower electron densities than Eq. 6.11, the ground state is overpopulated. With the typical conditions of laser plasmas, i.e. $T_e > 10^4$, the electron density should be $> 10^{17}$ cm^{-3} at the time of observation (see below).

Fig. 6.3 shows the typical temporal sequence of a LIBS analysis. Due to the dominating emission from free-free(ff) transitions (bremsstrahlung) and free-bound(fb) transitions (recombination), the early phase of plasma emission cannot be used for atomic spectroscopy. So only after a delay, relative to the plasma ignition, the atomic or ionic emission can be integrated for a defined time and utilized for analytical purposes. Typically, electron densities between 10^{16}–10^{18} cm^{-3} and temperatures of 0.5–2 eV (6000–23000 K) are observed after 1–5 µs delay at 1 atm of air. Both parameters, delay time and integration time, have to be optimized together with the laser irradiance. Typically, a number of prominent analyte lines are analyzed under variable conditions to determine the best signal-to-noise ratio for a given parameter set. The observed emission intensity is not only related to the probability of the population of the specific state and the probability of this transition, but also to the probability of reabsorption of the emitted photon. Hence, a quantitative LIBS analysis depends strongly on the assumption of an optical thin plasma. This assumption can be verified through technical means, e.g. a variable mirror behind the plasma, or through the ratio of observed and expected line intensities of multipletts. Other diagnostic possibilities are related to the line width and profile.

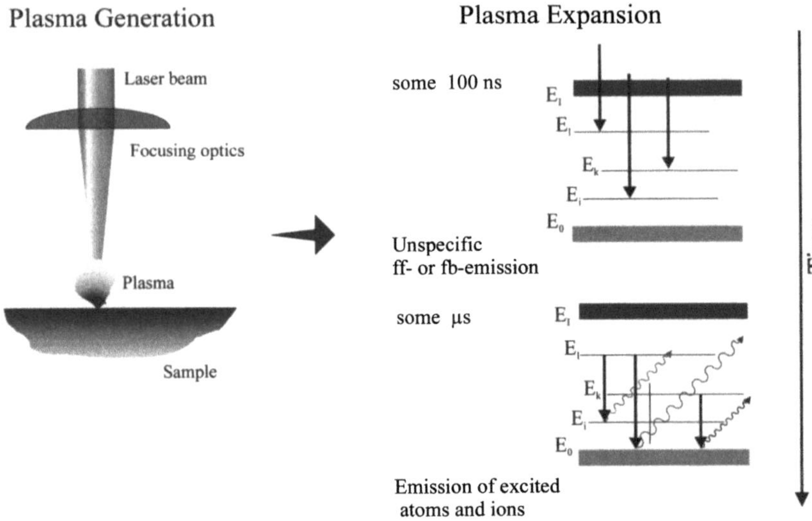

Fig. 6.3. Temporal sequence of a LIBS analysis.

For LTE, the characteristic times to establish the corresponding equilibrium is essential. Macroscopic deviations are observed when the excitation or deexcation is locally not compensated, i.e. either recombining or ionizing plasma conditions are found. Further, non-equilibrium situations can arise during the expansion of the plasma due to transport phenomena. In this case, n_e and T_e decrease faster than the corresponding ionization and population of excited states.

In a transient plasma the characteristic time τ_{plasma} for a change in the temperature or electron density

$$\tau_{Plasma} = \min\left(\frac{n_e}{\partial n_e / \partial t}, \frac{T_e}{\partial T_e / \partial t} \right) \tag{6.12}$$

should be significantly larger than the time τ_a

$$\tau_a = \frac{1}{n_e \langle \sigma v \rangle} \tag{6.13}$$

of atomic processes such as excitation or recombination. σ denotes here the cross section for the most frequent process and v is the electron velocity given through the Maxwell distribution and related thereby to T_e. If $\tau_{plasma} >> \tau_a$, a LTE can be safely assumed even in an transient plasma. For a delay time relative to the plasma ignition of 1–3 μs, a focal diameter (typically, $d_{laser} \approx 100$ μm), and an expansion velocity of the plasma ($v_{exp} \approx 5 \times 10^4$ cm s^{-1}) τ_{plasma} can be estimated via d_{laser}/v_{exp} ($\approx 5 \times 10^{-6}$ s). Although with τ_a in the order of 5×10^{-7} s, the LTE assumption seems to be verified, the difference of only an order of magnitude indicates that deviations from LTE in form of a recombining plasma can still occur in practice.

The oldest and frequently used approach for calibration in LIBS is based on the line ratio of the analyte and an internal standard, i.e. a main component of constant concentration and similar chemical properties in a series of calibration samples. For atomic emission spectroscopy in general, the energy levels of the upper states and the ionization potential of the internal standard and the analyte should be comparable. Then, variable electron temperatures can be corrected through the line ratio. However, as this a-priori knowledge on the sample is not always available, it seems more sensible to use the simultaneously recorded plasma characteristics for normalization. In general, the line intensity is related to the population of an excited state via

$$I_{ki} = \int I \nu \, d\nu = \frac{1}{4\pi} A_{ki} n_k h\nu l = \frac{1}{4\pi} A_{ki} n_k h\nu l \frac{g_k}{Z(T_e)} \exp\left(-\frac{E_k}{k T_e}\right) \tag{6.14}$$

A_{ki} denotes the Einstein probability coefficient for spontaneous emission, l the effective length of the plasma, and I_{ki} the intensity emitted in a 4π solid angle. In a multielement sample, the temperature is modified via the electron density and other plasma species. In practice, a constant F for the observation geometry and the plasma length has to be introduced, which allows to simplify Eq. 6.14 to

$$I_{ki} = F n_k \frac{g_k}{Z(T_e)} \exp\left(-\frac{E_k}{k T_e}\right). \tag{6.15}$$

From Eq. 6.14 also the classical linearization for determination of the plasma temperature from different lines (Boltzmann plot) is derived

$$\ln\left(\frac{g_k A_{ki} c}{I_{ki} \lambda}\right) = \frac{E_k}{kT_e} + \ln\left(\frac{4\pi Z(T_e)}{l h n_k}\right). \tag{6.16}$$

An important prerequisite to utilize Eq. 6.16 correctly, is a large difference between the energy levels taken into account. The error for the determination of T_e is also influenced by the error in the A_{ki}, which is typically between 10 and 40%. The estimation of the error in T_e through

$$\frac{\partial T_e}{T_e} = \frac{k T_e}{(E_1 - E_2)} \frac{\partial\left(I_1 / I_2\right)}{\left(I_1 / I_2\right)} \tag{6.17}$$

demonstrates the significance for the differences in the E_k. A more sophisticated calculation of T_e and n_e can be done by combining Eq. 6.16 with the Saha equation from Eq. 6.7 to

$$\ln\left(\frac{I_{ik}\, g_{aj} A_{a,ji} \lambda_i}{I_{aj}\, g_{ik} A_{i,ki} \lambda_a}\right) = \frac{\varepsilon - \Delta\varepsilon + E_{ik} - E_{aj}}{kT_e} + \ln\left(\frac{2\left(2\pi m_e kT_e\right)^{3/2}}{n_e h^3}\right). \tag{6.18}$$

I_{ik} and I_{aj} denotes here the line intensity of ionic and atomic lines. From the slope of Eq. 6.18 the plasma temperature can be determined, while the ordinate section gives the electron density.

After determination of T_e resp. n_e, the constant F in Eq. 6.15 can be roughly approximated via the line intensity of a known internal standard and the determined plasma temperature. However, without considering the Saha equation, this allows only an imprecise estimation of the concentration. More successful is the determination of the plasma temperature through Eq. 6.16 and subsequent evaluation of the electron density via Eq. 6.7 and Eq. 6.8 for the line ratio of successive ionization levels or the stark broadening of a single line. The later is, however, only accurate for hydrogen lines, while an error of 10–25% has to be considered for heavy elements. Other more sophisticated diagnostic possibilities to determine T_e and n_e are discussed in (Griem 1997;Lochte-Holtgreven 1968).

In this context, a promising approach was described by Palleschi et al. (Ciucci et al. 1999) based on the hypothesis that for every element only a single line and T_e, and n_e have to be known to determine the concentration atomic or ionic species via the second term in Eq. 6.16. The constant F is calculated via normalization to 100%, assuming that all major elements in the sample are detected. The later assumption gives an error in the order of the concentration of the non-detected elements. The difference to the earlier described approach is that here in principle no calibration or other a-priori-knowledge from the sample is needed. Further, the method assumes that all elements in question can be detected simultaneously which in principal requires a sophisticated echelle system (see below).

Further calibrations methods described in the literature are rather of a phenomenological kind, e.g. the acoustic signal from the plasma ignition (Chaleard et al. 1997;Chen and Yeung 1988). An approach termed ‘absolute calibration’ by Schechter et al. (Xu et al. 1997) based on a normalization to the overall plasma intensity was disproved by Gornushkin et al. (Gornushkin et al. 1999).

6.4 Instrumentation

A distinct advantage of LIBS is not only the conceptual simplicity, but also the versatility and simplicity of the experimental set-up. Most present systems are based on either a Q-switched Nd:YAG-laser or a gas laser such as excimers and nitrogen lasers for plasma ignition. An advantage of the solid-state Nd:YAG-lasers is the low maintenance and minimum infra structure needed. In addition, the small dimensions of modern Nd:YAG-lasers permit the construction of transportable LIBS systems.

The detection system comprises a monochromator with a photomultiplier tube or a spectrograph with an intensified array detector, e.g. a linear photodiode array, or two-dimensional charge-coupled devices (CCD). Non-intensified detectors can be used, but permit only analysis of the main components due to insufficient discrimination against the early unspecific emission (see Fig. 6.3). Most conventional detection systems can yield either a wide spectral coverage or a high spectral resolution ($\lambda/\Delta\lambda > 10000$), which results in limited applicability for multielement analysis in complex matrices. Fig. 6.3 illustrates this spectral dilemma for a number of elements recorded with conventional 0.3 m Czerny-Turner spectrograph.

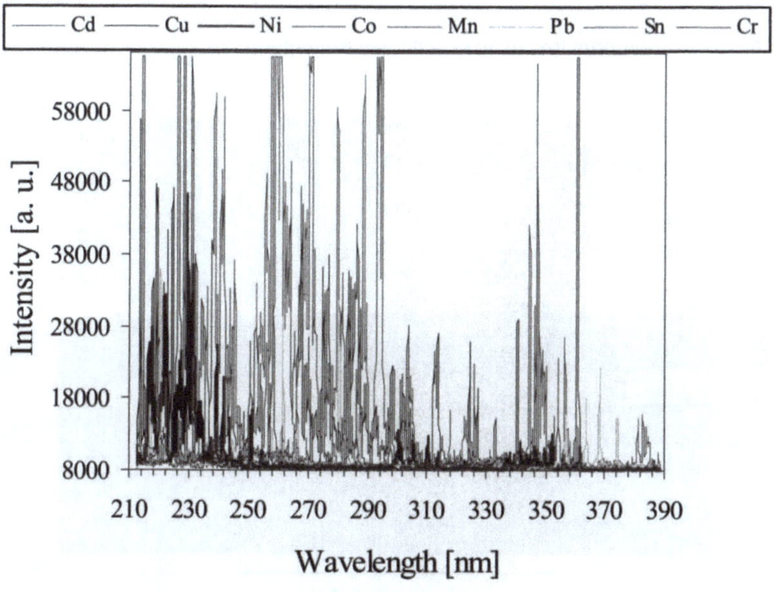

Fig. 6.4. Spectral interference of emission for some typical heavy metal analytes (0.3 m spectrograph, 1200 l mm^{-1} grating, spectral bandpass 0.4 nm)

For simultaneous detection of multiple lines over a large spectral range, polychromators or echelle spectrographs are the systems of choice. Both approaches offer the necessary spectral coverage and spectral resolution for minimizing spectral interferences in complex matrices. An echelle spectrograph employs two different elements for spectral dispersion: an echelle grating is utilized in higher orders (30-120) for high resolution, usually a prism in an orthogonal orientation is employed to resolve the overlapping orders in a second dimension

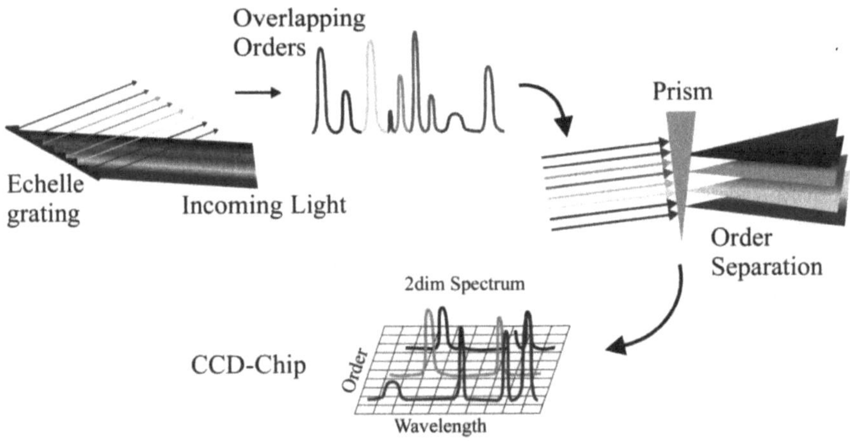

Fig. 6.5. Operation principal of an echelle spectrograph

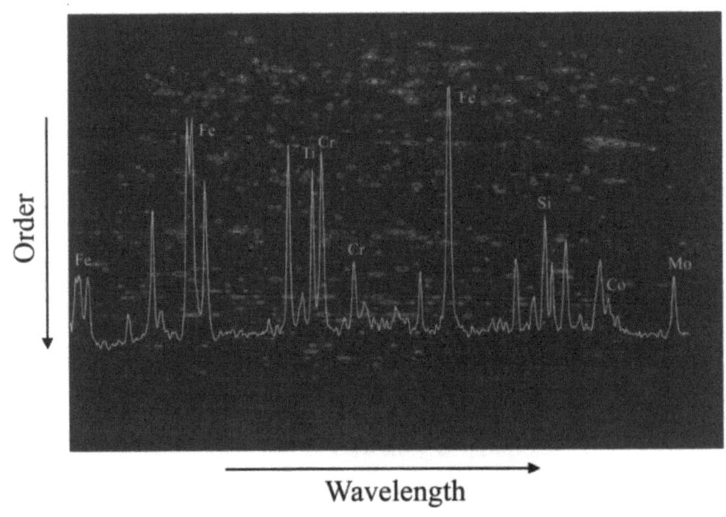

Fig. 6.6. Echelle spectrum from a tungsten mineral (inset: 128th order, 302.0–304.8 nm); adapted from ref. (Florek et al. 2001)

Fig. 6.5 shows the basic operation principal of an echelle system and illustrates the need for a two-dimensional detector. With intensified CCD cameras, current systems allow a spectral range of 200-780 nm with a spectral resolution above 10000. Fig. 6.6 shows exemplarily an echelle spectrum form a tungsten mineral in the wavelength range between 200–400 nm.

Polychromators are based on concave reflection gratings for spectrally resolving and focusing emission lines at the same time (see Fig. 6.7). Due to their horizontal focusing property, no other optical elements are needed in the spectrometer. Various designs are described in the literature, which have all in common that the entrance slit, the grating, and the detectors, typically photomultipliers interfaced to boxcar integrators, are located on a circle with a diameter equal to the radius of curvature of the diffraction grating. The performance of this mountings depend on the image quality, which is limited by astigmatism and coma: The first limits the useful spectral range, while the last restricts the spectral resolution. With focal length between 0.5 and 1 m, typically a spectra resolution above 10000 can be realized in the spectral range 200–600 nm (Neuhauser, Ferstl et al. 1999).

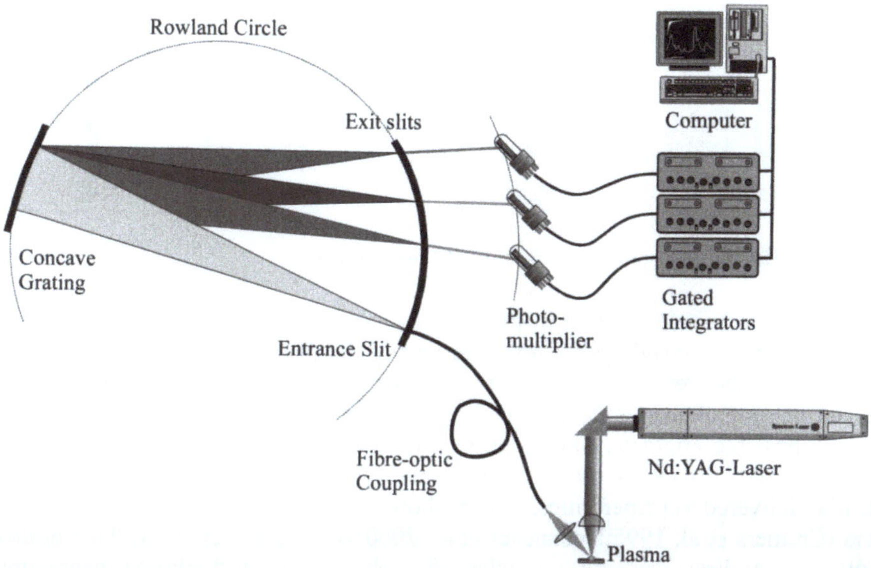

Fig. 6.7. Configuration of a Rowland polychromator and the corresponding experimental set-up for LIBS

In practice the combination of a polychromator with multiplexed boxcars permits to set the timing for each integrator (i.e. observed emission wavelength) independently. Each element can be individually tuned to the best signal-to-noise ratio via a specific delay to the plasma ignition and integration time. Further, with a the current boxcar technology fast read-out rates can be realized so that also Nd:YAG lasers with high repetition rates (e.g. 1 kHz) can be utilized for ablation. In contrast, the read-out speed of the current CCDs is still limited and the nature of the simultaneous detection of all elements allows only a compromise for the timing. However, due to the availability of the whole spectral information, especially for unknown samples with a complex matrix much more information can be gained (Bauer et al. 1998;Florek et al. 2001;Haisch et al. 1998). This includes informations about line width and shape as well as about the spectral background.

Several authors investigated the influence of the focusing optics and the observation geometry (Castle et al. 1998;Multari et al. 1996). Fig. 6.8 shows some of the possibilities utilizing fiber-optical elements which permit a very flexible set-up.

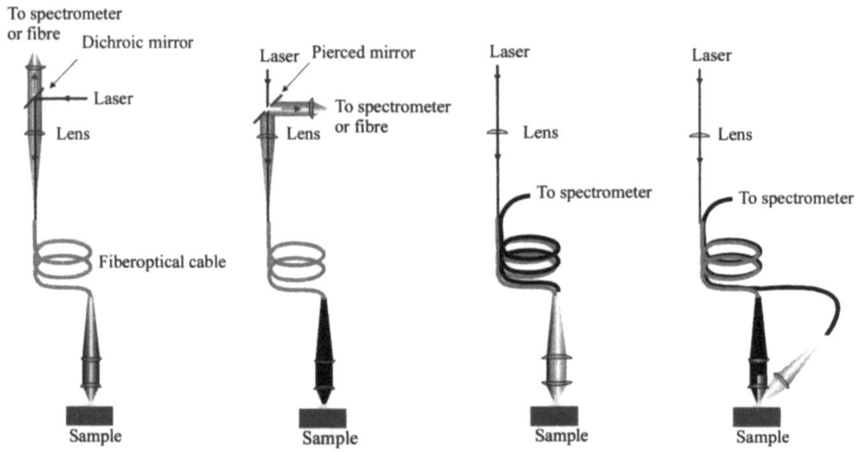

Fig. 6.8. Different fiber-optical set-up for LIBS experiments

Due to the transient nature of the plasma, changes in the observation height or angle lead to the imaging of a different part of the plasma on the entrance slit of the spectrometer resp. the optical fiber. Hence, the observation through a pierced mirror or a simultaneous fiber-optical observation of different plasma regions can reduce possible artefacts. Similar problems can also arise from a heterogeneous sample surface ('dancing plasma'). Not only the emission but also the laser beam can be delivered via fiberoptics, which allows a remote sensing over several meters (Cremers et al. 1995;Neuhauser et al. 2000;Whitehouse et al. 2001). Another option is a direct sensing via telescopic observation and plasma generation (Cremers 1987;Palanco et al. 2002).

6.5 Applications in Environmental and Process Analysis

6.5.1 Solid Samples

Solid sample analysis is probably the most successful area of application for LIBS. Due to the fact that metals have been the preferential samples for all kind of fundamental and analytical studies with LIBS since the first investigations of laser ablation, alloys and steel samples have been investigated in detail either in laboratory-based studies or process analysis work (Aguilera et al. 1992;Andre et al. 1994;Aragon et al. 1999;Bassiotis et al. 2001;Gonzalez et al. 1995;Kagawa et al. 1994;Kim et al. 1997;Niemax and Sdorra 1990;Sabsabi and Cielo 1995;St-Onge et al. 1998;Sturm et al. 2000;Goode, 2000). Usually, the LIBS detection limits rival the classical spark analysers and additionally permit the analysis of non-conducting samples as slags. Besides the simple analysis of macroscopic samples, also the direct analysis of liquid steel was reported (Gruber et al. 2001;Lorenzen et al. 1992;Noll et al. 2001). Due to the multielement character of LIBS, also an unequivocal identification of steel and alloys is possible to avoid mix-ups during production (Goode et al. 2000;Noll et al. 2001). This approach can be also used to monitor further modification in secondary metallurgy, e.g. welding processes (Palanco et al. 2001).

Another very popular area of LIBS analysis is the fast screening of natural materials such as soils, sands, and waste materials such as sewage sludge (Barbini et al. 1997;Eppler et al. 1996;Jensen et al. 1995;Multari et al. 1996;Wisbrun et al. 1993;Yamamoto et al. 1996). Although several problems were reported due to the heterogeneity of the matrix both spatially and chemically, water content, or density of the sample, usually detection limits in the order of 10-100 ppm are obtained if the sample can be conditioned through drying, mixing with a reference material, or pressing of pellets.

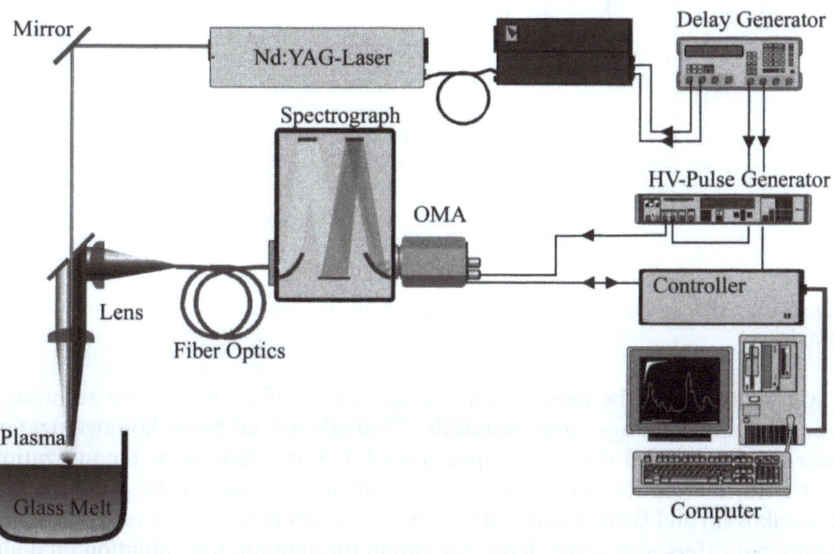

Fig. 6.9. Experimental set-up for remote analysis of glass melts via LIBS (adapted from (Panne et al. 1998))

Other materials of industrial importance investigated by LIBS are glass and ceramics (Kurniawan et al. 1996;Y.I. Lee and Sneddon 1994;Panne et al. 1998;1998;Su et al. 2000), which can be also analysed in remote fashion (Panne et al. 2002). Fig. 6.9 shows the experimental set-up for a remote LIBS analysis of hot glass melts (Panne et al. 1998), while Fig. 6.10 illustrates that no difference between LISB spectra at room temperature and 1200°C can be found. This allows to calibrate the system with various samples at 20°C, before transferring the calibration to the process analysis.

Further LIBS applications can be found for ores and coals (Castle et al. 1998;Chadwick and Body 2002;Gornushkin et al. 2000;Grant et al. 1991;Ottesen 1992;Sun et al. 2000;Wallis et al. 2000), paints especially for lead containing pigments (Castle et al. 1998;Marquardt et al. 1996;Yamamoto et al. 1996), recycling materials such as wood impregnated with inorganic salts (Uhl et al. 2001) and plastics with different kind of fillers and pigments (Anzano et al. 2000;Fink et al. 2001;Sattmann et al. 1998).

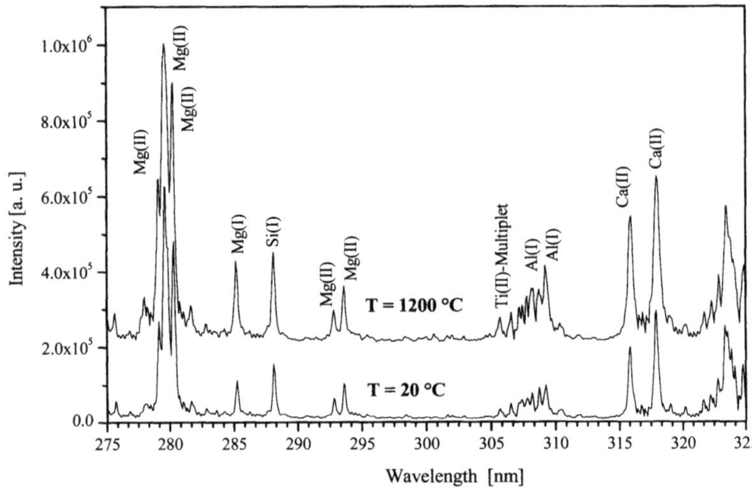

Fig. 6.10. LIBS spectrum for a typical glass melt in coparison with the same solid glass sample.

Certainly one of the most promising applications for LIBS, is the scanning microanalysis of heterogeneous materials. Through special beam homogenization, a laser can be focused down to a spot size of 3–5 µm, fs-laser or the utilization of near field fiber optics allows even smaller ablation craters in the order of 0.5 µm (Kossakovski and Beauchamp 2000). Further, LIBS provides not only spatial resolution, but offers also some depth resolution through multiple ablation on a single spot. In comparison to other beam techniques, LIBS needs only minor sample preparation and no vacuum. This approach was demonstrated by several authors for different heterogeneous and layered materials such a paper coatings, catalysts, or micro inclusions in steel (Garcia et al. 2000;Geertsen et al. 1996;Romero and Laserna 1997;Vadillo et al. 1998). The full potential of this technique is expected for future application with fast laser repetition rates in the order of 0.5–1 kHz and corresponding fast echelle spectrographs.

6.5.2 Liquid Samples and Colloids

The detection limits for the direct analysis of dissolved analytes in water or colloidal materials in water significantly decrease (in the order of 0.1%) due to the fast deexcitation (see above). Fig. 6.11 demonstrates a fast plasma quenching for a colloidal SiO_2 sample: even at elevated concentrations not significant Si(I) signal could be detected.

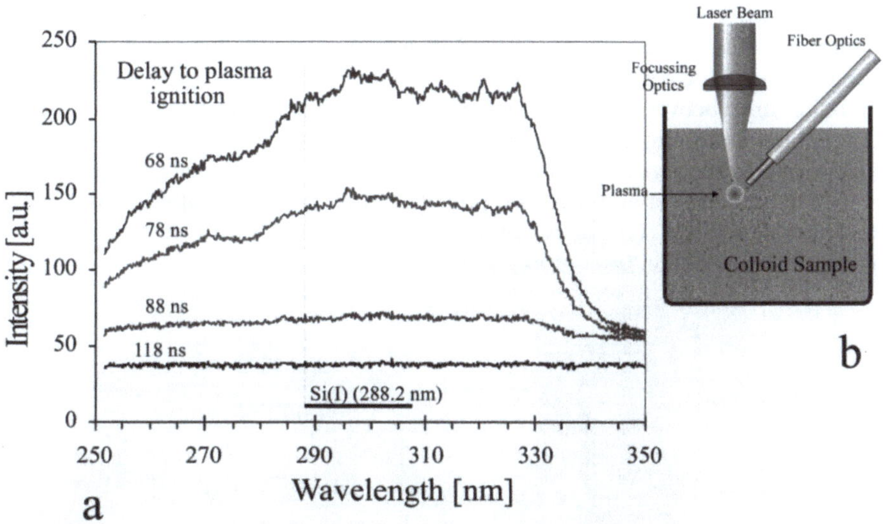

Fig. 6.11. Fast decay of plasma emission from a LIBS analysis of colloidal material (SiO_2 d_p = 250 nm, c = 220 mg l^{-1}) in water (a); inset (b) shows the experimental set-up (adapted from (Haisch et al. 1997))

The only exception are the alkali and alkaline earth metals, due to their low ionization potential (C. W. Ng and Cheung 2000;C . W. Ng et al. 1997). Not surprisingly, the most successful approach for analysis of liquids via LIBS is the ablation form the surface. In this case the plasma is expanding into a buffer gas, which permits ppm detection limits (Arca et al. 1997;Fichet et al. 1999;Fichet et al. 2001). Another approach is to use a second laser pulse to ignite a second plasma in the gas bubble generated after the first breakdown in water (Stratis et al. 2000). This second pulse can be from a second laser or a modified Nd:YAG-laser in a multi-pulse mode. Experimental alternatives to the plasma generation on the sample surface are the generation of a liquid jet (Ito et al. 1995;Nakamura et al. 1996) or a stream of liquid aerosols with µm-diameter (Parigger and Lewis 1993). However, both methods also lead to a further dilution of the original sample, i.e. a decrease in sensitivity. An interesting off-line approach is to absorb dissolved ions on ion-exchange membrane and to ablate the membrane. The corresponds to the ultrafiltration of colloids with a membrane filter with an appropriate cut-off diameter (Haisch et al. 1997;Haisch et al. 1998).

6.5.3 Gaseous Samples and Aerosols

Many of the early laser plasma experiments where performed in gases. However, this fundamental investigations were more focused on the mechanism of plasma formation (Morgan 1975;Nordstrom 1995) than analytical applications. Gas analysis with LIBS was mainly studied for volatile metals especially mercury and halogenated compounds (Cremers and Radziemski 1983;Haisch et al. 1996;Lancaster et al. 1999;Williamson et al. 1998) with detection limits in the ppm range. The detection of F, Cl, and Br are hampered by the fact that all resonance lines are located in the VUV range < 200 nm, so that for LIBS analysis NIR emission lines between 700–850 nm have to be utilized. However, for this wavelength range most currently available detectors have a rather low efficiency. Developments in the area of avalanche photodiode could improve the situation.

For the analysis of aerosols, two principal experimental set-ups, shown in Fig. 6.12, can be realized: (i) a direct, on-line detection in the focal volume or (ii) sampling of the aerosol on filter or impactor.

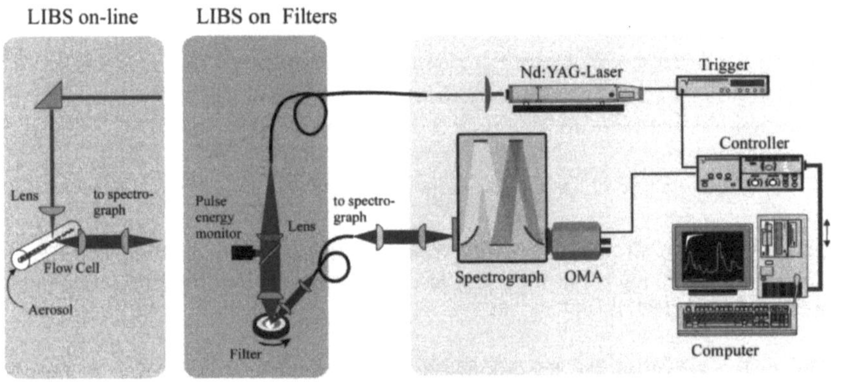

Fig. 6.12. Experimental set-ups for the detection of aerosols via LIBS

While the first approach permits an on-line detection, it is hampered by the low sampling volume and the corresponding decreased sensitivity (detection limits in the μg m^{-3} range). Due to the threshold dependence of the plasma ignition (Zuev et al. 1984), the pulse energy can be adjusted in such a way that only with an aerosol in the focal volume a plasma is generated (Essien et al. 1988;Martin and Cheng 2000;Neuhauser et al. 1997;NeuhauserPanne et al. 1999;Ottesen et al. 1991;Ottesen et al. 1989;Singh et al. 1997). This allows a dedicated single pulse analysis which can improve the detection limits considerable (Hahn and Lunden 2000). The analysis on filter samples allows to integrate larger volumes of air so that analysis of lower aerosols concentration (in the ng m^{-3} range), e.g. from remote areas, is possible. Aerosols were analyzed via LIBS not only for industrial hygiene and anthropogenic emissions, but also for climate change related studies (Arnold and Cremers 1995;Cremers and Radziemski 1985;NeuhauserPanne et al. 1999;Panne et al. 2001).

6.6 Conclusion and Outlook

For some analytical problems, LIBS represents the only reasonable solution to in situ and/or remote sensing. Despite the initial investment cost, the number of specialized applications will probably increase in the future due to the minimum sample preparation, the reduced cost for a single measurement, and the potential for an extensive automation. Quantitative analysis is currently still hampered by the apparent complexity of the transient plasma and laser-matter interactions. If more details of the ablation mechanism are uncovered, improved calibration algorithms taking the plasma and sample characteristics into account can be employed. Undoubtedly, this will be an area of active research also in the future. The application of solid sample analysis, especially microanalysis techniques for nanotechnology or ultrasmall volumes, will grow due to inherent blank limits in ultratrace analysis. In this area, laser ablation in combination with optical emission spectroscopy, i.e. LIBS, or mass spectrometry will have a major roll in the future. Due to the fact that LIBS is technology-driven, advances in the laser technology and optoelectronics will have direct impact on the design and miniaturization of LIBS systems. Applications in process analysis will further profit from fiber-optical systems delivering the laser to the point of demand and guiding the emission to a centralized detection system. The combination with other types of spectroscopy providing molecular information such as Raman spectroscopy will open further applications for LIBS.

References

Aguilera JA, Aragon C, Campos J (1992) Determination of carbon content in steel using laser-induced breakdown spectroscopy. Appl. Spectrosc. 46:1382-1387

Amoruso S, Bruzzese R, Spinelli N, Velotta R (1999) Characterization of laser-ablation plasmas. J. Phys. B 32:R131-R172

Andre N, Geertsen C, Lacour JL, Mauchien P, Sjostrom S (1994) UV Laser ablation-optical emission spectrometry on aluminium alloys in air at atmospheric pressure. Spectrochim. Acta B 49:1363-1372

Anzano JM, Gornushkin IB, Smith BW, Winefordner JD (2000) Laser-induced plasma spectroscopy for plastic identification. Polym. Eng. Sc. 40:2423-2429

Aragon C, Aguilera JA, Penalba F (1999) Improvements in quantitative analysis of steel composition by laser-induced breakdown spectroscopy at atmospheric pressure using an infrared Nd:YAG laser. Appl. Spectrosc. 53:1259-1267

Arca G, Ciucci A, Palleschi V, Rastelli S, Tognoni E (1997) Trace element analysis in water by the laser induced breakdown spectroscopy technique. Appl. Spectrosc. 51:1102-1105

Arnold SD, Cremers DA (1995) Rapid determination of metal particles on air sampling filters using laser-induced breakdown spectroscopy. Amer. Ind. Hyg. Assn. J. 56:1180-1186

Barbini R, Colao F, Fantoni R, Palucci A, Ribezzo S, vanderSteen HJL, Angelone M (1997) Semi-quantitative time resolved LIBS measurements. Appl. Phys. B 65:101-107

Bassiotis I, Diamantopoulou A, Giannoudakos A, Roubani-Kalantzopoulou F, Kompitsas M (2001) Effects of experimental parameters in quantitative analysis of steel alloy by laser-induced breakdown spectroscopy. Spectrochim. Acta B 56:671-683

Bauer HE, Leis F, Niemax K (1998) Laser induced breakdown spectrometry with an échelle spectrometer and and intensified charge coupled device detection. Spectrochim. Acta B 53:1815-1825

Bettis JR (1992) Correlation among the laser-induced breakdown thresholds in solids, liquids, and gases. Appl. Opt. 31:3448-3452

Brech F, Cross L (1962) Optical microemission stimulated by a ruby maser. Appl. Spectrosc. 16:59

Castle BC, Knight AK, Visser K, Smith BW, Winefordner JD (1998) Battery powered laser-induced plasma spectrometer for elemental determinations. J. Anal. At. Spectrom. 13:589-595

Chadwick BL, Body D (2002) Development and commercial evaluation of LIBS chemical analysis technology in the coal power generation industry. Appl. Spectrosc. 56:70-74

Chaleard C, Mauchien P, Andre N, Uebbing J, Lacour JL, Geertsen C (1997) Correction of matrix effects in quantitative elemental analysis with laser-ablation optical-emission spectrometry. J. Anal. At. Spectrom. 12:183-188

Chen G, Yeung ES (1988) Acoustic signal as an internal standard for quantitation in laser-generated plumes. Anal. Chem. 60:2258-2263

Ciucci A, Corsi M, Palleschi V, Rastelli S, Salvetti A, Tognoni E (1999) New procedure for quantitative elemental analysis by laser-induced plasma spectroscopy. Appl. Spectrosc. 53:960-964

Cremers DA (1987) The analysis of metals at a distance using laser-induced breakdown spectroscopy. Appl. Spectrosc. 41:572-578

Cremers DA, Barefield JE, Koskelo AC (1995) Remote elemental analysis by laser-induced breakdown spectroscopy using a fiber-optic cable. Appl. Spectrosc. 49:857-860
Cremers DA, Radziemski LJ (1983) Detection of chlorine and fluorine in air by laser-induced breakdown spectrometry. Anal. Chem. 55:1252-1256
Cremers DA, Radziemski LJ (1985) Direct detection of beryllium on filters using the laser spark. Appl. Spectrosc. 39:57-63
Eppler AS, Cremers DA, Hickmott DD, Ferris MJ, Koskelo AC (1996) Matrix effects in the detection of Pb and Ba in soils using laser-induced breakdown spectroscopy. Appl. Spectrosc. 50:1175-1181
Essien M, Radziemski LJ, Sneddon J (1988) Detection of Cd, Pb and zinc in aerosols by laser-induced breakdown spectrometry. J. Anal. Atom. Spectrom. 3:985-988
Fichet P, Mauchien P, Moulin C (1999) Determination of impurities in uranium and plutonium dioxides by laser-induced breakdown spectroscopy. Appl. Spectrosc. 53:1111-1117
Fichet P, Mauchien P, Wagner JF, Moulin C (2001) Quantitative elemental determination in water and oil by laser induced breakdown spectroscopy. Anal. Chim. Acta 429:269-278
Fink H, Panne U, Niessner R (2001) Analysis of recycled thermoplasts from consumer electronics by laser-induced plasma spectroscopy. Anal. Chim. Acta 440:17-25
Florek S, Haisch C, Okruss M, Becker-Ross H (2001) A new, versatile echelle spectrometer relevant to laser induced plasma applications. Spectrochim. Acta B 56:1027-1034
Garcia CC, Corral M, Vadillo JM, Laserna JJ (2000) Angle-resolved laser-induced breakdown spectrometry for depth profiling of coated materials. Appl. Spectrosc. 54:1027-1031
Geertsen C, Lacour JL, Mauchien P, Pierrard L (1996) Evaluation of laser ablation optical emission spectrometry for microanalysis in aluminium samples. Spectrochim. Acta B 51:1403-1416
Gonzalez A, Ortiz M, Campos J (1995) Determination of sulfur content in steel by laser-produced plasma atomic emission spectroscopy. Appl. Spectrosc. 49:1632-1635
Goode SR, Morgan SL, Hoskins R, Oxsher A (2000) Identifying alloys by laser-induced breakdown spectroscopy with a time-resolved high resolution echelle spectrometer. J. Anal. At. Spectrom. 15:1133-1138
Gornushkin IB, Ruiz-Medina A, Anzano JM, Smith BW, Winefordner JD (2000) Identification of particulate materials by correlation analysis using a microscopic laser induced breakdown spectrometer. J. Anal. At. Spectrom. 15:581-586
Gornushkin IB, Smith BW, Potts GE, Omenetto N, Winefordner JD (1999) Some considerations on the correlation between signal and background in laser-induced breakdown spectroscopy using single-shot analysis. Anal. Chem. 71:5447-5449
Grant KJ, Paul GL, O'Neill JA (1991) Quantitative elemental analysis of iron ore by laser-induced breakdown spectroscopy. Appl. Spectrosc. 45:701-705
Griem HR (1997) Principles of Plasma Spectroscopy. Cambridge University Press, Cambridge
Gruber J, Heitz J, Strasser H, Bauerle D, Ramaseder N (2001) Rapid in-situ analysis of liquid steel by laser-induced breakdown spectroscopy. Spectrochim. Acta B 56:685-693
Hahn DW, Lunden MM (2000) Detection and analysis of aerosol particles by laser-induced breakdown spectroscopy. Aerosol Sci. Technol. 33:30-48
Haisch C, Liermann J, Panne U, Niessner R (1997) Characterization of colloidal particles by laser-induced plasma spectroscopy (LIPS). Anal. Chim. Acta 346:23-35
Haisch C, Niessner R, Matveev OI, Panne U, Omenetto N (1996) Development of a sensor for element-specific determination of chlorine in gases by laser-induced-breakdown spectroscopy (LIBS). Fresenius J. Anal. Chem. 356:21-26

Haisch C, Panne U, Niessner R (1998) Combination of an intensified charge coupled device with an echelle spectrograph for analysis of colloidal material by laser-induced plasma spectroscopy. Spectrochim. Acta B 53:1657-1667

Ito Y, Ueki O, Nakamura S (1995) Determination of colloidal iron in water by laser-induced breakdown spectroscopy. Anal. Chim. Acta 299:401-405

Jensen LC, Langford SC, Dickinson JT, Addleman RS (1995) Mechanistic studies of laser-induced breakdown spectroscopy of model environmental-samples. Spectrochim. Acta B 50:1501-1519

Kagawa K, Kawai K, Tani M, Kobayshi T (1994) XeCl Excimer laser-induced shock wave plasma and its application to emission spectrochemical analysis. Appl. Spectrosc. 48:198-205

Kim DE, Yoo KJ, Park HK, Oh KJ, Kim DW (1997) Quantitative-analysis of aluminum impurities in zinc alloy by laser-induced breakdown spectroscopy. Appl. Spectrosc. 51:22-29

Kossakovski D, Beauchamp JL (2000) Topographical and chemical microanalysis of surfaces with a scanning probe microscope and laser-induced breakdown spectroscopy. Anal. Chem. 72:4731-4737

Kurniawan H, Kagawa K, Okamoto M, Ueda M, Kobayashi T, Nakajima S (1996) Emission spectrochemical analysis of glass containing Li and K in high concentrations using a XeCl excimer laser-induced shock wave plasma. Appl. Spectrosc. 50:299-305

Lancaster ED, McNesby KL, Daniel RG, Miziolek AW (1999) Spectroscopic analysis of fire suppressants and refrigerants by laser-induced breakdown spectroscopy. Appl. Opt. 38:1476-1480

Lee YI, Sneddon J (1994) Direct and rapid determination of potassium in standard solid glasses by excimer laser ablation plasma atomic emission spectrometry. Analyst 119:1441-1443

Lee YI, Sneddon J (1999) Laser-induced breakdown spectrometry. In: Sneddon J (ed.) Advances in Atomic Spectroscopy. Jai Press Inc, Hampton Hill, pp. 235-288

Lochte-Holtgreven W (1968) Evaluation of Plasma Parameters. In: Lochte-Holtgreven W (ed.) Plasma Diagnostics. North-Holland Publ. Company, Amsterdam, pp. 135-213

Lorenzen CJ, Carlhoff C, Hahn U, Jogwich M (1992) Applications of laser-induced emission spectral analysis for industrial process and quality control. J. Anal. At. Spectrom. 7:1029-1035

Marquardt BJ, Goode SR, Angel SM (1996) In situ determination of lead in paint by laser-induced breakdwon spectroscopy using a fiber optic probe. Anal. Chem. 68:977-981

Martin M, Cheng MD (2000) Detection of chromium aerosol using time-resolved laser-induced plasma spectroscopy. Appl. Spectrosc. 54:1279-1285

Miller JC, ed. (1994) Laser Ablation. Springer Verlag, Berlin

Miller JC Haglund RF, eds. (1998) Laser Ablation and Desorption. Academic Press, San Diego

Moenke-Blankenburg L (1989) Laser Micro Analysis. John Wiley & Sons, New York

Morgan CG (1975) Laser-induced breakdown of gases. Rep. Prog. Phys. 38:621-665

Multari RA, Foster LE, Cremers DA, Ferris MJ (1996) Effect of sampling geometry on elemental emissions in laser-induced breakdown spectroscopy. Appl. Spectrosc. 50:1483-1499

Nakamura S, Ito Y, Sone K, Hiraga H, Kaneko K (1996) Determination of an iron suspension in water by laser-induced breakdown spectroscopy with two sequential laser pulses. Anal. Chem. 68:2981-2986

Neuhauser RE, Ferstl B, Haisch C, Panne U, Niessner R (1999) Design of a low-cost detection system for laser-induced plasma spectroscopy. Rev. Sci. Instrum. 70:3519-3522

Neuhauser RE, Panne U Niessner R (2000) Utilization of fiber optics for remote sensing by laser-induced plasma spectroscopy (LIPS). Appl. Spectrosc. 54:923-927

Neuhauser RE, Panne U, Niessner R, Petrucci GA, Cavalli P, Omenetto N (1997) On-line and in situ detection of lead aerosols by plasma spectroscopy and laser-excited atomic fluorescence spectroscopy. Anal. Chim. Acta 346:37-48

Neuhauser RE, Panne U, Niessner R, Wilbring P (1999) On-line monitoring of chromium aerosols in industrial exhaust streams by laser-induced plasma spectroscopy (LIPS). Fresenius J. Anal. Chem. 364:720-726

Ng CW, Cheung NH (2000) Detection of sodium and potassium in single human red blood cells by 193-nm laser ablative sampling: A feasibility demonstration. Anal. Chem. 72:247-250

Ng CW, Ho FW, Cheung NH (1997) Spectrochemical analysis of liquids using laser-induced plasma emissions: Effects of laser wavelength on plasma properties. Appl. Spectrosc. 51:976-983

Niemax K, Sdorra W (1990) Optical emission spectrometry and laser-induced fluorescence of laser produced sample plumes. Appl. Opt. 29:

Noll R, Bette H, Brysch A, Kraushaar M, Monch I, Peter L, Sturm V (2001) Laser-induced breakdown spectrometry - applications for production control and quality assurance in the steel industry. Spectrochim. Acta B 56:637-649

Nordstrom RJ (1995) Study of laser-induced plasma emission spectra of N_2, O_2, and ambient air in the region 350 nm to 950 nm. Appl. Spectrosc. 49:1490-1499

Ottesen DK (1992) Laser Spark Emission Spectroscopy of Individual Coal Particles. In: Meuzelaar HLC (ed.) Advances in Coal Spectroscopy. Plenum Press, New York, pp. 91-118

Ottesen DK, Baxter LL, Radziemski LJ, Burrows JF (1991) Laser spark emission spectroscopy for in situ, real-time monitoring of pulverized coal particle composition. Energ. Fuel 5:304

Ottesen DK, Wang JCF, Radziemski LJ (1989) Real-Time laser spark spectroscopy of particulates in combustion environments. Appl. Spectrosc. 43:967-976

Palanco S, Baena JM, Laserna JJ (2002) Open-path laser-induced plasma spectrometry for remote analytical measurements on solid surfaces. Spectrochim. Acta B 57:591-599

Palanco S, Klassen M, Skupin J, Hansen K, Schubert E, Sepold G, Laserna JJ (2001) Spectroscopic diagnostics on CW-laser welding plasmas of aluminum alloys. Spectrochim. Acta B 56:651-659

Panne U, Haisch C, Clara M, Niessner R (1998) Analysis of glass and glass melts during the vitrification of fly and bottom ashes by laser-induced plasma spectroscopy. Part 1: Normalization and plasma diagnostics,. Spectrochimica Acta B 53:1957-1968

Panne U, Haisch C, Clara M, Niessner R (1998) Analysis of glass and glass melts during the vitrification of fly and bottom ashes by laser-induced plasma spectroscopy. Part 2: Process analysis. Spectrochim. Acta B 53:1969-1981

Panne U, Neuhauser RE, Haisch C, Fink H, Niessner R (2002) Remote analysis of a mineral melt by laser-induced plasma spectroscopy. Appl. Spectrosc. 56:375-380

Panne U, Neuhauser RE, Theisen M, Fink H, Niessner R (2001) Analysis of heavy metal aerosols on filters by laser-induced plasma spectroscopy. Spectrochim. Acta B 56:839-850

Parigger C, Lewis JWL (1993) Measurements of sodium chloride concentration in water droplets using laser-induced plasma spectroscopy. Opt. Comm. 12:163-173

Radziemski LJ (1994) Review of selected analytical applications of laser plasmas and laser ablation, 1987-1994. Microchem. J. 50:218-234

Radziemski LJ, Cremers DA, eds. (1989) Laser-Induced Plasmas and Applications.Marcel Dekker, New York

Ready JF (1971) Effects of High-Power Laser Radiation. Academic Press, New York
Romero D, Laserna JJ (1997) Multielemental chemical imaging using laser induced breakdown spectrometry. Anal. Chem. 69:2871-2876
Rusak DA, Castle BC, Smith BW, Winefordner JD (1998) Recent trends and the future of laser-induced plasma spectroscopy. TrAC, Trends Anal. Chem. 17:453-461
Russo RE (1995) Laser-Ablation. Appl. Spectrosc. 49:A14-A28
Sabsabi M, Cielo P (1995) Quantitative analysis of copper alloys by laser-produced plasma spectrometry. J. Anal. Atom. Spectrom. 10:643-647
Sacchi CA (1991) Laser-induced eletric breakdown in water. J. Opt. Soc. Am. B 8:337-345
Sattmann R, Monch I, Krause H, Noll R, Couris S, Hatziapostolou A, Mavromanolakis A, Fotakis C, Larrauri E, Miguel R (1998) Laser-induced breakdown spectroscopy for polymer identification. Appl. Spectrosc. 52:456-461
Singh JP, Yueh FY, Zhang HS, Cook RL (1997) Study of laser induced breakdown spectroscopy as a process monitor and control tool for hazardous waste remediation. Process Cont. Qual. 10:247-258
Song K, Lee YI, Sneddon J (1997) Applications of laser-induced breakdown spectrometry. Appl. Spectrosc. Rev. 32:183-235
St-Onge L, Sabsabi M, Cielo P (1998) Analysis of solids using laser-induced plasma spectroscopy in double-pulse mode. Spectrochim. Acta B 53:407-415
Stratis DN, Eland KL, Angel SM (2000) Dual-pulse LIBS using a pre-ablation spark for enhanced ablation and emission. Appl. Spectrosc. 54:1270-1274
Sturm V, Peter L, Noll R (2000) Steel analysis with laser-induced breakdown spectrometry in the vacuum ultraviolet. Appl. Spectrosc. 54:1275-1278
Su CF, Feng S, Singh JP, Yueh F-Y, Rigsby JT, III, Monts DL, Cook RL (2000) Glass composition measurement using laser induced breakdown spectrometry laser spectroscopy. Glass Technol. 41:16-21
Sun Q, Tran M, Smith BW, Winefordner JD (2000) Determination of Mn and Si in iron ore by laser-induced plasma spectroscopy. Anal. Chim. Acta 413:187-195
Uhl A, Loebe K, Kreuchwig L (2001) Fast analysis of wood preservers using laser induced breakdown spectroscopy. Spectrochim. Acta B 56:795-806
Vadillo JM, Vadillo I, Carrasco F, Laserna JJ (1998) Spatial distribution profiles of magnesium and strontium in speleothems using laser-induced breakdown spectrometry. Fresenius J. Anal. Chem. 361:119-123
Vogel A, Busch S, Parlitz U (1996) Shock wave emission and cavitation bubble generation by picosecond and nanosecond optical breakdown in water. J. Acoust. Soc. Am. 100:148-165
Vogel A, Nahen K, Theisen D, Noack J (1996) Plasma formation in water by picosecond and nanosecond Nd:YAG laser pulses - part I: optical breakdown at threshold and superthreshold irradiance. IEEE J. Sel. Top. Quantum Electron. 2: 847-860
Wallis FJ, Chadwick BL, Morrison RJS (2000) Analysis of lignite using laser-induced breakdown spectroscopy. Appl. Spectrosc. 54:1231-1235
Whitehouse AI, Young J, Botheroyd IM, Lawson S, Evans CP, Wright J (2001) Remote material analysis of nuclear power station steam generator tubes by laser-induced breakdown spectroscopy. Spectrochim. Acta B 56:821-830
Williamson CK, Daniel RG, McNesby KL, Miziolek AW (1998) Laser-induced breakdown spectroscopy for real-time detection of halon alternative agents. Anal. Chem. 70:1186-1191
Wisbrun R, Niessner R, Schröder H (1993) Laser-Induced breakdown spectrometry as a fast screening sensor for environmental analysis of trace amounts of heavy metals. Anal. Methods Instrum. 1:17-22

Xu L, Bulatov V, Gridin VV, Schechter I (1997) Absolute analysis of particulate materials by laser-induced breakdown spectroscopy. Anal. Chem. 69:2103-2108

Yamamoto KY, Cremers DA, Ferris MJ, Foster LE (1996) Detection of metals in the environment using a portable laser-induced breakdown spectroscopy instrument. Appl. Spectrosc. 50:222-233

Zuev VE, Zemlyanov AA, Kopytin YD, Kuzikovskii AV (1984) High-Power Laser Radiation in Atmospheric Aerosols. D. Reidel Publishing Company, Dordrecht

7 Intracavity-, Laser-Desorption- and Cavity Ring-Down Techniques as Detection Devices for Samples in Condensed Phases

J. Lauterbach, H. Bettermann, R. Steinert, P. Hering and K. Kleinermanns

This contribution presents three different experimental approaches for studying molecules in condensed phases: an intracavity Raman technique which is used to identify isomeric samples which could not be separated by standard HPLC settings, a laser desorption technique which provides the transport of large molecules into the gas phase under mild conditions and the cavity-ring-down spectroscopy for condensed phases. The latter method developed recently has been applied to questions in the field of nano-technology.

7.1 An Intracavity Laser Raman Detection Device for HPLC Chromatography

Over many decades the Raman spectroscopy has been successfully applied to investigate molecular properties and structures. The success of the Raman spectroscopy based on its simple instrumentation which basically consists of a powerful light source, a well-dispersing monochromator and a sensitive detection device. Furthermore, no special efforts have to be made to prepare the samples for recording the spectra.

In addition, a comprehensive interpretation of Raman spectra can be carried out by quantum mechanically assisted vibrational analyses which are rapidly provided by modern computers.

On the other hand, the Raman spectroscopy has an intrinsic disadvantage. Since the beginning of measuring Raman spectra, multifarious attempts have been made to obtain larger signals in order to record weaker transitions or to study signals from small sample concentrations. Compared to molecular fluorescence emissions, Raman signals are notoriously weak. The magnitude of scattered radiation depends linearly on the intensity of the excitation laser light source, the number of exposed molecules, the fourth power of the absolute scattering wavenumber, and the sum of the squared elements of the scattering tensor, respectively. The smallness of the latter quantity determines the small yield of Raman photons. As a rule of thumb, one million laser photons generate one Raman photon only. The magnitude of scattering tensor elements can be increased by matching the frequency of the laser to an optically allowed electronic transition of the molecules which are under study. In this case, certain Raman signals of the molecule can be strongly enhanced by resonance effects. Beside this property, the increase of the frequency and the power of the excitation source as well as the extension of the solid angle of the scattered light are further tools to enhance the scattering intensity.

In this contribution, we are concerned with a better detectability of Raman signals by improving the instrumental components, especially the imaging of the scattered light (Steinert et al. 1997, 1998). The improved imaging conditions of our setup enable the record of scattering signals from small-sized sample volumes usually present in chromatographic separations and make it possible to apply Raman spectroscopy as detection device for HPLC (Iriyama et al. 1983; Koizumi and Suzuki 1987).

7.1.1 Experimental Approach

Since the intensity of the scattered light depends linearly on the intensity of light exposing the sample, the sample cell was placed inside a modified laser resonator. In the case of liquid samples the fluctuations within the liquid may cause considerable intensity fluctuations of the laser light. Considering HPLC-flow conditions, these possible fluctuations can not be compensated by longer detection times. To avoid this, the sample was placed into a separated cavity which was optically coupled to the original laser resonator.

Fig. 7.1 presents the experimental setup. The excitation unit consists of an argon ion laser, a band-pass filter (F_1), a focusing lens (L_f), a pinhole (P), and the feedback mirror (M_3). After passing the sample cell (S), the laser beam is coupled back into the laser by the feedback mirror (radius 60 cm, reflectivity 99.9%). This enhances the laser power to a factor of 2.7 inside the sample compared to ordinary laser excitation. To achieve this increase in power, the feedback mirror must be adjusted in such a way that its radius of curvature fits the radius of the phase front of the laser beam.

The band-pass filter, which consists of a holographic diffraction grating between two quartz prisms, reduces the spontaneous emission lines from the argon discharge. This task is achieved by effectively directing the exciting laser beam in an angle of 90 degrees. The adjacent spectral lines, now angular-dispersed, are further rejected by a pinhole (P). The hole (diameter 3 mm) is drilled into a reflector of stainless steel (R). The reflector R is part of the light collecting unit likewise. The lens L_f produces a sharp focus of the laser beam with a diameter of about 10 μm in the sample.

The sample cell has an optical path length of 1 cm and width of 4 mm. Due to the imaging conditions, the main part of the scattered light origins from a sample volume of less than 1 nL.

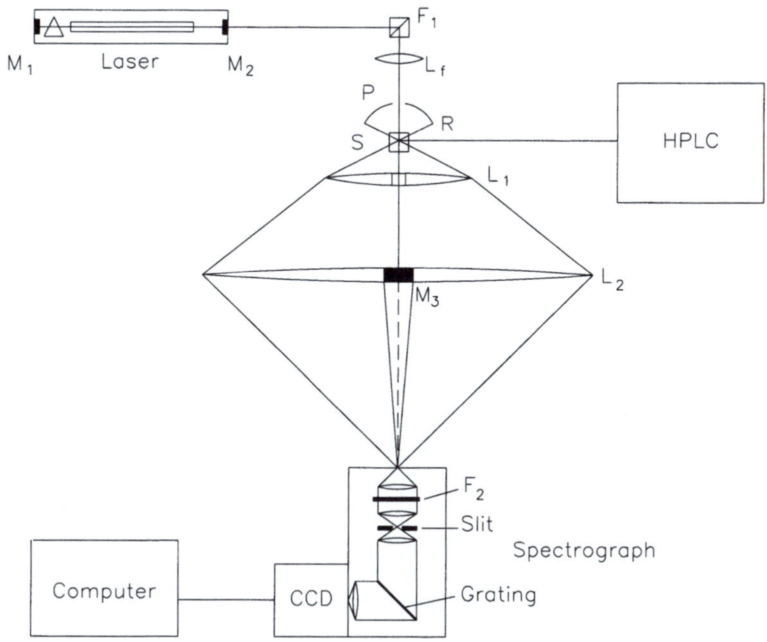

Fig. 7.1. Schematic setup of the HPLC-Raman detector

For HPLC separations, the sample cell was optionally replaced by a flow cell and an injection volume of 100 μL was chosen for each measurement. The flow velocity of the mobile phase water/acetonitrile (50 %) was set to the commonly selected value of 1 mL/min.

The light collection unit consists of two aspheric multi-element lenses ("Fresnel lenses"), made of acrylic plastic, and the spherical reflector (R). The first lens (L_1) has a pinhole of 3 mm diameter in its centre to let the laser beam pass without focusing. The second lens (L_2) holds the feedback mirror M_3, which is fitted into a bore in its centre. This construction reduces the shadowing of the scattered light by the mirror and its holder to a minimum. Furthermore, the part of the light which is scattered in the backside direction of the imaging lenses is collected by the reflector R which contains a hole (P) for letting through the laser beam.

The multi-element lenses are excellent for imaging a point-like light source into a point-like image, i.e., for illuminating the entrance slit of a spectrometer. Moreover, they give better ratios of focal length to diameter, hence larger solid angles of scattered light can be achieved and they are better corrected for spherical aberration than usual lenses.

The light detection unit contains the spectrograph with a large entrance ratio (f/1.8), a CCD camera cooled with liquid nitrogen (-90° C) and equipped with a 1024x256 pixel chip. The spacing of the pixels and the dispersion of the monochromator provide a spectral resolution of about 10 cm^{-1}. Inside the housing of the spectrograph, a holographic notch filter is inserted between the entrance plane of the spectrograph and the internal slit (SL, 100 μm). This filter suppresses the Rayleigh scattering that would generate a large amount of stray light. The reduction of stray light enhances the signal to background ratio considerably.

The sensitivity of the measurements is limited by the amount of background signal. This depends on the number of pixel columns in the selected pixel area and on the total number of incoming photons in the same area. For obtaining the best results, the spectral range must be set to a region which is characteristic for the examined molecules and not congested by the Raman signals of the solvent. To avoid saturation effects of the CCD, the total measuring time (typically 150 s) is divided into a sequence of several single exposures (typically 10 to 50 s each).

Near the detection limit, the evaluation of the data consists of several steps. The measured spectrum is corrected for background signals by subtracting a properly scaled spectrum of the pure solvent. The remaining background may contain fluorescence signals, which have a much larger spectral bandwidth than the Raman signals, and may show broad artificial structures that arise from imperfect corrections. Both contributions to the background signal are removed to obtain a flat baseline. This is carried out by subtracting a cautiously smoothed graph of the background. Finally, there are remaining periodic structures in the spectra caused by the CCD camera. These structures show a larger standard deviation than the Raman signal and the background and in this way are limiting the sensitivity of the measurements. The resulting spectrum is Fourier filtered.

We define the detection limit by the ratio of the height of Raman signals to the height of signals originated from adjacent residual structures. The detection limit is reached, if the ratio is unity.

7.1.2 Results

The presented experimental setup was first tested by measuring low-concentrated solutions of benzene. Acetonitrile has been chosen as solvent since it provides suitable spectral ranges in which vibrations of the solvent are missing. Restricting the measuring time for one spectrum to a value of about 150 s, the detection limit can be set to $3 \cdot 10^{-6}$ mol/L with respect to the above made definition. As a second test, spectra were recorded from a mixture of m-xylene and p-xylene diluted in acetonitrile, too. Both xylenes could still be identified by their main transitions at a concentration of $1 \cdot 10^{-5}$ mol/L (1 μg/mL, 430 ppb) for each isomer.

The third example may demonstrate the efficiency of Raman spectroscopy with regard to the analysis of multi-component systems. Fig. 7.2 shows Raman spectra of dichloro-phenols at 10^{-4} mol/L for each isomer dissolved in acetonitrile as well as the spectrum of the mixture including all dichloro-phenols. The dichlorophenols differ only in their substitution pattern of the two chlorine atoms. Using standard HPLC conditions, the structural isomers have nearly the same retention times. Thus, it can hardly be decided which isomer or how many compounds do generate a signal in a chromatogram. Since electronic transitions from compounds in liquids have large halfwidths and show only little differences by slight changes of the chromophoric system, fluorescence spectroscopy and absorption spectroscopy as conventional detection methods in HPLC are overtaxed to identify the compounds.

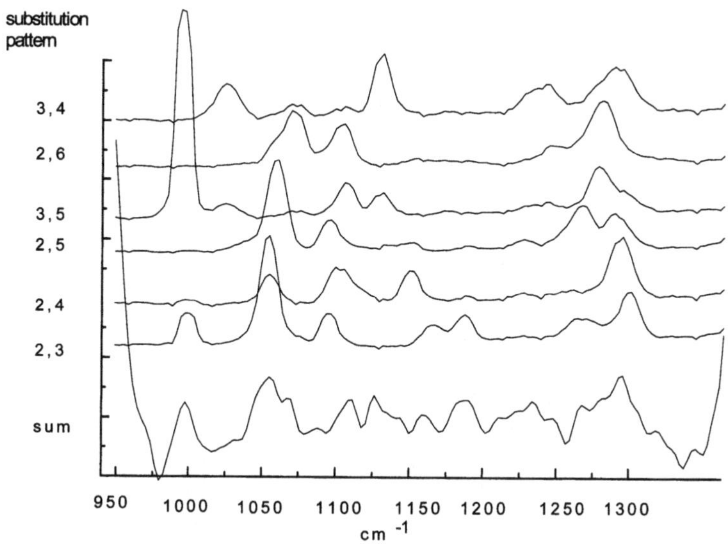

Fig. 7.2. Raman spectra of the single dichloro-phenols and the sum spectrum. The spectra are indicated by the substitution patterns of the compounds on the ordinate axis. The strong increases within the sum graph results from the signal processing procedure.

Following the approach of normal vibrations each atom of the molecule moves during a vibration. By comparing a large number of organic compounds it has been observed that certain vibrations can be attributed to specific types of bonds. These group frequencies change slightly in going from one compound to another. Group frequency ranges are mostly observed for stretching vibrations but they are rather invalid for deformation modes in which dominantly bond angles and dehedral angles are altered (Sverdlov et al. 1974). These latter fundamental mode frequencies depend sensitively on changes in the distribution of atomic masses and the internal force constants so that spectra of isomers differ considerably. Therefore the frequency pattern of the C-H in-plane and out-of-plane deformation modes is very characteristic for each dichloro-phenol. As can be seen from Fig. 7.2, this property enabled us to identify each dichloro compound among all the remaining isomers and this example shows the possibilities in solving problems of trace analysis by the combination of HPLC with Raman spectroscopy.

For considerably lower concentrations and the analysis of trace samples, it is surely necessary to make use of resonance Raman effects. For this the principal experimental scheme has not to be altered. Since most of the analytically relevant molecules absorb ultraviolet light, the laser light source has to preferably emit light in the ultraviolet range. Because of the fact that to our knowledge large multi-element quartz lenses are not available, standard lenses have be chosen. This will presumably account for only half an order of magnitude of signal, while the enhancement by resonance will improve the sensitivity of the Raman spectrometer by a factor of an estimated hundred or more.

7.2 Laser Desorption Spectroscopy

We used laser desorption spectroscopy as a wavelength selective mass spectroscopic method to analyse thin layer (TL) chromatograms. The main advantage of this setup is the transport of large molecules from the TL-plate into the gas phase without fragmentation. Additionally, this method provides a very low detection limit and a high quality in identifying molecular structures using time of flight mass spectroscopy (TOF) as detecting device. Furthermore, the implemented Resonant Two Photon Ionisation (R2PI) spectroscopy enables the analysis of isomers. R2PI has been applied by a number of working groups in order to determine various classes of chemical and biological substances.

Indolamine and peptides as representatives of thermally unstable biological molecules were desorbed with nanogram sensitivity from silica-gel covered TL-plates and analysed subsequently. The method is comprehensively described in (Liang and Lubman 1989). Additionally, this reference gives a good overview over the analytical applications. Fig. 7.3 presents our experimental setup of laser desorption spectroscopy. The progress with regard to the former publication is the adjustability of the TL-plate carrier. This advantage enables the analysis of various local areas in the plate.

A molecular beam is generated by expansion of Ar, He or CO_2. The desorption laser at 266 nm is used to ablate the TL-plate which is moveable by a xyz translator. The molecular beam passes the skimmer and the desorbed and jet cooled molecules are ionized by a tunable UV-laser. Finally, the ion masses are selected by a time of flight spectrometer.

We were able to detect benzotriazole on thin layer chromatographic plates with our apparatus. The TOF mass spectrum of benzotriazole obtained from resonant ionisation agrees well with the electron impact mass spectrum so that the Wiley library could be used to identify the substance. The nanogram sensitivity necessary to analyse typical TL chromatograms was, however, not achieved yet with the apparatus shown in Fig. 7.3. More sensitive is the desorption of molecules from a graphite substrate with a 1.06 μm Nd-YAG laser. With this method it was possible to detect dimers of DNA bases and obtain their UV spectra (Nir et al. 2000). We propose to sensitise the TL-plates by spraying graphite powder on their surfaces and desorb the chromatographically separated compounds with a Nd-YAG laser at 1.06 μm.

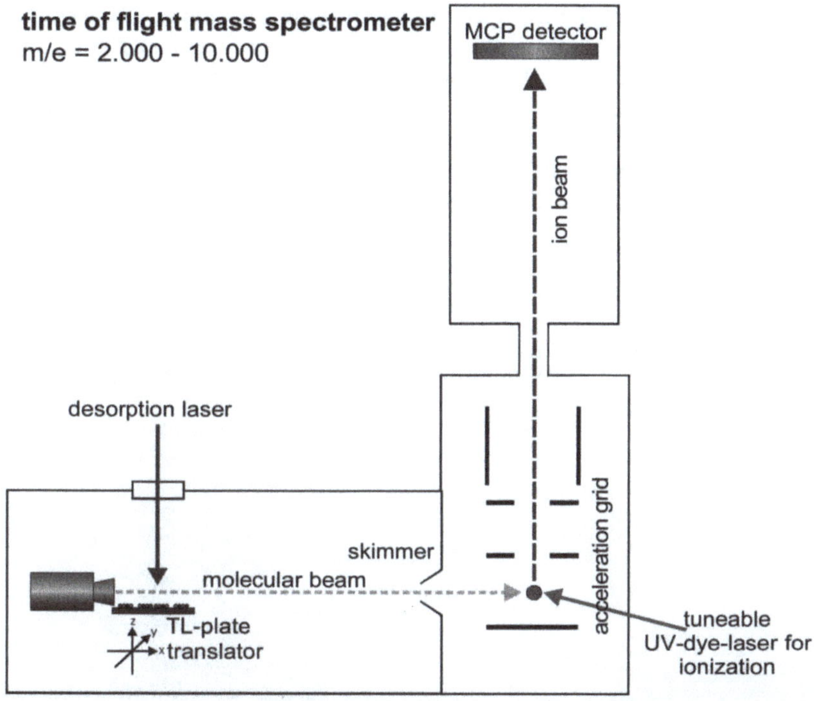

Fig. 7.3. Our experimental setup of laser desorption, laser ionisation spectroscopy

7.3 Cavity Ring-Down Spectroscopy of the Condensed Phase

7.3.1 Introduction

Cavity ring-down spectroscopy (CRDS) holds also great promise for the analysis of very thin solid layers and liquid films. This will be shown in this part of the article. But first of all the principles of pulsed CRDS are described briefly in this introduction. The details of the mathematical foundations and working principles of CRDS can be found in (Romanini and Lehmann 1993; Zalicki and Zare 1995; Zalicki et al. 1995; Lee et al. 1999; Hodges et al. 1996).

Cavity ring-down spectroscopy (CRDS) is a highly sensitive laser absorption method and was introduced in 1988 by O'Keefe and Deacon (O'Keefe and Deacon 1988). Often it is used in trace gas measurements for quantitative analysis of molecular species on the sub-ppb level. But also many other applications of CRDS like flame diagnostics, kinetic measurements, overtone spectroscopy and others are practised (Scherer et al. 1997; Paul and Saykally 1997; Ruth 1999).

The idea of Cavity-Ring-Down-Spectroscopy is quite easy but highly inspired. In using the decay time of an excited high reflectance optical resonator many different advantages come together. For example the absorption length can reach several kilometres, the results are not affected by intensity fluctuations of the laser and the absorption coefficients are obtainable directly without any calibration. Fig. 7.4 shows the schematic set-up of a pulsed cavity ring-down experiment. It consists of a pulsed tuneable laser, lenses for mode matching, a resonator (cavity) and a detector.

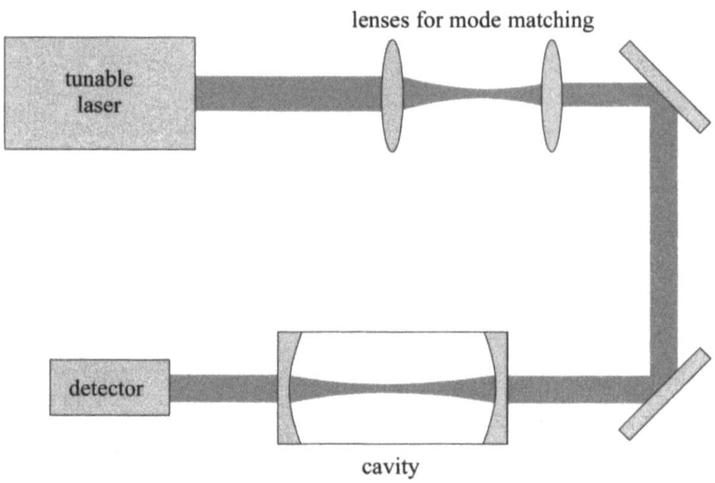

Fig. 7.4. Principle set-up of a pulsed CRDS experiment

The laser beam is mode matched to the cavity with lenses so that preferentially the TEM_{00} of the cavity is excited. The cavity is filled with the absorbing medium and is sealed on both sides by highly reflecting mirrors to enhance the absorption length. For our gas phase experiments we typically use mirrors with a reflectivity of 99.98 % and a cavity length of nearly 0.5 m which results in an effective absorption path-length of about three kilometers. Due to this high absorption length CRDS is very sensitive. After coupling a laser pulse into the cavity the light intensity decreases with every pass within the resonator because of reflection losses of the mirrors and absorption losses of the medium. Therefore an exponential decay of intensity is obtained at the detector. This decay is specified by the decay time τ, which normally is in the order of magnitude of µs. Of course a stronger absorption in the cavity causes a faster decay and therefore a shorter decay time results. It can be deduced [see (Zalicki and Zare 1995; Ruth 1999)] that the decay time of the CRDS signal is inversely proportional to the linear absorption coefficient k_ν:

$$k_{\nu} = \frac{d}{L_{Abs} \cdot c} \cdot \left[\frac{1}{\tau} - \frac{1}{\tau_0} \right] \tag{7.1}$$

where τ_0 is the decay time of the empty cavity, τ is the decay time of the cavity filled with an absorbing medium, d is the cavity length, L_{Abs} is the thickness of the absorbing medium and c is the velocity of light. Additionally, d and L_{Abs} are mostly equal in gas phase measurements.

The cavity signal is measured with a detector and visualised with an oscilloscope, which is read out by a personal computer. The decay times are fitted by the computer using a nonlinear fit.

By recording the decay times of the empty and the filled cavity for the interesting wavelength range an absorption spectra is received.

7.3.2 Current Status of the Development of Condensed Phase CRDS

In 1999, Engeln et al. (Engeln et al. 1999) integrated a solid ZnSe plate in their cavity and coated the plate with C_{60} to measure an IR-absorption line of this fullerene. Because of the presence of the 3 mm thick ZnSe plate there were massive losses due to scattering.

In order to avoid these losses and be able to measure broad-band spectra, Pipino et al. developed the evanescent CRDS technique (Pipino et al. 1997, 1997; Pipino 1999). Despite several years of research, this technique has not achieved a major breakthrough to date because of its very high experimental complexity.

We developed the method of direct coating the cavity mirrors in order to minimize both the complexity of the experimental set up and the scattering effects. This method allowed us to record the spectrum of a monomolecular iodine layer without any additional scattering. The iodine spectrum has been published in (Kleine et al. 2001) together with a discussion of the details of our experimental setup and possible future applications of condensed phase CRDS.

7.3.3 Further Developments of Mirror Coated Thin Film CRDS

Our measurements showed that the surface properties of the dielectric CRDS mirrors can affect the spectra of thin layers, especially in the case of small, polar molecules. Since the CRDS mirrors are manufactured in a not very controlled manner and the same materials and methods (e.g. coated versus sputtered) are not always used, the reproducibility of the spectra obtained with different CRDS mirrors is limited. The dependence of the spectra on the surface properties of the mirrors is lost when large, essentially unpolar molecules (such as polyaromatic hydrocarbons) are investigated. For molecules of this type, spectra are obtained that are largely independent of the thickness of the layer and the mirror properties. An example is shown in Fig. 7.5, in which the position and shape of the CRDS spectrum are largely independent of the increase in anthracene concentration. The peak and shape of the anthracene band agrees well with the conventional UV-VIS absorption spectrum.

Further disadvantages of the dielectric mirrors are the lack of reflectivity over a broad range of wavelengths, rotation of the polarization plane of the light by the mirrors, lack of reproducibility in mirror surface production and difficulties to characterize the surface, as well as the high costs of the mirrors.

A new generation of mirrors developed by 3M is made of polymers (Weber et al. 2000) and seems to avoid most of these disadvantages. Aside from their broadband reflectivity, the polymer surfaces of these mirror films can be manufactured more reproducibly and the surface constitution can be characterized. The polymer surface is expected to have little, if any, effect on the spectra, and should be easy to standardize. The only shortcoming of these films at the moment is related to their comparatively low reflectivity. The reflectivity of the 3M broadband film in the wavelength range from approximately 350 to 850 nm typically exceeds 98 % and approaches 99 %.

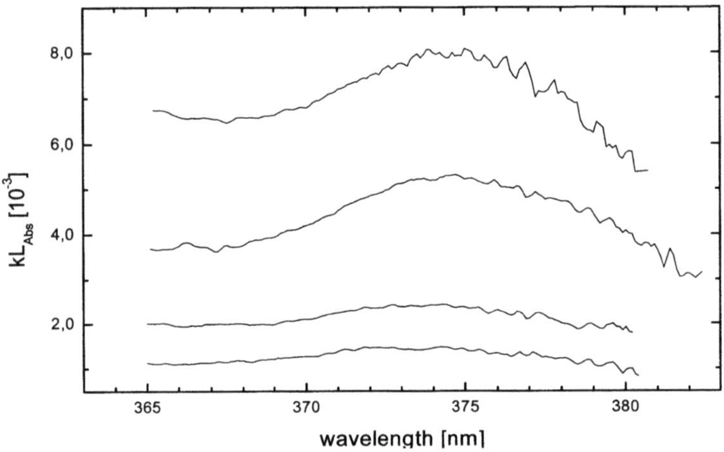

Fig. 7.5. CRDS spectra of solid anthracene deposited on the CRDS mirror by spin coating from a solution containing $5 \cdot 10^{-5}$ mol/l anthracene in isopropanol. The amount of anthracene applied to the mirror covers the range from approximately 1-4 monolayers (spectra from bottom to top). The mirrors from Laseroptik Garbsen had 99.8 % reflectivity; DMQ (2-Methyl-5-t-butyl-para-quarterphenyl) in dioxane was used as the laser dye. The laser step width was 0.1 nm.

In some preliminary experiments, we were able to show that good CRDS decay signals can be recorded with the 3M mirror films. The relative uncertainty of the cavity decay time τ is: $\Delta\tau/\tau = 7.5 \cdot 10^{-3}$, averaged over 10 decay signals. Most of our attention has been devoted to the 500 nm mirror film (product name: 3M Radiant Color Film CM 500), which, mounted in our conventional CRDS mirror holder, looked like a normal dielectric mirror and could be aligned alike with the consequence that no change in the normal experimental setup was required. The broadband mirror film was also investigated (product name: 3M Radiant Mirror Film VM 2000). In order to generate a stable resonator, a high reflectivity dielectric mirror (Layertec) with 99.98 % reflectivity and a radius of 2 meter was mounted to serve as the cavity end mirror. The known reflectivity of the Layertec mirrors and the mean reflectivity derived from the measured decay times were used to obtain the reflectivity of the mirror films for different wavelengths.

We were able to demonstrate that the transmission of the 3M mirror films is sufficient for CRDS measurements and that the typical CRDS decay signals are obtained. Moreover, in our experiments, the mirror films remained undamaged with no effect on the CRDS decay signal after exposure of the films to the pulsed laser beam for 15 minutes and more. The films proved to be resistant to mild solvents. We find a high degree of correlation between the reference spectra supplied by 3M and the spectra of the experimental films obtained with our UV-VIS spectrometer and with CRDS.

Fig. 7.6 shows part of the spectrum of the 500 nm-film obtained with our UV-VIS spectrometer and the reflectivity values derived from the CRDS measurements. Both the absolute reflectivity values and the wavelength dependence of the reflectivity are well reproduced in the CRDS results. This demonstrates the suitability of these films for use in CRDS. Combined with our method of direct coating, these mirrors can be expected to circumvent many of the difficulties of condensed phase CRDS.

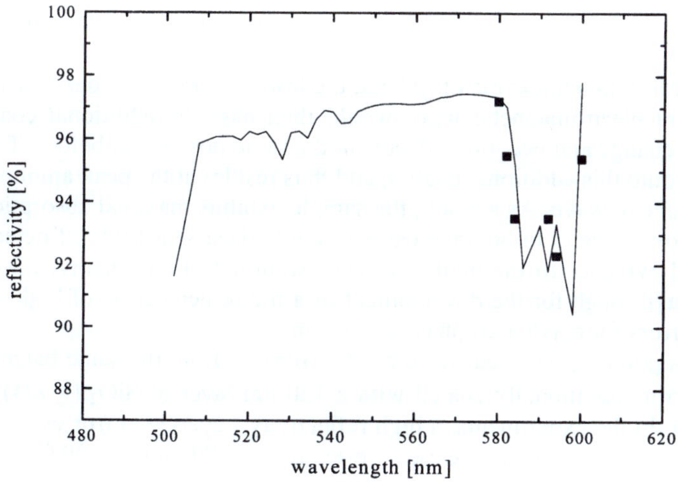

Fig. 7.6. Part of the reflectivity curve of the 500 nm-mirror film (3M) recorded with our UV-VIS spectrometer. The dots correspond to reflectivity values determined by CRDS measurements. The laser dye in the experiments was rhodamine 6G.

It should be noted that good CRDS decay signals have been obtained also with the broadband mirror film. This film is quite interesting because of its especially broad wavelength range of good reflectivity. It was not possible, however, to mount this less flexible film in our mirror holder without very tiny undulations of the film surface. This caused the laser light to be not ideally reflected during the many CRD round trips. By mounting the film in the mirror holder more tightly, we succeeded to reduce the surface undulations, but caused a distortion of the polymer structure of the film which results in the loss of the CRD signal. However, the broadband film can be expected to keep its full reflectivity at CRDS-measurements if it is coated directly on a glass substrate. Direct coating is also required to provide for the necessary radii of curvature for a stable resonator in future CRDS-broadband measurements.

7.3.4 Improved Sensitivity of Detection by Utilizing the Dependence of Absorption on the Spatial Position of the Ultrathin Layer in the Electromagnetic Wave

In order to enhance the sensitivity of detection in the CRDS measurements of thin layers and for correct interpretation of the results, it is important to keep in mind that the thickness of the investigated layers is substantially smaller than the length of the electromagnetic wave used in the experiments (compare to Fig. 7.5). Since the electromagnetic wave forms a node directly at the mirror surface, the measurements on ultrathin coated mirror surfaces are performed at spatial positions where the electromagnetic wave has a small amplitude. The maximum of the wave amplitude is at a distance of $\lambda/4$ from the mirror surface and the measurements should be performed there. This can be achieved by applying an additional coating to the mirror with a thickness of $\lambda/4$ and the lowest refractive index possible, since the incident electromagnetic wave would then pass the additional coating with only little change and become reflected at the same position as before. The sample is applied onto this additional coating and thus resides at the peak amplitude of the electromagnetic wave. As a result, the sample exhibits maximal absorption, which in turn allows spectra to be obtained with the highest sensitivity of detection. Experimental evidence of the predicted increase in detection sensitivity would be a major breakthrough for the development of a future generation of high sensitivity CRDS mirrors for condensed phase applications.

For this purpose, we used two dielectric mirrors from the same batch with one of the mirrors additionally coated with a 150 nm layer of SiO_2 (= $\lambda/4$) by vapor deposition. In the experiments, a high reflectivity Layertec mirror was used as the cavity end mirror, such that the mean reflectivity at 580 nm was 99.94 %. The loss in reflectivity due to the presence of the additional coating was observed to be negligible in the investigated wavelength range. The experiments were performed on a ultrathin pentacene layer. The absorption spectrum of this substance can be presumed to be largely independent of the mirror surface (Ta_2O_5 or SiO_2), because pentacene is a completely non-polar, polyaromatic hydrocarbon (see also the results at the beginning of the previous chapter).

Some difficulties were encountered in reproducibly preparing pentacene layers of equal thickness and identical constitution for the mirrors with and without SiO_2 layer – both factors are indispensable requirements for a correct comparison of absorption values. The results obtained indicate however, that coating with an additional layer of a thickness of $\lambda/4$ leads to an increase in sensitivity (J. Lauterbach, Doctoral Thesis).

7.3.5 Extension of Our Method to Liquids

The typical approach to get absorption spectra of liquids involves the use of a cuvette. In the CRDS experiment, this would be a microcuvette between the cavity mirrors. The decay time of the CRDS signal, however, is too small with this arrangement. We instead used our routine CRDS setup and installed a specially prepared CRDS mirror: 30 µl of solution is applied to the planar mirror or the one with the smaller curvature and is covered with a commercially available cover slip (Roth) with a thickness of 150 µm. Because of the adhesive forces between the mirror surface, the liquid, and the glass surface of the cover slip, this mirror is easy to install in the cavity without any loss of liquid. In addition, the packing ring of the mirror-holder carefully and evenly presses against the cover slip. The usual precise adjustment of the mirrors follows.

Reflections from the surface of the cover slip and from the liquid will be captured again, and will not be noticed as overall losses from the cavity [see (Engeln et al. 1997, 1999)]. The resulting (scattering) losses are sufficiently low, so that sensitive absorption measurements can be performed. Fig. 7.7 shows the losses resulting due to the presence of the cover slip and the solvent (in this case: 30 µl ethanol). The losses are shown directly as decrease of reflectivity. The remaining reflectivity of R = 99.8 % is > 99 % which is sufficient for CRDS-measurements of liquids.

Fig. 7.8 shows – to our knowledge - the first CRD liquid phase spectrum. The spectrum was obtained from a solution of 9-methylanthracene in ethanol ($2 \cdot 10^{-4}$ mol/l). The pure solvent was used to obtain the blank spectrum – as is common procedure in measurements of the liquid phase.

The spectral curve of Fig. 7.8 agrees very well with the conventional UV absorption spectrum (Clar 1949) as is expected from the low degree of interaction between the polyaromatic hydrocarbon and the polar surfaces of the cover slip and the dielectric mirrors. This relationship has been already discussed.

It should be noted that the method presented here can in principle also be used to determine absolute concentrations, provided the cross-section of absorption is known from the literature. The thickness of the liquid film (order of magnitude is 50 µm) can be calculated from the very precisely defined volume of the liquid used and the dimensions of the cover slip. Of course, well defined spacers could be used also.

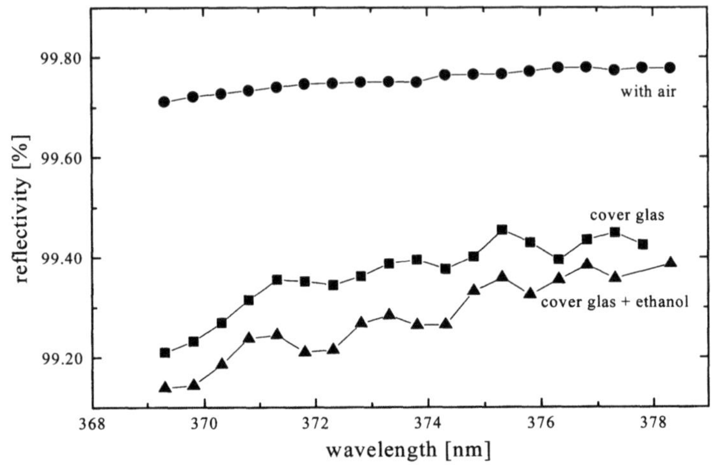

Fig. 7.7. Losses resulting due to the presence of the cover slip and 30 μl of solvent.

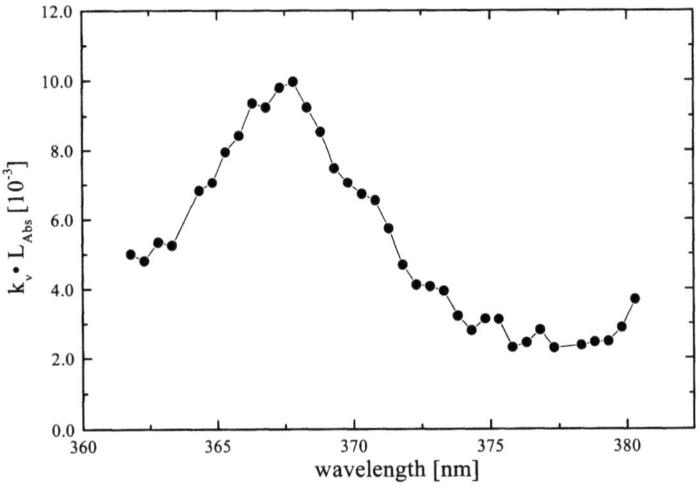

Fig. 7.8. CRDS spectrum of a solution of 9-methylanthracene in ethanol recorded with the 99.8 % reflectivity mirrors. The laser dye used for the displayed wavelength range was DMQ in dioxane.

Acknowledgements

We wish to thank Dr. Daniel Kleine for his essential contributions to achieve the CRDS results and Mr. A. Hicks (3M-UK) for supplying the 3M mirror films.

References

Clar E (1949) Das Anellierungsprinzip und die Resonanz in aromatischen Kohlenwasserstoffen. Chem. Ber. 82: 495-514

Engeln R, Berden G, v. d. Berg E, Meijer G (1997) Polarization dependent cavity ring down spectroscopy. J. Chem. Phys. 107 (12) : 4458-4467

Engeln R, v. Helden G, v. Roij AJA, Meijer G (1999) Cavity ring down spectroscopy on solid C_{60}. J. Chem. Phys. 110 : 2732-2733

Hodges JT, Looney JP, van Zee RD (1996) Response of a ring-down cavity to an arbitrary excitation. J. Chem. Phys. 105 : 10278-10288

Iriyama K, Ozaki Y, Hibi K, Ikeda T (1983) Raman spectroscopic detection of haemoproteins in the eluate from high-performance liquid chromatography. J. Chromatgr. 254 : 285-288

Kleine D, Lauterbach J, Kleinermanns K, Hering P (2001) Cavity ring-down spectroscopy of molecularly thin iodine layers. Appl. Phys. B 72 : 249-252

Koizumi H, Suzuki Y, (1987) Micro high performance liquid chromatography of aliphatic amines by means of resonance Raman detection. J. High Resolut. Chromatogr. Chromatogr. Commun. 10 : 173-176

Lee JY, Lee HW, Hahn JW (1999) Time domain study on cavity ring-down signals from a Fabry-Perot cavity under pulsed laser excitation. Jpn. J. Appl. Phys. 38 : 6287-6297

Liang L, Lubman DM (1989) Resonant two-photon ionization spectroscopic analysis of thin-layer chromotography using pulsed laser desorption/volatilization into supersonic jet expansions. Anal. Chem. 61 : 1911-1915

Nir E, Kleinermanns K, de Vries MS (2000) Pairing of isolated nucleic-acid bases in the absence of the DNA backbone. Nature 408 : 949-951

O'Keefe A. and Deacon D.A.G. (1988) Cavity ring-down optical spectrometer for absorption measurements using pulsed laser sources. Rev. Sci. Instrum. 59 (12) : 2544-2551

Paul J. B. and Saykally R. J. (1997) Cavity Ringdown Laser Absorption Spectroscopy Anal. Chem. News and Features A : 287-292

Pipino ACR, Hudgens JW, Huie RE (1997) Evanescent wave cavity ring-down spectroscopy with a total-internal-reflection minicavity. Rev. Sci. Instrum. 68 : 2978-2989

Pipino ACR, Hudgens JW, Huie RE (1997) Evanescent wave cavity ring-down spectroscopy for probing surface processes. Chem. Phys. Lett. 280 : 104-112

Pipino ACR (1999) Ultrasensitive surface spectroscopy with a miniature optical resonator. Phys. Rev. Lett. 38 : 3093-3096

Romanini D, Lehmann KK (1993) Ring-down cavity absorption spectroscopy of the very weak HCN overtone bands with six, seven, and eight stretching quanta. J. Chem. Phys. 99 : 6287-6301

Ruth A. A. (1999) Cavity-Ring-Down-Spectroscopy Phys. Bl. 55 (2) : 47-49

Scherer J.J., Paul J.B., O'Keefe A. and Saykally R. J. (1997) Cavity Ringdown Laser Absorption Spectroscopy: History, Development and Applications to Pulsed Molecular Beams. Chem. Rev. 97 : 25-51

Steinert R, Bettermann H, Kleinermanns K (1997) Identification of xylene isomers in high-pressure liquid chromatography eluates by Raman spectroscopy. Appl. Spectrosc. 51 : 1644-1647

Steinert R, Bettermann H, Kleinermanns K (1998) Identification of compounds in HPLC-eluates by Raman spectroscopy. Int. J. Vibr. Spec. 2 : 27 (feature article)

Sverdlov LM, Kovner MA, Krainov EP (1974) Vibrational Spectra of Polyatomic Molecules. Wiley, New York Toronto

Weber MF, Stover CA, Gilbert LR, Nevitt TJ, Ouderkirk AJ (2000) Giant birefringent optics in multilayer polymer mirrors. Science 287 : 2451-2456

Zalicki P, Zare RN (1995) Cavity ring-down spectroscopy for quantitative absorption measurements. J. Chem. Phys. 102 : 2708-2717

Zalicki P, Ma Y, Zare RN, Wahl EH, Dadamio JR, Owano TG, Kruger CH (1995) Methyl radical measurements by cavity ring-down spectroscopy. Chem. Phys. Lett. 234 : 269-274

8 Application of Two-Dimensional LIF for the Analysis of Aromatic Molecules in Water

F. Lewitzka, M. Niederkrüger and G. Marowsky

8.1 Introduction

The majority of environmentally relevant substances can be quantitatively analyzed in various matrices – provided that the samples may be sent to an analytical laboratory. The analysis there involves an extraction and a clean-up step prior to the –mostly chromatographic – separation and analysis. Thus this procedure is quite time-consuming. Moreover, these laboratory based techniques are very costly if continuous monitoring is necessary or if hundreds of soil samples have to be analysed to assess the hazard potential of a particular site. For these screening and monitoring applications optical methods are much more suitable.

A very sensitive optical technique is fluorescence spectroscopy, especially with lasers as the excitation source. Many environmental pollutants show fluorescence when illuminated with ultra-violet (UV) light. The most important – and therefore regulated – substances are monocyclic compounds referred to as BTEX (benzene, toluene, ethyl-benzene, xylene) and polycyclic aromatic hydrocarbons (PAH). Due to their high mutagenic and/or carcinogenic potential, the US Environmental Protection Agency (US-EPA) has assigned 16 PAH as priority pollutants. Mineral oil products also contain BTEX and PAH (beside other fluorescent compounds) so they can always be detected by fluorescence spectroscopy.

In the past few years, a variety of laboratory and field instruments based on laser-induced fluorescence (LIF) spectroscopy have been introduced. Some instruments use a nitrogen laser (337 nm excitation wavelength) as the excitation source while others use a flash-lamp pumped Nd:YAG laser (3rd or 4th harmonic, excitation wavelength is 355 nm and 266 nm respectively). For more details we refer the reader to the review article from U. Panne (Panne 1997).

The design considerations of the LIF systems presented here can be summarized as follows. The excitation wavelength should be well below 300 nm to allow one to detect also monocyclic and small polycyclic compounds. The pulse length should be less than 10 ns. The system should be mobile and it should be possible to operate it with a 12 V battery. With the lasers commercially available in the mid nineties, these specifications could not be met. Thus a laser with the required specifications was developed.

A general concept for in-situ analysis is to increase the dimensionality of the measured data. The higher information content is crucial for a subsequent analysis since there is no clean-up step and no separation involved; the substances have to be analysed in a complex matrix and in the presence of other interfering compounds. The realized LIF systems are described below. All systems can record two-dimensional (2-D) fluorescence data, i.e. fluorescence intensity vs. emission wavelength and time.

After developing and testing the laser-fluorimeters, the research work was directed to improve the calibration procedure since quantitative and selective results are regarded as a prerequisite for the acceptance of the new technique.

8.2 Hardware

8.2.1 Overview

The general set-up of the LIF systems is shown in Fig.8.1. The light from a pulsed UV laser is focused on to an optical fibre and guided to the sensor head which is submerged in a water sample. The pollutants present there are excited and emit fluorescence light. This light is guided back through the detection fibres to the detection system. There its spectral and temporal behaviour is registered. A control unit serves for the synchronisation of the laser and the detector and transfers data and assures communication with a computer.

There are different ways to realize the 2-D fluorescence detection. Spectra at a number of time intervals can be recorded or, equivalently, decay curves at a number of wavelengths can be captured to build up the 2-D data. Both techniques have been realized and are described in more detail below.

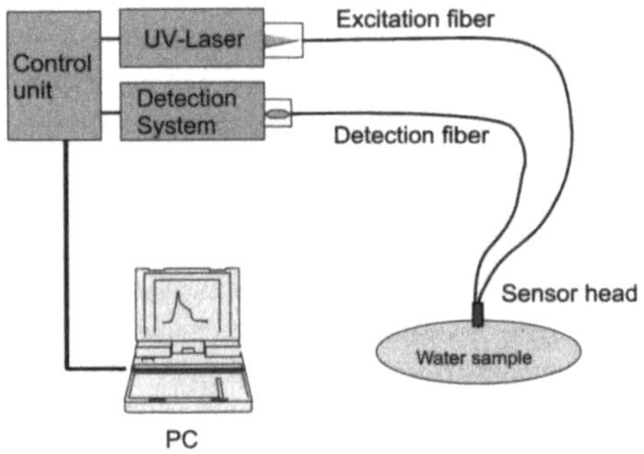

Fig. 8.1. General set-up of the LIF-systems

8.2.2 Laser

The requirements for a short wavelength and a short pulse width in conjunction with small system size and low power consumption could be fulfilled at best with a diode-pumped, frequency converted solid-state laser. Lasers with the desired specifications were not commercially available at that time, therefore two models were developed at the LLG.

8.2.2.1 Model I

The first laser employed a 4 bar 240 W quasi-cw diode array that pumped a Nd:YAG crystal in a longitudinal geometry (see Fig.8.2). To achieve the nanosecond pulses, a Pockels cell was used as the active Q-switch. With a pump duration of 300 μs, pulses of 2.5 mJ energy and 9 ns length at the fundamental wavelength at 1064 nm were obtained in a multi-mode beam.

For the generation of UV light, the IR pulses were first converted to 532 nm by a KTP crystal, the fundamental wavelength still being present. Then two alternative optical paths could be selected by moving a mirror (HR 532 and 1064, 45°) into the beam. With the mirror positioned in the beam, 532 nm and 1064 nm light is focused into a BBO crystal for frequency mixing. Without the mirror, the beam is reflected by a HR 532 mirror and focused on to a second BBO crystal for further frequency doubling. Finally, both paths are combined so that the same filter (to remove residual 532 nm light) and optics can be used to couple both alternative laser wavelengths beam into an optical fibre. The movable mirror is mounted on a motor driven translation stage. More details can be found in (Karlitschek 1998).

At 266 nm, the laser generated pulses with a maximum energy of 140 μJ at a maximum repetition rate of 100 Hz. At 355 nm pulse energies of 180 μJ were available. In both cases the pulse width was 7 ns.

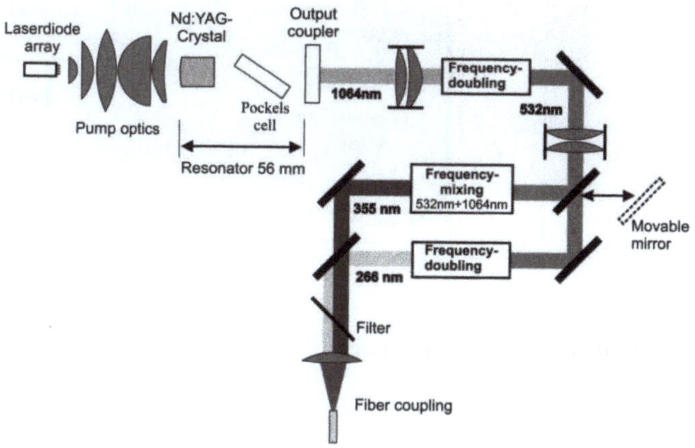

Fig. 8.2. Set-up of the longitudinally pumped Nd:YAG laser

8.2.2.2 Model II

A second model was developed to further decrease the size and costs. With an innovative pump geometry (see Fig.8.3), the efficiency could be increased. The wavelength selection feature was omitted to keep the size and cost of the laser as low as possible.

The cylindrical Nd:YAG rod (2 mm diameter, 12 mm length) is placed in a glass block (BK 7). The laser is pumped by two perpendicular positioned quasi-cw laser diode bars, each with 100 W power. The two backsides of the glass block are coated with gold to reflect the pump radiation back into the laser crystal. The glass block serves as a mount for the Nd:YAG crystal, as a heat conductor (it is attached to a thermo-electric cooler) and as a beam shaper for the pumping light; it actually works as a cylindrical lens.

The nanosecond pulses are obtained by using a Pockels cell again. Frequency conversion to 532 nm and 266 nm is done by a KTP crystal and a BBO crystal, respectively. Residual 532 nm radiation is removed by two mirrors (HR 266, 45°) that are transmitting at 532 nm.

The maximum ratings at 266 nm were: repetition rate: <100 Hz, pulse energy: up to 200 µJ; pulse width: 7 ns. The laser is typically operated at 50 Hz repetition rate and at pulse energies of 100 µJ. The size of the laser head is 26 cm x 15 cm x 11 cm.

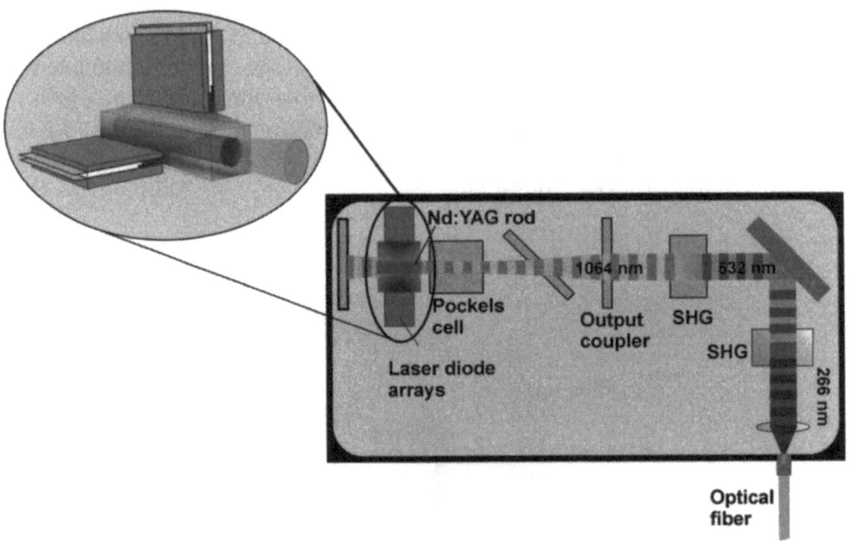

Fig. 8.3. Transversally pumped Nd:YAG laser

8.2.3 Detection System I

The first detection system developed is realized as an optical multi-channel analyser. The fluorescence light collected by the optical fibres is coupled to a spectrograph (Jobin-Yvon CP 140). Four detection fibres are arranged vertically in front of the entrance slit (0.25 mm width). With this set-up, an optical resolution of 7 nm was achieved, which was regarded as sufficient for the analysis of organic fluorophores in water.

At the exit focal plane, an image intensifier allows short gating (5 ns) and variable amplification of the fluorescence light. The intensified spectrum is imaged by two 85 mm lenses on to a CCD-camera (Hamamatsu C5809). It is a thermo-electrically cooled slow-scan camera with 512 x 64 pixels and is operated in the line-binning mode. Data are converted by a 16-bit analogue-to-digital converter, stored and transferred to a PC via the parallel (printer) port. The maximum read-out frequency is 100 Hz.

Another circuit controls the synchronisation of the gate-electronics with the laser. Gate widths ranging from 5 ns to 250 ns can be set and also the delay relative to the laser pulse (recorded by a fast photo-diode) be set in 1 ns steps using 3 delay generators. One generator is used to compensate the transit time in the 15 m long optical fibres, the other two provide the start and stop pulses for the gate-electronics for the image intensifier. The energy of the laser is monitored by another photo-diode and stored together with the data. Thus energy fluctuations during the recording of a time-resolved spectrum can be compensated. A photo of the laser-fluorimeter with the CCD detection is depicted in Fig.8.4.

Fig. 8.4. Photo of laser-fluorimeter with detection system I (CCD), 15 m sensor cable, and notebook computer for data acquisition and analysis

8.2.4 Detection System II

The second system developed should replace the costly optical multi-channel analyser by a more cost effective module. The concept is to record the fluorescence decay curves sequentially at a number of wavelengths. Thus the detection system consists of a miniature monochromator (CVI Digikrœm CM 110) and a fast (0.7 ns rise-time) compact photomultiplier module (Hamamatsu H5773) with integrated high-voltage supply. The photomultiplier signal is digitised by a low-cost digital storage oscilloscope (Tektronix TDS 220, 1 GSample/s, 100 MHz analogue bandwidth) from which only the mainboard and the interface board is used. The time-resolution of the detection system fits well to the pulse length of the laser.

In contrast to the above mentioned CCD-detection system, the computer is integrated in the fluorimeter. It is an AT96-Bus industrial PC that is equipped with an 8-fold RS-232 interface board. With these interfaces, the computer communicates with the oscilloscope board and controls the monochromator and the laser (start/stop). One RS-232 interface is connected to an I^2C-Bus where additional functions are available: an analogue-to-digital converter digitises the signal from a energy monitor diode and a digital-to-analogue converter controls the amplification of the photomultiplier.

For easy field operation the fluorimeter is equipped with a 6.5" TFT touch-screen display. The whole system is mounted in a 19" case and is depicted in Fig.8.5. The opened top shows the monochromator and the photomultiplier on the left, and the laser (model II) on the right. The lower part of the case contains the computer, the DSO mainboard, the power supply and the display.

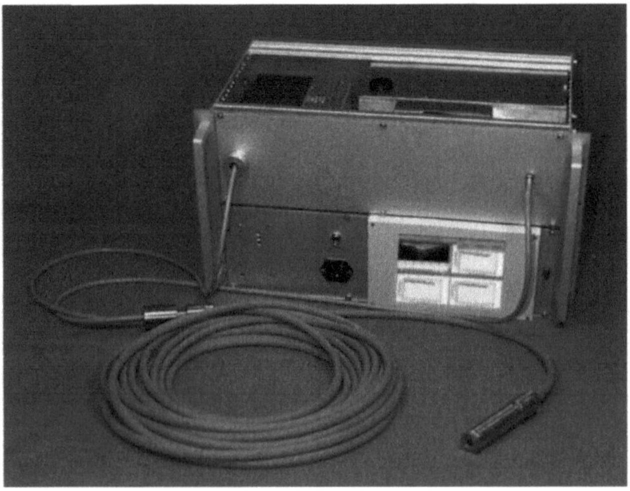

Fig. 8.5. Photo of the laser-fluorimeter with solid-state laser (model II) and detection system II (monochromator/ photomultiplier/ DSO)

8.2.5 Current development

To further decrease the costs of the laser-fluorimeter, the most cost-intensive component – the actively Q-switched Nd:YAG laser – had to be replaced. The laser of choice is a passively Q-switched, so-called microchip laser (Crystal FQS 266), see (Zayhowski 1998). It delivers pulses of 1 ns length with repetition rates between 5 to 15 kHz and pulse energies of 0.1 nJ per pulse. With the low pulse energies and relative high repetition rates, it is a nearly ideal source in conjunction with time-correlated single-photon-counting (TCSPC) detection, see also (O'Connor and Phillips 1984).

The detection system is made up of a monochromator, a photomultiplier module (Hamamatsu H5773 P) and the PC card (SPC 300, Becker&Hickl), which is incorporated in a standard PC. A small portion of the laser light is coupled into a photodiode that generates a synchronisation signal for the photon-counting electronics. Another photo-diode delivers an energy monitor signal.

The time resolution of the system is about 1 ns, the spectral resolution can be set to either 2 nm or 5 nm. The fluorimeter will be commercialised by Analytik Jena AG, Germany. In the commercial version, the controlling computer will be linked to the system via USB interface. Fig.8.6 shows the laboratory set-up of the prototype system.

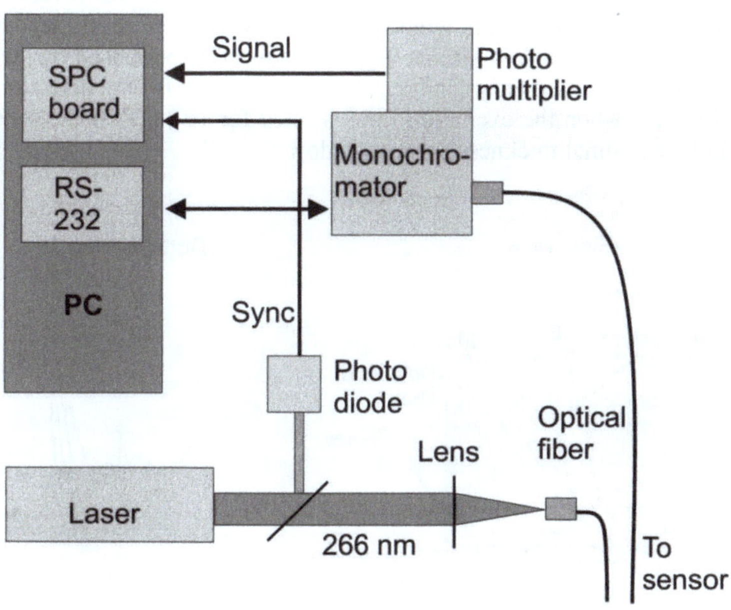

Fig. 8.6. Laboratory set-up of the laser-fluorimeter with microchip laser and time-correlated single-photon-counting detection

8.2.6 Optimised Sensor Geometry

The fibre-optical sensor head is a crucial component for the overall performance of the laser-fluorimeter. Extensive calculations have been carried out to find the best geometry for excitation and detection fibres for measurements in water (Bünting and Karlitschek 1998). The calculations considered the interaction volume and the efficiency with which the light from each volume element inside the interaction volume can be captured by the detection fibre. The optimal angle between the excitation and detection fibres was 20° and was also confirmed experimentally. To increase the detectable fluorescence signal, four detection fibres (400 µm core diameter) were positioned around one excitation fibre (600 µm core). At the detection end, the four detection fibres were arranged in-line to match the dimensions of the polychromator slit and the height of the CCD chip.

The system with the sensor head performed well. However, there was only one disadvantage. The sensor was in direct contact with the medium so that it could be easily soiled or contaminated. This hindered calibration measurements. Hence a new sensor head should be developed with a window to protect the fibres. Preliminary experiments showed that a quartz window, mounted perpendicular to the excitation fibre, generated a huge background signal. It was assumed that the signal is already generated in the excitation fibre and reflected to the detection fibre by the outer window surface. Simple geometrical considerations reveal that this effect can be avoided when the window is tilted by 10°. The basic layout of the head is depicted in Fig.8.7. A similar layout was recently published by (Wright et al. 1999). The goal was then to find out the best geometry for the detection fibres, i.e. the tilt angle when the excitation fibre is tilted for 10° against the sensor axis and to find the optimal thickness for the window.

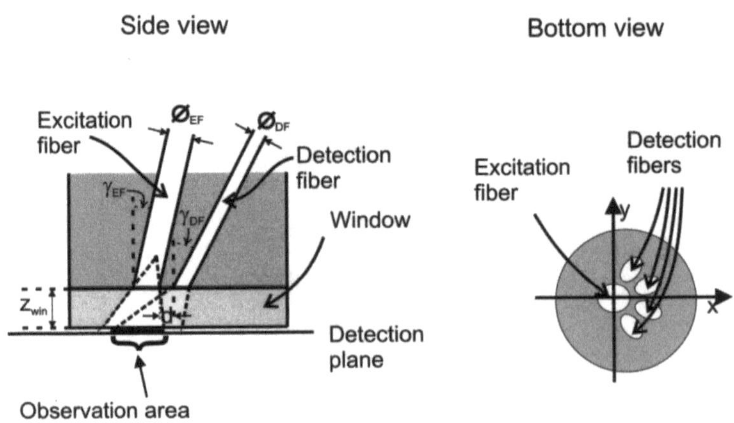

Fig. 8.7. Schematic drawing of the fibre optic sensor head with protection window

It is not possible to build an analytical model of the sensor head that reveals the sensor's sensitivity under different geometric conditions. A ray-tracing program was thus developed in which all relevant sensor parameters could be varied and the sensitivity could be calculated (Bünting et al. 1999). A typical result is shown in Fig.8.8. Here the sensitivity was calculated in dependence of the tilt angle of the detection fibre (γ_{DF}) and the thickness of the (quartz) window (z_{win}) and is displayed as a contour plot. From this graph, an optimum tilt angle of 35° is deduced when the thickness of the quarz window is 1 mm.

Similar calculations have been carried out with sapphire as window material. Due to its higher index of refraction (1.80 instead of 1.49 for quartz), the maximum sensitivity is reached with a window thickness of 1.4 mm, but yielding only one half of the sensitivity as with a quartz window.

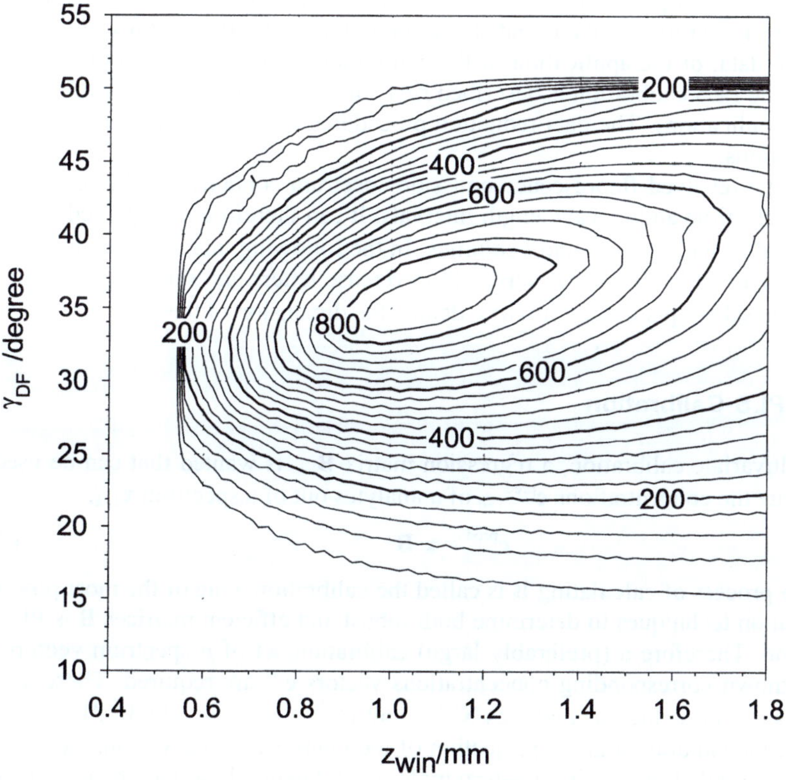

Fig. 8.8. Contour plot of the sensor sensitivity versus the tilt angle of the detection fibre (γ_{DF}) and the thickness of the protection window (z_{win})

8.3 Data processing and Calibration

8.3.1 Introduction

The goal of the data analysis is to get quantitative results, i.e. concentrations of the substances of interest. An established technique to derive concentrations in multi-component mixtures from spectral data is partial least squares (PLS) regression (Martens and Naes 1998). We used different approaches to reduce the 2-D spectra to conventional (1-D) spectra that can be processed with classical PLS. The use of two-way-PLS, that can handle 2-D data, was not regarded as appropriate since the additional dimension, the temporal profile, is only a superposition of exponentially decreasing functions.

Thus the data processing is a two-step procedure. The first step is the data reduction of 2-D to 1-D data, and the second step is the PLS calibration with the spectral data, or the application of the calibration for the prediction of concentrations. The PLS calibration is described first since it remains unchanged for the different applications. The data reduction procedures are described together with the applications.

A time-resolved fluorescence emission spectrum contains the fluorescence intensity at t time and w wavelength intervals. It can therefore be regarded as a data matrix $\mathbf{M}_{w\times t}$ with w rows and t columns. In the following, matrices are symbolised as bold capital letters, vectors as bold lowercase letters and scalars as regular letters. The subscripts in the form $a\times b$ denote the dimension of the data.

8.3.2 PLS-Calibration

In multivariate calibration, a regression matrix $\mathbf{B}_{k\times a}$ is wanted that can be used to calculate the concentrations $\mathbf{c}^{\mathrm{pred}}{}_{1\times a}$ of a analytes out of a spectrum $\mathbf{x}_{1\times k}$,

$$\mathbf{c}^{\mathrm{pred}} = \mathbf{x}\ \mathbf{B} \tag{8.1}$$

The process of calculating $\mathbf{B}$ is called the calibration. One of the most powerful calibration techniques to determine both robust and efficient matrices $\mathbf{B}$ is PLS regression. Therefore a (preferably large) calibration set of p spectrum vectors $\mathbf{x}^{\mathrm{cal}}$ with known corresponding concentrations vectors $\mathbf{c}^{\mathrm{cal}}$ are required. These sets of vectors can be written as matrices $\mathbf{X}^{\mathrm{cal}}{}_{p\times k}$ and $\mathbf{C}^{\mathrm{cal}}{}_{p\times a}$, respectively. In PLS, the first step of the calibration is the projection of the many variables $\mathbf{X}^{\mathrm{cal}}$ onto a few new variables (called *factors* in PLS-terminology) and using these factors as regressors for $\mathbf{C}^{\mathrm{cal}}$. By using the intermediate step of the factors, the common structures in the spectra are compressed into a stabilised set of less variables, leaving out much of the noise as residuals. The major advantage of PLS is its ability not only to describe the known components but also unknown (but systematic) phenomena in the spectra by the factors, e.g. background from natural organic compounds.

8.3.3 Applications

8.3.3.1 Analysis of fluorescence tracers in water

One application where the in-situ analysis capabilities of the laser-fluorimeter have proven to be advantageous is the analysis of fluorescence tracers in waters. These substances are water soluble, non-hazardous, persistent fluorescence dyes that are employed to track surface and subsurface water flows. In the experiment described below, Sulforhodamine G (SRG) was used. Typically 1 kg is introduced into the upper part of a river and monitored as it flows downstream. The movement and distribution of the dye gives the necessary information for a mathematical model for the description of the behavior of the rivers current. Such a model is of great importance for hydraulic engineering and flood prevention.

The laser-fluorimeter was slightly modified for the SRG analysis. The laser was operated at 532 nm (second frequency doubler removed) and a 532 nm holographic notch filter was placed in front of the spectrograph to remove scattered light.

The preliminary experiments showed that the fluorescence lifetime of SRG does not differ much from the instrument response function and the background (humic substances). Therefore only an one dimensional spectra with a 20ns gatewidth were recorded, what is equivalent with integration of a 2-D spectrum along the time axis. Hence for this application, the spectra were taken for calibration without any processing.

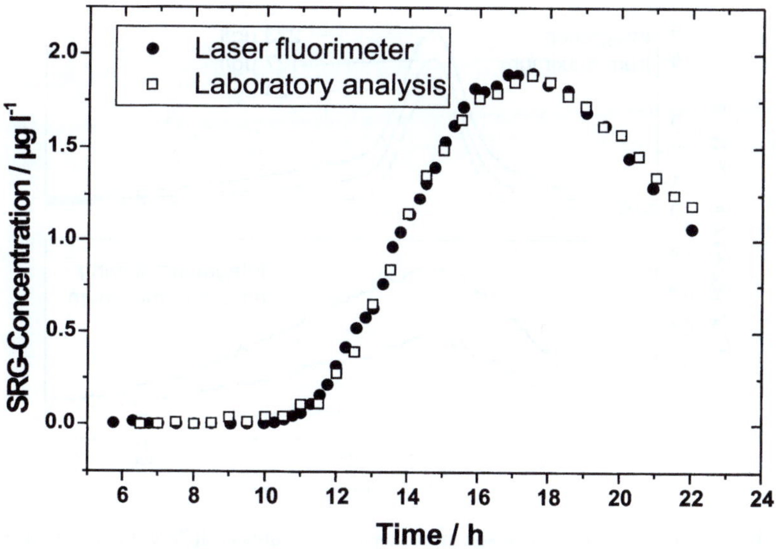

Fig. 8.9. Comparison of the results of an SRG analysis with the laser-fluorimeter (CCD-system) and with classical analysis

For the calibration, a number of samples with SRG in the concentration range between 0 and 1000 ng/l in tap water and in two different surface waters were prepared. Concentrations below 200 ng/l were taken for calibration, samples with higher concentrations were taken for validation and showed excellent agreement with the actual concentrations.

In a tracer experiment of the German Federal Institute of Hydrology (BfG) on the river Elbe in April 1999, the tracer SRG was monitored with the laser-fluorimeter. Fig.8.9 shows the concentration of the tracer near Dresden, two days after the introduction into the river 70 miles upstream. The curve illustrates the increase of the concentration when the SRG tracer cloud arrives and its decrease. Comparisons with parallel reference measurements, available a few weeks later, yield a detection limit for SRG of 10 ng/l (ppt) and a detection inaccuracy of less than 5%, which was consistent with the results from calibration.

8.3.3.2 Calibration for xylene in water

The PLS calibration method was used for a one-substance ($a = 1$) calibration for xylene. Within the BTXE group, xylene is the compound with the highest fluorescence intensity when excited with 266 nm and the longest fluorescence lifetime (10 ns). Thus it is easy to distinguish from other BTXE molecules and the Raman scattering signal of water.

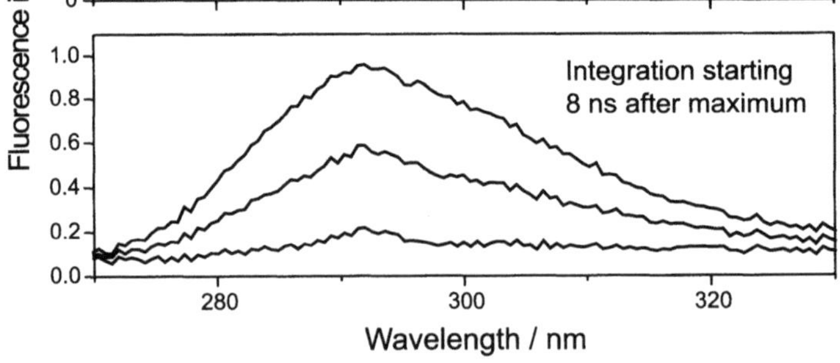

Fig. 8.10. Integrated fluorescence spectra of xylene in water at different concentrations, recorded with the CCD detection system. Top: Integration starts at the maximum of the Raman peak, bottom: integration starts 8 ns after the peak intensity

The experiments were carried out with an isomeric mixture of ortho-, meta-, and para-xylene and the CCD-detection system. A more detailed investigation later on revealed that the above mentioned properties can be mainly attributed to para-xylene.

The main problem of the xylene analysis is the strong overlap with the Raman signal of water (at 293 nm), when 266 nm light is used for excitation. This is shown in the top panel of Fig.8.10 for different concentrations of xylene. Here the time-resolved spectra were integrated for 30 ns starting at the maximum of the Raman signal. The Raman signal clearly dominates the spectra.

Since the xylene fluorescence decays much slower than the Raman signal it can be better discriminated from the Raman signal, when it is observed at a later time period. This is demonstrated in the lower panel of Fig.8.10. The spectra have been integrated for 30 ns starting 8 ns after the maximum of the peak intensity, when the Raman signal is diminished. Now only the xylene spectra are observed. The drawback of this method is also obvious in the figure: the signal-to-noise ratio is decreased.

With this 'late' integration as pre-processing procedure a number of calibration measurements have been carried out. Spectra from synthetic samples (prepared in the lab) and from field samples were recorded. Apart from the xylene the field samples contained plenty of other substances, among them toluene, xylene and ethylbenzene in high concentrations.

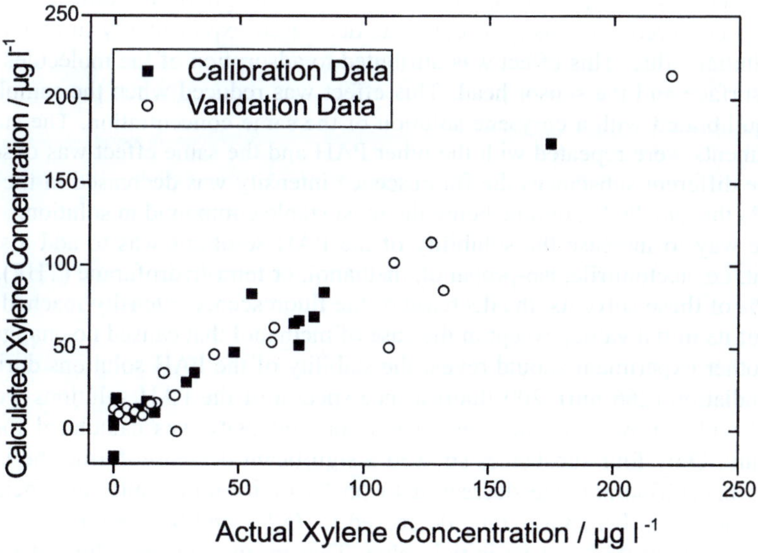

Fig. 8.11. Calculated vs. actual concentration for calibration and validation data for xylene after PLS calibration

One half of the data was used for the PLS calibration. This calibration was tested by applying it to the other half of the data, what is called validation. The result is shown in Fig.8.11. Since the error in the results are similar for both the calibration and the validation data set, it can be concluded that the calibration can be applied to unknown samples, and that the data are not over-fitted. From the calibration, a prediction error of 15 µg/l was calculated.

8.3.3.3 Analysis of PAH in water

Preliminary Investigations

Similar calibration experiments as for BTEX were planned for PAH in water. Due to their low solubility in water, some preliminary experiments concerning the sample preparation and handling were carried out. The substances of interest were the 16 PAH from the EPA priority list, except acenaphthylene, that does not fluoresce and dibenz(a,h)anthracene, indeno(1,2,3-c,d)pyrene, and benzo(g,h,i)-perylene that are only slightly (< 1 µg/l) soluble. Chrysene was selected as a model substance due to its moderate solubility (2 µg/l) and its large fluorescence signal.

A solution of 1.5 µg/l chrysene was prepared and spectra with 150 ns gate-width and 30 averages were recorded in 10 minute intervals over 60 minutes. The samples were prepared and handled in the dark. To compensate the variations in the laser energy, the spectra were normalised to the height of the Raman signal.

The fluorescence intensity of chrysene decreased exponentially to about 75 % of its initial value. This effect was attributed to adsorption of the molecules at the glass surface and the sensor head. This effect was reduced when the sample was pre-equilibrated with a chrysene solution of the same concentration. The stability experiments were repeated with the other PAH and the same effect was observed. For the different substances the fluorescence intensity was decreased in the range of 30 % through 90 %; pyrene being the most stable compound in solution.

One way to increase the solubility of the PAH solutions was to add a second solvent, i.e. acetonitrile, iso-propanol, methanol, or tetra-hydrofurane (THF). With 4 vol % of these solvents, the decrease of the fluorescence intensity reached about 80 % of its initial value, except in the case of methanol that caused no change.

Another experiment should reveal the stability of the PAH solutions during laser irradiation (266 nm). 200 fluorescence spectra of the PAH solutions were recorded with 30 averages and the fluorescence intensity was calculated for each spectrum. Only four substances showed a significant decrease during the irradiation, benzo(b)fluoranthene decreased to 55 % of its initial intensity, benzo(k)-fluoranthene and benzo(a)pyrene decreased to 75 %, and benzo(a)anthracene decreased to about 95 % of its initial value. That means, for recording of time resolved spectra of these five-ring aromatics, a much lower number of time intervals and/or averages is recommended.

As a consequence of these stability problems, only freshly prepared solutions were analysed for the calibrations described in the following.

Concept of data analysis

To understand the following data processing steps, it is reasonable to envision the structure of multi-component time-resolved emission spectra. The time-resolved fluorescence spectrum $\mathbf{M}_{w\times t}$ of a single substance can be described by the product of a row vector $\mathbf{s}_{w\times 1}$, that contains the spectral information and a column vector $\mathbf{d}_{1\times t}$, that describes the temporal (decay) behaviour of the fluorescence, i.e. $\mathbf{s}\,\mathbf{d} = \mathbf{M}$. For mixtures with n components this expression is extended to:

$$\mathbf{S}_{w\times n}\,\mathbf{D}_{n\times t} = \mathbf{M}_{w\times t} \tag{8.2}$$

Here the n individual vectors $\mathbf{s}_{w\times 1}$ and $\mathbf{d}_{1\times t}$ are simply concatenated to the matrices $\mathbf{S}_{w\times n}$ and $\mathbf{D}_{n\times t}$. While the product **S** and **D** give a unique solution (**M**), the decomposition of the matrix **M** into spectral and time-profile matrices has infinite solutions. One way to find physical meaningful solutions is to restrict the matrix **D** (the temporal behaviour of the fluorescence) to exponential decay functions. Convolution with the instrument response function can be neglected, when the analysis is restricted to a time interval where the excitation pulse has already decayed:

$$d_{ij} = \exp(-j\Delta t/\tau_i) \tag{8.3}$$

where $i = 1..n$, $j = 1..t$, d_{ij} are the elements of the matrix **D**, Δt is the time interval between two time slices and τ_i are the decay constants. With given **D** the matrix **S** can be calculated – in a least squares sense – by multiplying the data matrix **M** with $\mathbf{D}^+$, the pseudo-inverse of **D**, leaving a matrix of residuals **R**:

$$\mathbf{S} = \mathbf{M}\,\mathbf{D}^+ + \mathbf{R} \tag{8.4}$$

This is the first step of a technique described by (Knorr and Harris 1982). They reproduced **M** by iteratively optimising the decay constants τ_i in eq. 8.3 which – at the end – also yielded an optimised **S**. Their algorithm works quite well with binary mixtures without the presence of background fluorescence, and under these conditions the spectra $\mathbf{s}_i$ resemble the spectra of the pure analytes.

To test the procedure with a more complex sample, a mixture of 16 PAH in water (5 µg/l)[1] was prepared by adding a 10 ng/µl standard solution to the water, and the time resolved fluorescence spectrum was recorded. For the Knorr-Harris-analysis, the number of components was set to 3. A higher number yielded very noisy spectra, a lower number did not fit the data very well. The results, three spectra and the corresponding time constants, are shown in Fig.8.12 (upper panel). It is interpreted as follows: The spectrum belonging to the 4.9 ns decay time can be identified as fluorene in the short wavelength range, the signal at the longer wavelengths can be attributed to benzo(k)fluoranthene. The spectrum with 107 ns decay time belongs to pyrene, which has a lifetime of 125 ns.

[1] This concentration is above the solubility limit of some PAH, so the concentration might be lower for a few of them

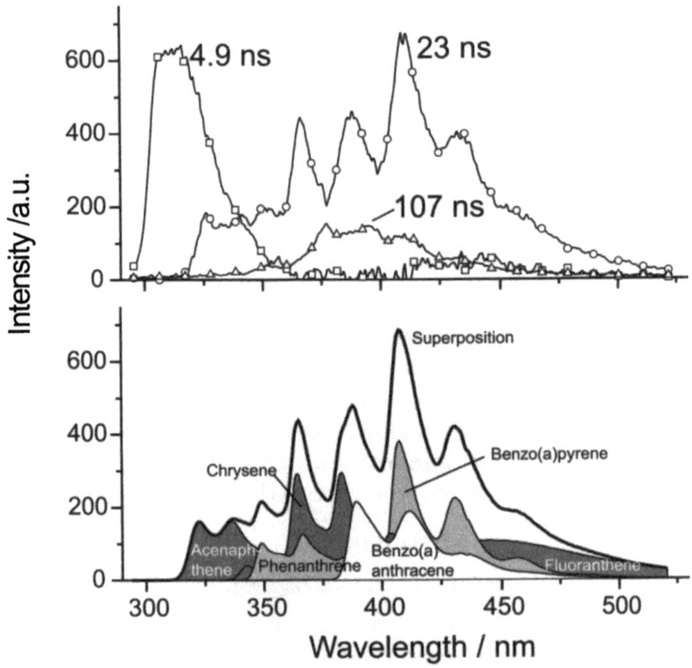

Fig. 8.12. Upper panel: 3-component Knorr-Harris analysis of a 2-D fluorescence spectrum of a sample containing the 16 EPA PAH in water (5 µg/l each). Lower panel: linear combination of 6 PAH (least squares fit) to reproduce the 23 ns spectrum

In the lower panel of Fig.8.12, the fitted 23 ns-spectrum is shown together with the spectra of the substances it is made up. The scaling factors for the single spectra have been optimised by a least-squares routine to reproduce the mixture spectrum. It comprises the remaining PAH, five of them can already be visually identified owing to their characteristic peaks. Two sub-groups exist which can not be separated due to their similar spectra, fluoranthene/ benzo(b)fluoranthene and acenaphthene/ naphthalene.

The result of this analysis can be summarised as follows: when a multi-component mixture is analysed by the Knorr-Harris procedure with only a few spectra (and time constants), then the calculated spectra are superpositions of the spectra of the single substances, provided their decay constants are similar. Fig.8.13 makes the situation clear. Here the spectral range and the decay times of 12 PAH are plotted. There is one group (fluorene, anthracene and benzo[k]fluoranthene) with decay times between 4 ns and 10 ns, one *group* (only pyrene) with a decay time above 100 ns and a large group of the remaining 8 PAH with decay times between 26 ns and 37 ns. These are the same groups that the Knorr-Harris-analysis identified.

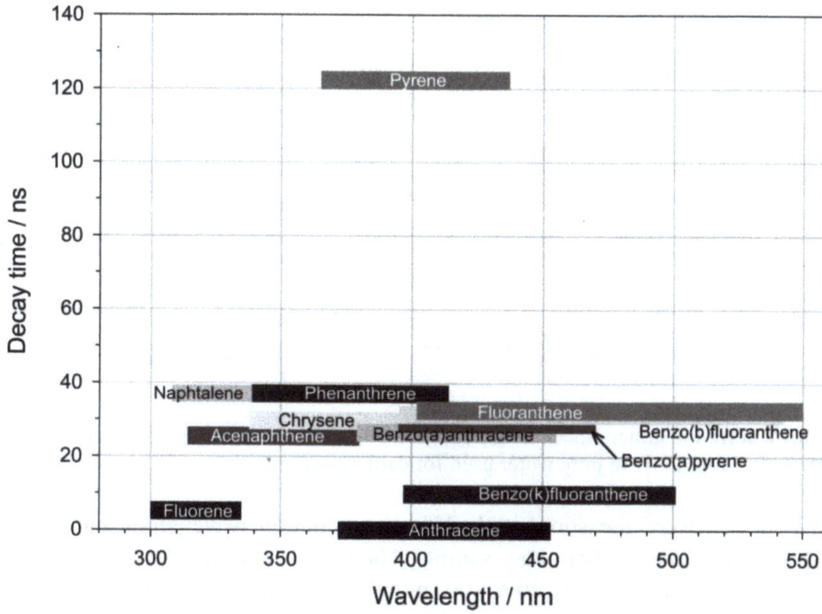

Fig. 8.13. Decay-times and spectral ranges (> 10% of peak intensity) of 12 PAH

Hence it seemed to be reasonable to use the Knorr-Harris-analysis for data pre-processing and to use the optimised spectra for the later PLS calibration. However, a more detailed investigation on different PAH mixtures revealed that the separation of the spectra and the recovery of the time-constants depends very much on the signal-to-noise ratio and the concentration ratios of the substances in the mixture. The solution of this problem was to not optimise the time-constants and spectra but to use a carefully selected set of typically three decay profiles, and to calculate the corresponding spectra according to eq. 8.4. After this factor decomposition, the n vectors $\mathbf{s}_{1\times w}$ are transformed into one vector $\mathbf{x}_{1\times k}$ with $k = nw$ by simply concatenating the individual vectors $\mathbf{s}$. The resulting data now have the data structure required for the classical PLS calibration and the spectra still contain the temporal information.

Sample selection

For the PAH calibration, 354 time-resolved fluorescence spectra of natural water samples were taken and artificially contaminated with PAH. The water samples were taken from 31 different locations all over Germany, including water from the North-Sea, some major rivers, several lakes and brooks and one bog. The natural samples were assumed to be free of PAH, and to each sample, one to five of the 12 above mentioned EPA PAH were added in the ppb- and sub-ppb-range.

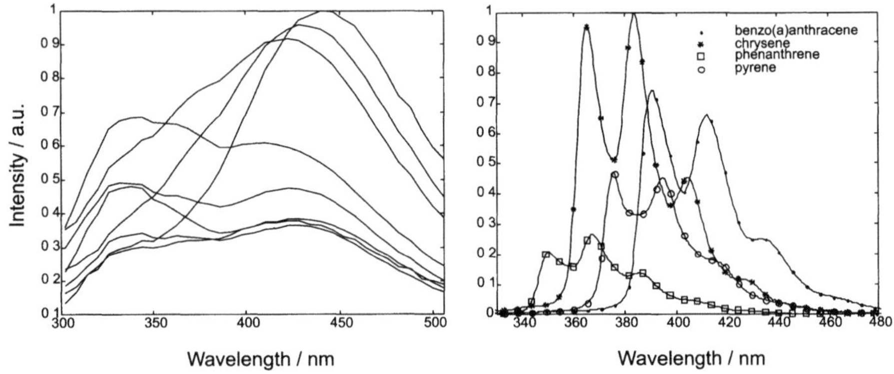

Fig. 8.14. left: Integrated spectra of eight natural water sample from different locations; **right:** and of four PAH in pure water used for calibration

All spectra were measured with 266 nm excitation wavelength and with the CCD detection system. For every sample, 100 time slices with $\Delta t = 2$ ns were collected, covering a range of -20 to 180 ns relative to the maximum of excitation. Therefore, the resulting data matrices $\mathbf{M}_{w\times t}$ consisted of $w = 512$ rows and $t = 100$ columns.

To give an impression about the data, we have plotted integrated spectra of eight different waters (Fig.8.14, left) and of four PAH (benzo(a)anthracene, chrysene, phenanthrene and pyrene, Fig.8.14, right). The background spectra show two broad maxima around 330 nm and 430 nm, probably due to amino acids, i.e. tryptophane, and humic material respectively. The PAH spectra show more structure but strongly overlap with each other and with the background spectra.

Calibration

To investigate and improve the calibration, additional spectral data $\mathbf{X}^{\mathrm{val}}{}_{q\times k}$ with known concentrations $\mathbf{C}^{\mathrm{val,is}}{}_{q\times a}$ are needed, called the validation data. Therefore the natural water samples were divided into three sets. The first set #1 with samples from five locations was contaminated with PAH in steps of 0.4, 1 and 2 µg/l. The second set with water from 13 other locations was contaminated in the same way. Set #3 of natural samples from the remaining 13 locations was contaminated with PAH in concentrations between 0 and 7 µg/l (the upper limit depended on the substance). Though the samples were contaminated with 12 different PAH, in the calibration we focused on 9 PAH. The calibration was carried out as described above using set #1 for the primary calibrations and set #2 to calculate the prediction error for optimising these calibrations.

Good results were obtained when the factor decomposition was carried out with decay matrices **D** made up with decay constants of 7 ns, 27 ns, and 125 ns (see eq. 8.3 and 8.4).

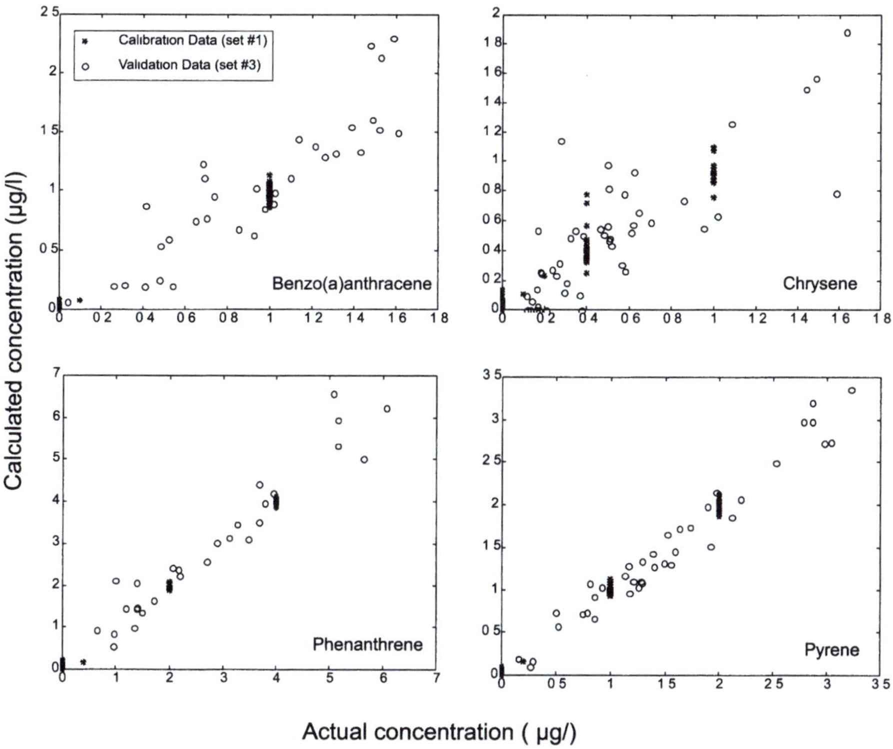

Fig. 8.15. Result of a calibration for 4 PAH

Since set #2 influenced the calibration during the optimisation step, set #3 was used to assess the final calibration. The result of a calibration of four PAH is shown in Fig.8.15. The predicted concentration is plotted against the actual concentration for the calibration data set (#1) and the validation data set (#3). The prediction error in the calibration data is obviously not lower than in the validation data set. This is an indication that the data are not over-fitted.

The calculated prediction errors for all of the 9 PAH are summarized in Fig.8.16. They range from 0.07 µg/l for benzo(a)pyrene to 1.47 µg/l for anthracene. These remaining errors are at least partly due to the fact that the three sets of calibration and validation data contained different background signals, because the water samples from different locations were restricted to one of the sets only.

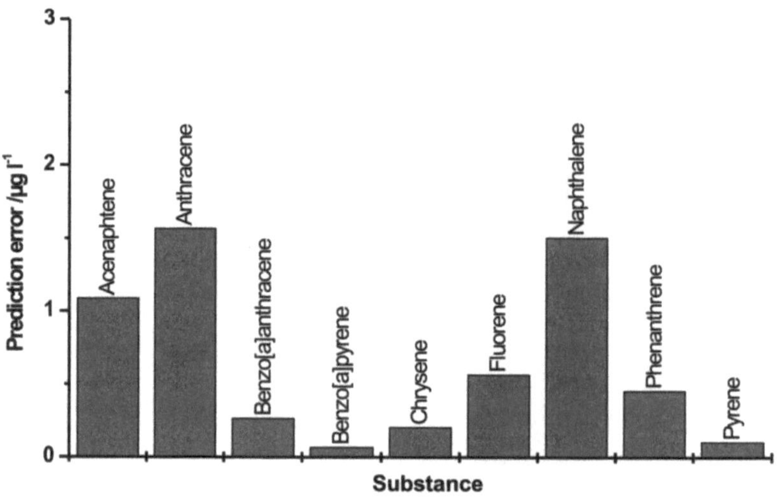

Fig. 8.16. Prediction errors for the analysis of the 9 PAH under investigation

Acknowledgements

The authors thank Peter Karlitschek for his work on the diode-pumped solid-state lasers, Tobias Nörthemann for the hardware development of the LIF-systems and Uwe Bünting for sensor optimisation, measurement software and the optimisation of the calibration procedure. Financial support by the "Deutsche Bundesstiftung Umwelt" (Grant No. # 01989) and the German "Bundesministerium für Bildung, Wissenschaft, Forschung und Technologie" (Grant # 16 SV 558/0 and # 16 SV 1112/5) is gratefully acknowledged.

References

Bünting U and Karlitschek P (1998) Mathematical model for optimum fibre optic probe design and characterisation. Spectrochim Acta A 54: 1369–1374

Bünting U, Lewitzka F and Karlitschek P (1999) Mathematical Model of a Laser-Induced Fluorescence Fibre-Optic Sensor Head for Trace Detection of Pollutants in Soil. Appl Spectr 53: 49–56

Karlitschek P, Lewitzka F, Bünting U, Niederkrüger M, and Marowsky G (1998) Detection of aromatic pollutants in the environment using UV-laser-induced fluorescence. Appl Phys B 67: 497–504

Knorr FJ and Harris JM (1981) Resolution of Multicomponent Fluorescence Spectra by an Emission Wavelength-Decay Time Data Matrix. Anal Chem 53: 272–276

Martens H and Naes T (1998) Multivariate Calibration. John Wiley & Sons, Chichester

O'Connor DV and Phillips D (1984) Time-Correlated Single Photon Counting. Academic Press, London

Panne U, Niessner R (1997) Laserverfahren in der Umweltanalytik, Analytiker Taschenbuch Bd. 16, Springer Verlag, Berlin, 155–270

Zayhowski JJ (1998) Passively Q-switched microchip lasers and applications. Rev Laser Eng 26: 841–846

Part III

Applications for Gaseous Substances and Aerosols

9 Chemical Analysis with Multi-Dimensional and On-Line Selectivity Using Laser Spectroscopy Combined with Mass or Species Separation

Ulrich Boesl

9.1 Introduction

Chemical trace and ultra-trace analysis has reached an exceptionally high technological level during the last decades. This is mainly due to the combination of two or even more analytical methods which resulted in a "multi-dimensional" selectivity. The leading "two-dimensional" analytical technique is gas chromatography-mass spectrometry. The price for this high technological standard is a time-consuming sample preparation. Thus, species-selective detection of traces of organic pollutants (e.g. dioxins or pesticides) may take days or even weeks. Therefore, these conventional methods of trace analysis are not adapted to special problems such as:

- rapid measurement for fast counter measures in the case of chemical accidents
- trace analysis of dynamic chemical processes, e.g. combustion processes
- large amounts of samples, e.g. close-meshed spot checks of polluted areas or biochemical and medical or atmospheric analytics

A prerequisite to solve these problems is the availability of rapid on-line or mobile on-site methods of selective detection. Their development is a challenge of modern technology and subject of modern research.

In particular laser-based techniques (Letokhov 1986; Andrews 1990; Kompa, Sick et al. 1993; Andrews 1994) promise to open new doors. Absorption spectroscopy with semiconductor lasers, laser-induced fluorescence, laser-induced desorption of large molecules for fast detection of substances adsorbed on surfaces or particles, or LIDAR (and many other techniques presented in this book) are examples, where laser-based methods already became established in chemical analysis. Many of these methods use laser spectroscopy for species selection. However, in the case of highly complicated mixtures of components (e.g. organic pollutants in soil, water, or air) even highly resolved laser spectroscopy often is not sufficient for species selective, time-resolved analysis. Having in mind the above mentioned problems of conventional analytical methods, the combination of a laser-based with a second selective detection method is the way of choice for developing fast on-line (and nevertheless highly selective and sensitive) analytical methods. Primarily the combination with mass spectrometric techniques (see chapter 10) carries a large variety of new analytical methods (for a survey of lasers in mass spectrometry see (Cotter 1984; Lubman 1990; Vertes, Gijbels et al. 1993; Boesl 2000; Boesl, Heger et al. 2000)).

Many conventional analytical methods work by separation of species in time (e.g. chromatographic techniques, ion mobility spectrometry, time-of-flight mass analysis) and therefore molecular detection or excitation by pulsed lasers is particularly favorable. Furthermore, the combination of laser-based methods with techniques which do not select molecular species, but microscopic particles, open up many fascinating possibilities to study chemical and physical processes of and on such particles. One example is the study of single microdroplets (Stockel et al. 2002). To detect and analyze particles with environmental relevance, lasers can be used to vaporize solid samples which then are analyzed by laser based (Zhan, Voumard et al. 1995; Morrical, Fergenson et al. 1998) or conventional means. On the other hand, On-line and On-site analysis ask for miniaturized sensors. Diode lasers are the laser types of choice for this purpose (see chapter 11). The goal of this article is to give an insight in the possibilities of combining laser-based molecular excitation or detection with conventional methods of separating chemical species in a mixture. This insight will be given by discussing selected examples rather than giving an exhaustive review. The emphasis will be on analytical methods with "two-dimensional" selectivity and/or "On-line" capability.

9.2 Resonant Laser Mass Spectrometry

The combination of two selective detection methods to achieve one method with "two-dimensional selectivity" is realized by so-called resonant laser mass spectrometry (Boesl, Neusser et al. 1978; Letokhov 1987) with the two selective methods UV-spectroscopy and mass spectrometry. The link between these two techniques is resonance enhanced multiphoton ionization (REMPI) which consists of two absorption steps: the first one is absorption to a bound excited molecular state (resonant intermediate state), the second one absorption up to the ionization continuum. The first absorption step strongly depends on the UV-spectroscopy of the molecule and therefore on the wavelength of the exciting light source representing one of the two species selective parameters of resonant laser-MS. The second absorption results in molecular ions and thus generates the connection to mass spectrometry; the molecular mass is the second species selective parameter. In conventional analytical laboratories, mostly low resolution UV-spectroscopy of solutions and magnet sector or quadrupole mass spectrometers are used. Resonant laser-MS involves UV-spectroscopy in the gas phase and time-of-flight mass analyzers. UV-spectroscopy in the gas phase can be highly selective (e.g. allowing separation of structural isomers or isotopomers) due to molecular cooling in a supersonic gas beam which strongly simplifies molecular spectra (Smalley, Wharton et al. 1977; Levy 1981; Hayes 1987). On the other hand, time-of-flight mass spectrometry is a pulsed technique delivering a whole mass spectrum for every single ionization pulse. Since REMPI is mostly performed by pulsed lasers, this combination of UV-gas phase spectroscopy and time-of-flight mass spectrometry via REMPI allows high speed of measurement combined with exceptionally high selectivity (Meijer, deVries et al. 1990; Weickhardt et al. 1994), and high power of detection. In addition, the typical REMPI ion source is compatible with many different types of sample inlet techniques.

As for all two-dimensional techniques, several working modes are possible depending on which of the selection parameters is kept constant and which one is tuned. With a fixed excitation wavelength UV-selected mass spectra are obtained, within a fixed mass window a mass selected UV-spectrum (e.g. of one isotopomere out of the natural isotopic mixture) can be measured. From a historically point of view, two-dimensional UV/mass spectra of diatomic molecules (Herrmann, Leutwyler et al. 1977), of benzene (Boesl, Neusser et al. 1978) and of other medium sized molecules (Antonov, Knyazev et al. 1978; Zandee and Bernstein 1979) represent the beginning of resonant laser-MS. This was preceded by the development of REMPI-UV-spectroscopy in a gas cell (Johnson 1980) and by non-resonant multiphoton ionization of molecular iodine and heavy water in low resolution mass filters. Nowadays, resonant laser-MS is introduced in many research laboratories so well that there exists extensive review literature (Johnson 1980; Bernstein 1982; Gedanken, Robin et al. 1982; Lubman and Kronick 1982; Fassett, Moore et al. 1985; Gobeli, Yang et al. 1985; Letokhov 1987; Lubman 1990; Vertes, Gijbels et al. 1993; Boesl 1999; Boesl 2000).

However, it turned out that for some molecular species fast internal molecular processes (in particular dissociation) reduces the ionization efficiency heavily or even suppresses any multi-photon ionization of parent molecules at all. Typical mass spectra then reveal intense fragment ion peaks; parent ion peaks often are missing completely. One way to cope with this problem is to use sub-picosecond laser pulses to rule out fast intramolecular processes. Several applications of fs-laser ionization are presented in chapter 10. In addition, combination of laser desorption of neutral molecules and resonant laser ionization allows application even to solid samples (see section 9.5, this chapter and chapter 10).

In figure 9.1, another example is presented, demonstrating the two-dimensional character of resonant laser-MS (Zimmermann, Heger et al. 1999). The chemical compound of interest is monochlorobenzene representing the class of chlorinated aromatic compounds. In particular polychlorinated dibenzodioxins, dibenzofurans, and biphenyls are highly toxic environmental pollutants and the need for their online monitoring (e.g. during industrial combustion processes) is obvious. In principle, laser spectroscopy would allow selective excitation (Weickhardt, Zimmermann et al. 1993; Weickhardt, Zimmermann et al. 1994), even in the case of the 22 isomers of tetrachlorinated dioxins. However, fast intramolecular relaxation

tion processes, unfavorable shifts of intermediate level energies (Zimmermann, Lenoir et al. 1995), the large number of relevant species, and extremely low concentrations in the sub-ppt range would complicate the necessary laser mass spectrometer and make it impractical for routine application. Nevertheless, there are research groups who try to develop a highly sensitive and selective On-line dioxin detector based on resonant laser-MS (Oser, Coggiola et al. 2001). On the other hand, a good correlation has been found between the toxicity equivalent of polychlorinated dioxins and the concentration of monochlorobenzene. It is more ambient by typically three orders of magnitude and shows more favorable conditions for resonance enhanced multiphoton ionization than dioxins. Therefore, monochlorobenzene is well suited as an indicator of toxicity due to chlorinated dioxins.

The goal of the experiment presented in figure 9.1 was to check if traces of monochlorobenzene can be detected on-line in the flue gas of a waste incinerator. Details of the experimental set-up are described in (Heger, Boesl et al. 1999). The spectra have been taken on-site and on-line at a sampling flange of the gas cleaning unit of an hazardous waste incinerator. The laser mass spectrometer has been transported to this incinerator and lifted by a crane to the sampling point for these measurements. At the top of figure 9.1, the room-temperature REMPI spectrum of the $S_1 \leftarrow S_0$ transition of monochlorobenzene is shown. Two vibronic bands associated with a system of hot bands can be seen. The one of them lies in the red of the 266.04 nm wavelength (4th harmonic of Nd:YAG laser) by just a few tenth of a nm. With other words: no strong absorption will take place in monochlorobenzene at this frequently used wavelength. The second vibronic band has its maximum absorption at 269,82 nm. At this wavelength, monochlorobenzene should show a strongly enhanced ionization efficiency: a detection limit (S/N = 3) of 75 pptv has been determined for this wavelength from the mass spectrum of a gas standard (whereas the detection limit at 266.04 nm is only 10 ppb).

Figure 9.1 also shows a mass spectrum taken at a wavelength of 266.04 nm, where most substituted and/or polycylcic aromatic compounds are ionized, but not with ultimate sensitivity. Several aromatic compounds, which are present at the combustion flue gas (which is a complex mixture of hundreds of different species), give rise to peaks at several masses, but not at mass 112 and 114 where monochlorobenzene (MCB) and its isotopomere should appear (see also the inset in the upper right). In figure 9.1 below, the wavelength has been changed for optimum detection efficiency of MCB. Now, the isotopic mass pattern expected for monochlorinated aromatic compounds appears at the masses 112, 113 and 114 (see also inset). A concentration of 510 pptv of MCB can be deduced from the signal intensity. This is in good agreement with the expected MCB concentration of several 100 pptv which is typical for this special sampling point of the incinerator. In conclusion, resonant laser-MS is indeed able to monitor indicators for the overall toxicity of chlorinated dioxins on-site and on-line.

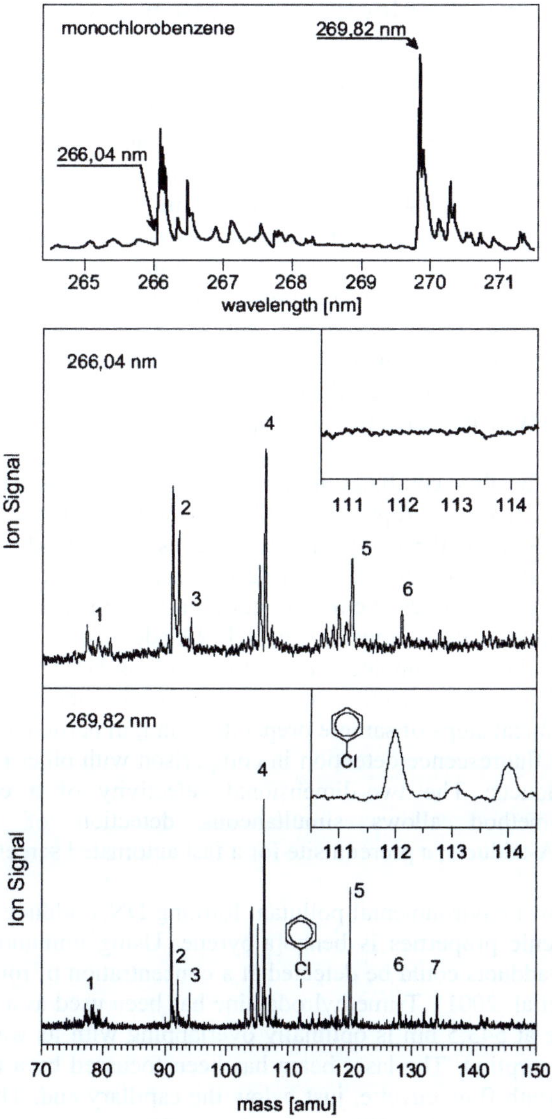

Fig. 9.1. Excitation spectrum of monochlorobenzene (MCB) (top) and optically selected mass spectra of mixtures containing traces of MCB with the excitation wavelength out of resonance (middle) and in resonance (bottom). All spectra are due to resonance-enhanced multiphoton ionization. (1: benzene, 2: toluene, 3: phenol, 4: xylene, 5: trimethyl-benzene, 6: naphthalene, 7: tetramethyl-benzene). Reprinted from (Heger, Boesl, Zimmermann et al. 1999) with permission from IM Publications, Chichester, United Kingdom

9.3 Laser-assisted Selective Detection in Chromatography

Mass spectrometers are high-level instruments of conventional chemical analysis. The price for this stage of development is a high technical and financial effort. A compromise between effort and selectivity are chromatographic selection methods, including capillary electrophoresis. In addition, many of these methods have features which are not available or difficult to achieve with mass analyzers, such as isomer selectivity (e.g. by gas chromatography), separation of very large molecules (e.g. by electrophoresis), investigation of a mixture of components directly out of a solution (e.g. by liquid or high pressure liquid chromatography or electrophoresis). The combination of laser-based and chromatographic selection methods is of particular interest, since low power lasers can be used and thus the advantage of reduced size and technological effort is preserved but selectivity and therefore sensitivity is increased considerably in comparison to non-laser assisted detection.

An example for this combination is described in chapter 17, namely the application of capillary electrophoresis and laser-induced fluorescence in an very actual field of research: the study of carcinogenesis and maybe even the early diagnosis of pre-stages of cancer. The role of laser excitation here is to induce selective fluorescence of dyes chemically attached to unmodified nucleotides and modified nucleotides (DNA-adducts). These DNA adducts are caused by reactive compounds attacking macromolecular systems such as DNA and are a hint for their damage. The article mentioned above nicely describes the role of these adducts, the chemical steps of sample preparation and, in particular, the advantage of laser induced fluorescence detection in comparison with other modern ways to detect DNS adducts: The two-dimensional selectivity of a combined electrophoresis/LIF-method allows simultaneous detection of modified and unmodified DNA-adducts, a prerequisite for a fast automated screening.

One well known environmental pollutant forming DNA-adducts and exhibiting strong carcinogenic properties is benzo[a]pyrene. Using immunoelectrophoresis with LIF, DNA adducts could be detected at a concentration of roughly 3 10^{-10} M (Tan, Carnelly et al. 2001). Trimethylrodamine has been used as a dye. The green He-Ne-laser line at 543.5 nm is optimally overlapping with its wavelength range of maximum absorption. The laser beam has been focussed by a microscope objective into a sheath flow cuvette, just below the capillary end. The emitted light was spectrally filtered with a 580 nm band-pass filter. Other recently published examples of combined capillary electrophoresis/ LIF are the study of hirudin which is a potential anticoagulant and antithrombotic agent (Ban, Nam et al. 2001) and the detection of biogenic amines which are accumulated in foodstuffs and beverages during improper storage (Male and Luong 2001).

For some analytes such as inorganic ions or carboxylic acids, derivatization with a fluorescent probe is not easily achieved. In this case, indirect laser-induced fluorescence is a good alternative. It is based on the displacement of a fluorescing species present in the background electrolyte by the analyte (Melanson, Boulet et al. 2001). This displacement results in a decrease of the fluorescence signal. The sensitivity of such a signal-decrease detection intrinsically depends on the power stability of the exciting light source. Necessary power stabilization of UV-gas lasers used up to now lead to increased instrumental complexity, cost and loss of laser power, however. The new generation of violet diode lasers overcomes this problems; in fact, diode lasers are especially attractive concerning size, costs, lifetime, and power stability. With such an approach, a detection limit of 10^{-7} M (9 ppb, 0.7 fmol injected) could be achieved for chemical warfare agent degradation products (Melanson, Boulet et al. 2001).

Gas (GC) and liquid (LC) chromatography are other conventional low cost (in comparison with mass spectrometry) selection techniques which are suited for combination with laser-assisted detection. A very popular GC-detection method is flame ionization detection (FID). A considerable increase of sensitivity and, in particular, selectivity can be achieved by laser-enhanced ionization in the flame. This has been demonstrated for element-specific detection of organotin which is a compound widely used as catalyst and additive in industry and as biocide in agriculture (Ke, Su et al. 2001). By double-resonance excitation of tin atoms using two wavelength-tunable dye lasers and subsequent collisional ionization, very high selectivity is achieved. An alternative approach is excitation with one laser wavelength and dispersion of the fluorescence in a monochromator. Of course, the use of two Nd:YAG-laser pumped dye lasers makes this set-up not attractive for practical applications. However, it has been shown by other research groups (see last section) that highly selective and sensitive laser-based detection of elements is possible using a multi-wavelength diode laser assembly; its size is believed to be reducible to that of a shoe box.

In the following, the ultratrace analysis of fluorescent dyes in environmental samples will be described. The technique of choice is the combination of high-performance liquid chromatography (HPLC), laser-induced fluorescence and multiwavelength emission detection (Kleimeyer, Rose et al. 2001). The goal was to lower the limits of detection of fluorescent tracers, a widely spread method to study migration and transport of environmental waters. Such studies may provide information for geothermal reservoirs, about the migration of species between surface and subsurface aquifers, about stream flows and the mixing of open ocean currents. Since the volume of water in which the fluorescent traces are diluted can be enormous (e.g. 10^{12} l) even sub-part-per-billion (ppb) detection limits may require hundreds of kilogram of dye. There is a significant benefit, therefore, in lowering the detection limit for such tracers.

The instrumentation used consists of a LC-pump, a manual injection valve with a large volume sample loop, and a modified 1/16 in O.D. HPLC T-fitting. This fitting represents a flow cell with a volume of 100 nL; it includes a fiber-optic system for exciting the analyte by an Ar-ion laser (488 nm, 25 mW) and for gathering the emitted fluorescence and coupling it into a f/1.2 spectrometer. A cooled CCD-camera is used for simultaneous recording of a spectral range from 500 to 600 nm. In this range, it exhibits a quantum efficiency of 80%. Instead of measuring within a two-dimensional selective window of parameters, here the full two-dimensional spectral information (elution time/wavelength) is used for the analysis. Special data analysis techniques (rank annihilation, self-modeling curve resolution, for further details see (Kleimeyer, Rose et al. 2001)) have been used to discriminate contributions of the mobile phase (Raman scattering) or interference of other fluorescing compounds (e.g. due to incomplete chromatographic separation).

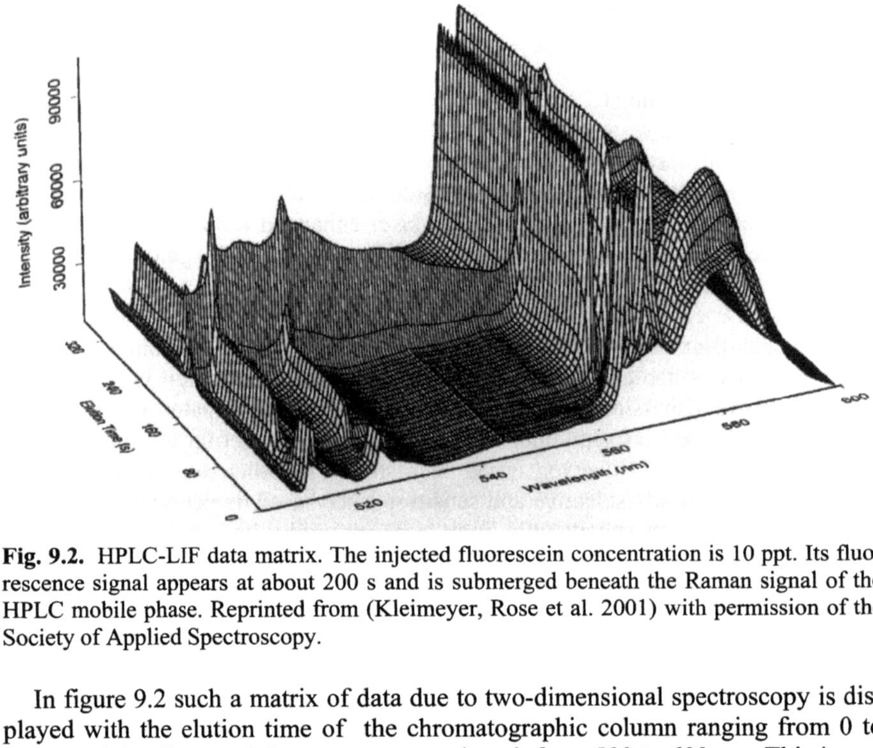

Fig. 9.2. HPLC-LIF data matrix. The injected fluorescein concentration is 10 ppt. Its fluorescence signal appears at about 200 s and is submerged beneath the Raman signal of the HPLC mobile phase. Reprinted from (Kleimeyer, Rose et al. 2001) with permission of the Society of Applied Spectroscopy.

In figure 9.2 such a matrix of data due to two-dimensional spectroscopy is displayed with the elution time of the chromatographic column ranging from 0 to 300 s and the dispersed fluorescence wavelength from 500 to 600 nm. This is one spectrum from a set of calibration solutions ranging in concentration from 10 ppt (10^{-11} g/mL) as in figure 9.2 to 10 ppq fluorescein standard in aqueous solution. The observed background spectrum is dominated by Raman scattering from methanol and water, which are the major components of the mobile phase and present during the whole elution time. At the lowest concentration a fluorescence has to be detected which is only 0.15 % of the total Raman scattering background at the corresponding elution time of the dye. The capability of this ultratrace detection method has been tested on geothermal well samples: about 40 ppq ($4\ 10^{-14}$ g/mL) of fluorescein could be detected in the presence of 15-fold greater fluorescence from unknown interferents.

This section about combined chromatography/ laser-assisted detection shouldn't be finished without mentioning the possibilities of three-dimensional selectivity by including mass spectrometry. Gas chromatography/mass spectrometry (GC-MS) may be considered as the method with highest selectivity under widely used conventional analytical tools. In contrast to conventional ionization techniques (e.g. electron ionization) laser-induced ionization brings to bear a third selective parameter, the excitation wavelength. One-way of laser assisted ionization combined with gas-chromatopgraphy has been described in this section (Ke, Su et al. 2001), another one is resonance enhanced multiphoton ionization (see section 9.2). Molecular cooling of the GC-efluent in a supersonic beam gas inlet (Hayes and Small 1982; Lubman 1987; Opsal and Reilly 1988; Köster, Grotemeyer et al. 1990; Zimmermann, Lermer et al. 1995) even allows the combination of high resolution UV-spectroscopy with high resolution GC-MS. Catalytic conversion of isomers and congeners to a common basic molecular structure after GC-separation (Zimmermann, Rohwer et al. 1999) or a new miniaturized supersonic beam inlet system for optimum coupling of gas inlet and laser ionization (Hafner, Zimmermann et al. 2001) are further promising developments of laser-induced ionization GC-MS for ultra-trace analysis of very complex mixtures of chemical compounds.

9.4 Two-Dimensional Selectivity by Absorption/Emission Spectroscopy

Absorption, as well as emission spectroscopy are species-specific, of course. A combination of both, e.g. by exciting at one absorption wavelength and detecting fluorescence at another emission wavelength provides two-dimensional selectivity. To achieve this, tunable lasers as well as spectral filters or a spectrograph are needed to be able to choose the wanted absorption wavelength and to disperse the fluorescence spectrum and set a wavelength window of detection. Dye lasers, optical parametric oscillators (OPO), tunable excimer lasers (tunable over a small wavelength range), or Raman shifters emitting several stokes and antistokes lines are available as tunable lasers. In the visible and near-UV-range even diode lasers can be used, if no high peak powers are needed. The approach of combined absorption/emission spectra has for instance been used to improve the low selectivity of UV-spectroscopy of PAH´s in the gas phase and in solution (Löhmannsröben and Roch 1996). A further enhancement of selectivity is possible by considering the fact that fluorescence lifetimes of different PAH´s may differ considerably.

Another advantage of selective laser excitation-induced fluorescence is that it provides the possibility of obtaining instantaneous two-dimensional images of absolute concentration fields without influencing the process of interest, e.g. combustion. For taking advantage of this feature, the laser excitation is performed in a plane sample area (e.g. cross section of a flame) and the fluorescence emitted vertically to this area is recorded by a spatially resolving CCD camera after spectral dispersion in a spectrograph. In the following, an experimental arrangement is presented, which allows the spatially resolved detection of NO in high-pressure flames (Dreizler, Sick et al. 1997; Schulz, Sick et al. 1997) and even in the combustion chamber of spark ignition engines (Schulz, Yip et al. 1995; Schulz, Sick et al. 1996). The problems to be solved are caused by strong Raman, Rayleigh, and LIF signals from O_2, OH and strong absorption by PAHs and partially oxidized hydrocarbons. Another problem is the need of a high-speed measurement to study highly dynamic chemical processes in combustion engines. This can only be achieved by single-shot experiments which provide the whole spectroscopic information for one single laser shot. High-power laser pulses are necessary for selective excitation to obtain fluorescence which is sufficiently strong to be spectrally dispersed and spatially resolved.

It turned out that KrF excimer lasers are suitable laser sources despite their narrow tuning range (which overlaps with the NO A-X(0,2) band, by accident). To determine the optimum wavelength windows for excitation and fluorescence detection a two-dimensional excitation/emission wavelength plot is helpful. Such a plot is displayed in figure 3. The spectrum has been measured at a laminar, premixed methane/air flame at pressures between 1 and 40 bar with NO concentrations of 50 parts in 10^6 which were increased up to 370 ppm by doping NO to the gas mixture (which is much less than in typical combustion engines). A KrF excimer laser with a linewidth of 0.003 nm has been focussed with a cylindrical telescope to form a 20 mm x 0.5 mm light sheet aligned parallel to the burner surface. The spectral intensity at 247.95 nm was 40 MW/cm^2. The fluorescence was collected at right angle to the exciting laser beam and focussed with a f=100 mm lens into a spectrometer equipped with a 300 grooves/mm grating. In addition, a narrow band dielectric mirror (248 nm) was used to reduce the Rayleigh signal by two orders of magnitude. An intensified CCD camera allowed spatial resolution of the dispersed fluorescence along the spectrometer slit.

The excitation-emission plot in figure 9.3 has been obtained by averaging over the spatial coordinate. For every step of the excitation wavelength by 0.001 nm the signal is averaged over 250 single laser shots. The spectra on the left (excitation spectra) show the total emission signal as function of the excitation wavelength (at the NO A-X(0,2) transition) for three different pressures. Single rotational lines are resolved in the 1 bar spectrum; due to pressure broadening the rotational fine structure is blurred at higher pressures and finally disappears at 40 bar. The emission spectrum (dispersed spectra) at the bottom was obtained at 1 bar after excitation at 247,95 nm. The NO A-X(0,v′′) progression is clearly resolved with the (0,0) and the (0,1) band in the blue of the strong Rayleigh peak (off scale) which overlaps with the (0,2) band. The small frames in the two-dimensional plot and the gray areas in the excitation and emission spectra indicate a set of excitation and detection windows which guarantee minimum interference from other gas components, here mainly O_2 and OH-radicals. The height of the small frame corresponds to the bandwidth of the exciting laser, its width to the detection filter function.

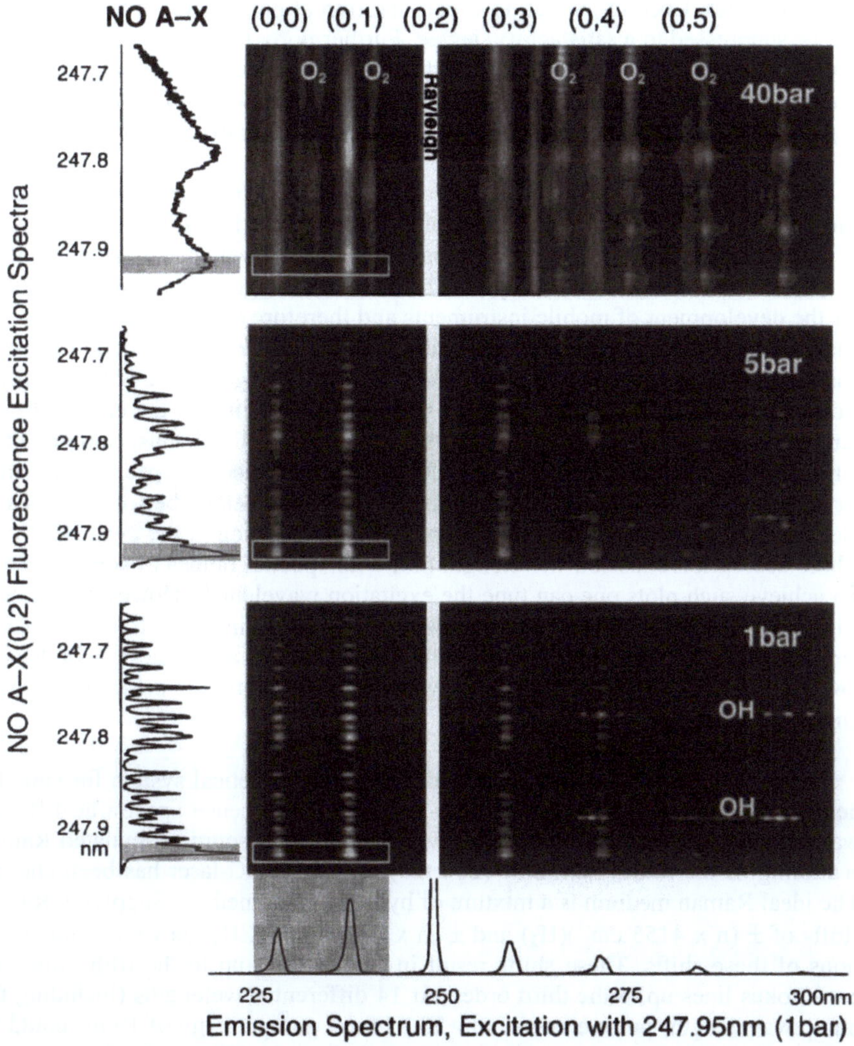

Fig. 9.3. Two-dimensional excitation-emission wavelength plot obtained in the lean CH_4/air flame at 1, 5, and 40 bar. On the left, excitation spectra are displayed, at the bottom a dispersed emission spectrum is shown. The white frame indicates a set of excitation and detection windows for maximum NO selectivity. Reprinted from (Schulz, Yip et al. 1995) with permission from the Optical Society of America.

Because of the anti-Stokes detection of NO, interference with OH LIF is not to be expected. In addition, as can be seen from Figure 9.3, the intensity of OH A-X(3-v′′) decreases with increasing pressure. H_2O, which could be excited by a two-photon process and contributes to the fluorescence via OH fragments, is not a problem either. Even O_2, which is a main interferent particularly at high pressure, can be suppressed to a satisfactory degree. Furthermore, it turned out, that the NO-signal is still in the linear range despite the high KrF excimer laser pulse energies. It could be shown that this NO detection scheme also works in a spark ignition engine fueled with isooctane (Schulz, Yip et al. 1995; Schulz, Sick et al. 1996).

As mentioned above, for the analysis of PAHs, the largest class of mutagenes and carcinogens known, multidimensional fluorescence including the dimension of time (i.e. lifetimes of excited electronic states) has been used as an alternative to conventional analytical techniques. The main advantage is that new possibilities for the development of mobile instruments and therefore of on-site and on-line (or quasi-on-line) measurements are opened up. In the case of NO-detection in combustion processes, the use of a single excitation wavelength and a different single emission wavelength window supplies sufficient selectivity. In the case of PAHs (and many other organic species), however, broad absorption bands, complex mixtures of different PAHs, and interference with many molecular species with similar spectroscopic features cause severe problems which cannot be solved by a single wavelength approach in a satisfying way. One solution is the evaluation of a whole absorption/emission wavelength plot with spectral ranges of some 100 nm. To achieve such plots one can tune the excitation wavelength. However, this only allows a sequential and therefore time-consuming recording, which is particularly problematic, if lifetime-resolved two-dimensional spectra are requested. In the following, a different approach for multi-wavelength excitation is presented (Panne, Dicke et al. 2000)].

The idea behind this approach is to develop a fiber-optical system for simultaneous excitation and detection of time-resolved fluorescence spectra at different wavelengths. As a simultaneous multi-wavelength light source stimulated Raman scattering of the fourth harmonic (266 nm) of a Nd-YAG-laser has been chosen. The ideal Raman medium is a mixture of hydrogen and methane supplying Raman shifts of ± (n x 4155 cm^{-1})(H_2) and ± (n x 2914 cm^{-1})(CH_4) as well as combinations of these shifts. These shifts result in Stokes lines up to the fifth order and anti-Stokes lines up to the third order. At 14 different wavelengths (including the pump wavelength) pulse energies of >50 µJ for a pump energy of 19 mJ could be achieved spanning a wavelength range from 240 nm to 400 nm. In figure 9.4 the experimental set-up is shown. The fourth harmonic of a Nd-YAG-laser (266 nm) had been focused into the Raman cell (stainless steel tube: length 1.17 m; inner diameter 20 mm equipped with a gas inlet system, pumping facility and pressure meter). The output beam, consisting of various spatially overlapping stimulated-Raman-scattering beams, has been collimated by a f=600 mm lens and then dispersed by a prism set. A lens (f=500mm) was used to couple eight selected dispersed beams with wavelengths between 250 and 400 nm into eight optical fibers simultaneously. These fibers had a length of 50 m and guided the excitation wavelengths to eight separated sensors positioned in a probe. One single sensor consisted of a pair of excitation and emission fiber with a distance of about 0.5 mm and an angle of 13° between their ends. The whole probe had an outer diameter of 48 mm. Cross-talk was minimized by a special mounting of the sensors.

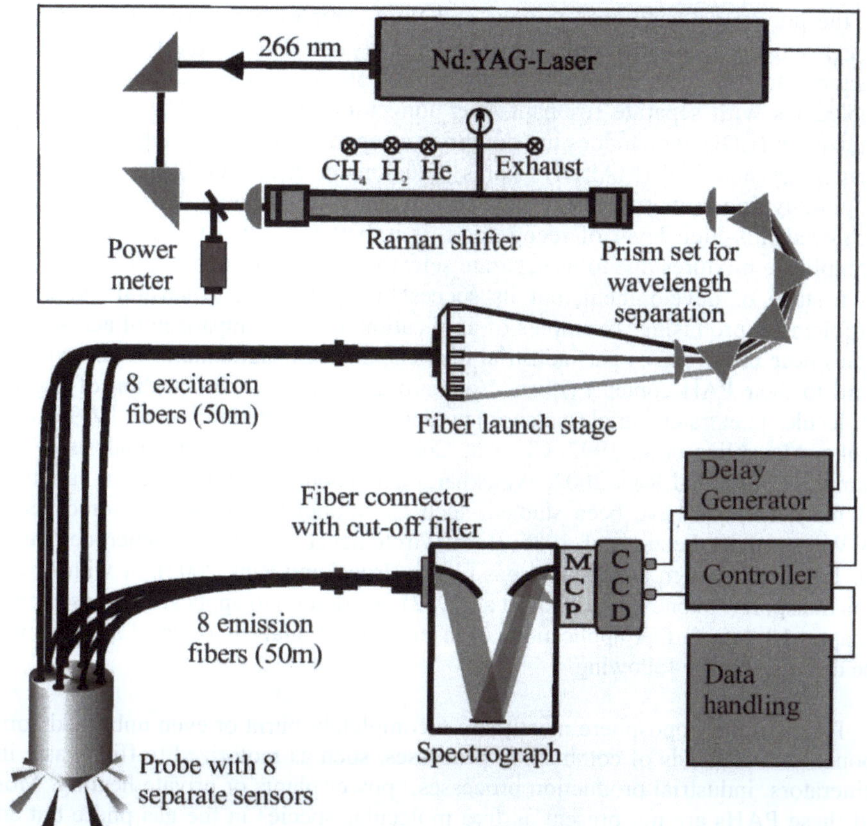

Fig. 9.4. Experimental set-up for simultaneously recording eight emission spectra induced by eight different excitation wavelengths. Each single laser shot supplies a two-dimensional excitation-emission wavelength plot. A third selection parameter is time (due to species selective lifetimes of excited electronic states). Reprinted from (Panne, Dicke et al. 2000) with permission from the Society of Applied Spectroscopy.

The emission fibers were coupled to a spectrograph. The height of the entrance slit, where the fiber ends were positioned in a single row, was imaged 1:1 onto a two-dimensional CCD chip. Thus, eight emission spectra corresponding to eight different excitation wavelengths could be recorded for every single laser pulse. For time-resolved spectra, the intensifier (MCP: multi channel plate) was gated and this gate was delayed in time. Special multi-way data analysis methods are used to allow quantitative pattern recognition of these three-dimensional (i.e. excitation wavelength, emission wavelength, and lifetime) data in complicate mixtures such as oil and petroleum contamination.

9.5 Laser-Assisted Analysis of Solid Samples

In the previous paragraphs, gaseous and fluid samples have been considered. In many situations, solid samples have to be analyzed. There exist two laser-based methods to cope with this problem: the combination of laser desorption of neutral molecules with separate resonant laser ionization (LD/LI) and laser desorption ionization (LDI), which does not comprise a separate ionization step. In particular matrix assisted LDI (MALDI) (Karas, Glückmann et al. 2000) has gained high popularity due to its success in biochemical analysis. Although LD/LI has not yet achieved this high level of recognition, it is better adapted to trace analysis of complicate mixtures due to its separate selective ionization step. LD/LI-MS is still in a stage of development, but its successful application to several interesting problems is promising. Examples of application are the comparison of aerosol particles near busy roads, near industrial districts, and near agricultural areas (with regard to their PAH content) (Zhan, Voumard et al. 1995), the detection of organic molecules in extraterrestrial (meteoric) particles (Zenobi, Philippoz et al. 1989; Kovalenko, Maechling et al. 1992; Clemett, Chillier et al. 1998), or the analysis of soil samples (Boesl and Rink 2002; Weickhardt and Tönnies 2002). Different variations of this technique have been studied, such as combination with supersonic beams (v.Weyssenhoff, Selzle et al. 1985; Boesl, Grotemeyer et al. 1987; Meijer, deVries et al. 1990), with room temperature gas beams (Boesl and Rink 2002), or without any gas transport (Antonov, Letokhov et al. 1982). In addition to applications presented in chapter 10, two further applications of laser-assisted analysis of solid samples will be discussed in the following.

PAHs in the troposphere mostly are incompletely burnt or even unburned components of all kinds of combustion processes, such as motorized traffic, waste incinerators, industrial production processes, power plants or private heating. Most of these PAHs are not present as free molecular species in the gas phase but adsorbed on particle surfaces. For environmental analysis of the atmosphere, therefore not only the gas phase, but also the aerosols have to be investigated. Since PAHs strongly differ in toxicity, some being well known as carcinogenic while others are more or less harmless, an adequate analytical technique should be outstanding not only in terms of its sensitivity but in particular in terms of its selectivity. In opposite to conventional techniques, LD/LI-MS enables one to study molecular species absorbed on surfaces without time-consuming and expensive chemical cleanup due to its high (two-dimensional) selectivity. Single samples may now be analyzed within a few minutes. For instance, this method has been applied to compare the PAH load of aerosols at different sites (Zhan, Voumard et al. 1995). For laser desorption, a pulsed CO_2 laser, and for ionization a XeCl excimer laser has been used. The latter emits at a wavelength of 308 nm where many PAHs absorb. For sampling aerosols in the atmosphere, air from a height of 45 cm above the ground was passed through filter material. The covered filter was cut, mounted onto the tip of a sample holder and introduced into the mass spectrometer through a vacuum interlock. No additional extraction, purification or preconcentration has been performed. The recorded mass spectra of aerosols from the countryside and an industrial zone show a different range of most intense mass peaks (namely 200 to 230 u) than the mass spectrum of a downtown road (mass 200 to 400 u). Furthermore, the integrated signal intensities of the former two spectra are substantially smaller than in the latter case. The few intense mass peaks in the rural site and industrial zone spectra (i.e. 202: pyrene or fluoranthene; 116, 230: methylated and ethylated mass-202 species) may be explained by aerosol transport over long distances from regions of high traffic or intense residential heating. Thus, traffic seems to be one of the main sources of PAH´s on aerosols.

In the measurement described above only crude spectroscopic selectivity has been achieved by using a wavelength which is specific for the class of PAHs, but which does not distinguish single PAHs, in particular not isomers. In fact, for a fast screening of the PAH load of particles or other solid samples a too high selectivity would be disadvantageous. However, there exist several isomeric PAHs which strongly differ in toxicity (e.g. benzo[a]pyrene, benzo[b]fluoranthene, and benzo[k]fluoranthene, all of them with mass 252) and therefore should be measured selectively. Unfortunately, the process of laser desorption gives rise to strong fluctuations of internal molecular energy (corresponding to strongly varying occupation of different molecular states) and therefore to mostly unstructured and unselective laser spectra (Haefliger and Zenobi 1998). This problem can be solved by cooling internal degrees of molecular motion in a supersonic beam. The result are exceptionally well structured molecular spectra (Meijer, deVries et al. 1990; Meijer, deVries et al. 1990), which even would allow to distinguish molecular isotopomers (Boesl, Neusser et al. 1978). The disadvantage of this method is a significant vacuum-technical effort which makes it impractical for analytical instruments, in particular if on-line and on-site analysis is requested. Nevertheless, there are researchers who try to develop a method for on-line separation of the 22 isomers of tetra-chloro-dibenzodioxins (Oser, Coggiola et al. 2001) using this method.

A different approach has been made in the following method which also tries to combine spectroscopy (although not as high resolving as in supersonic beams) and mass spectrometry and has been developed to detect traces of PAHs in soil (Boesl and Rink 2002). To perform a measurement, a soil sample is prepared by forming little pellets with a diameter of 2 mm in a small press. These are fixed on the surface of a movable sample support by gluing tape. The sample support is a thin rod; it has a specially designed tip and can be introduced into the working mass spectrometer via a vacuum lock system within about a minute. A pulsed laser desorbs neutral molecules (e.g. PAH) from the sample surface; these are carried away by a pulsed gas beam which is confined by a short cone-shaped opening in a solid block of Teflon. This beam guide has the advantageous feature to cause thermalisation of the internal motion of the desorbed molecules and of the direction and amount of their velocity. Due to well defined occupation of molecular states, fluctuations of the ionization yield are strongly reduced. Spectroscopy of the desorbed molecules and selective molecular excitation, respectively, now is much more reliable and reproducible than at laser desorption without any further means of achieving an internal equilibrium of state population. After the process of laser desorption, molecular ions are formed by resonance enhanced ionization and are analyzed and detected in a time-of-flight mass spectrometer. The whole procedure (from pressing pellets to obtaining a mass spectrum) takes about 5 minutes. The important advantage of gas beam thermalisation in comparison to supersonic beam cooling is that no extra effort for vacuum is necessary.

In figure 9.5 at the top, a part of the mass spectrum of a soil sample from the bottom of an American river is shown as an example for this method (with a desorption laser wavelength of 532 nm and a ionization laser wavelength of 266 nm, which is semiselective for the class of PAHs). There exists a EPA-certificate of this soil sample; the concentrations of the representative 16 PAHs have been determined by conventional routine analysis. In addition to the certified compounds, the resonant laser mass spectrum reveals a large number of further mass peaks due to numerous methylated PAHs, for example the series of methylated PAHs of mass 228 (242, 256, and 270) or of mass 252 (266 and 280). Typical concentrations of single PAHs were some ppm in this soil sample. From comparison with the background, a detectable level of concentrations of some 100 ppb can be deduced. At this state-of-the-art resonant laser-MS represents an ideal technique for fast PAH-screening. One should keep in mind that the whole mass spectrum in figure 9.5 has been obtained within 5 min including the preparation of the soil sample. An improvement of sensitivity is possible e.g. by tuning the laser wavelength in resonance of a wanted compound.

However, not only PAH-screening but also isomer-sensitive monitoring should be possible, since laser desorption does not disturb spectroscopy anymore. Figure 9.5 displays REMPI-spectra as well as conventional UV-gas phase spectra of the isomers Benzo[a]Pyrene, Benzo[b]Fluoranthene, and Benzo[k]Fluoranthene for comparison. The agreement of REMPI- and conventional absorption spectra is obvious. The increase of the REMPI-ion signal in the wavelength range 290 to 270 nm is caused by more and more favorable Franck-Condon factors of the second absorption step (causing ionization). The structural differences of the spectra of different species are significant enough to allow isomer enhanced detection in this laser-desorption / laser post-ionization assembly. From the ion signal at three different wavelengths, the contributions of these three isomers to the ion peak at mass 252 in an unknown sample then could be entangled within seconds. With modern tunable lasers a wavelength change of some 20 nm should be possible within much less than a minute. In summary, the combination of laser desorption, laser spectroscopy and time-of-flight spectrometry described above will make it possible to realize a very compact, fast, probably mobile and nevertheless highly specific trace analysis of solids. This might open interesting new aspects not only for soil analysis but, for instance, for aerosol and soot particle analysis, also.

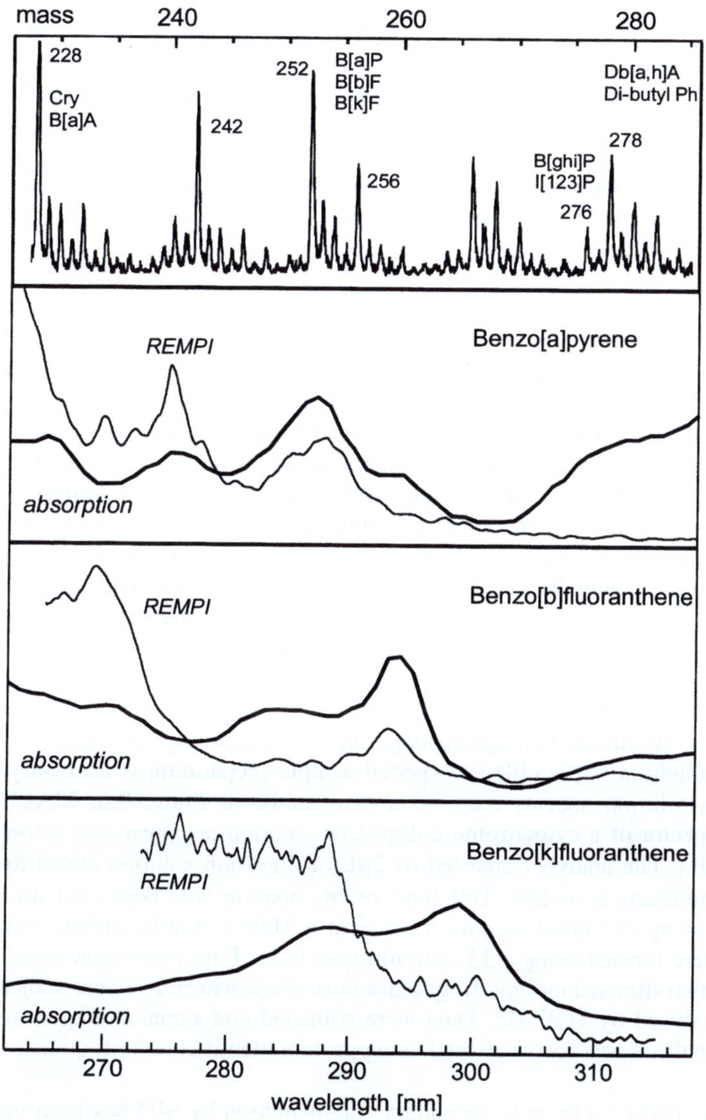

Fig. 9.5. Mass spectrum of a soil sample (certified by the US-EPA) from the bottom of an American river and resonant multiphoton ionization spectra (REMPI) of three isomeric PAHs with mass 252. The REMPI-spectra are compared with UV-gas phase spectra. The good correspondence of the spectral structure (differences in intensity are due to changing transition probabilities of the second absorption step to the ionization continuum) proofs that the population of internal states of the desorbed molecules has reached thermodynamic equilibrium roughly corresponding to room temperature. See also (Boesl and Rink 2002)

Much more popular than laser desorption laser postionization is the single step process laser desorption ionization (LDI), in particular matrix assisted LDI (MALDI). Since desorption and ionization process are not separated anymore, spectroscopy cannot be involved and therefore LDI and MALDI are non-selective methods to form molecular ions of involatile molecular compounds. To detect traces of compounds in a complicate mixture, another method of selection than spectroscopy has to be combined with mass spectrometry.

A fine example of such a two-dimensional analysis including MALDI is discussed in the following (Stone, Gillig et al. 2001). The experimental set-up is roughly as follows: After performing MALDI the desorbed molecular ions are separated by an ion mobility spectrometer. Optionally, several specially coated gold grids are used for surface induced dissociation in front of the extraction plates of the time-of-flight spectrometer. SID is performed at a grazing incidence angle and at collision energies ranging from 20 to 100 eV. Ion mobility spectrometry (Baumbach and Eiceman 1999) has some special features concerning isomer separation (Illenseer and Löhmannsröben 2001); in addition, secondary fragmentation is a well known method of mass spectrometry to characterize molecular ions. Since the time scale of the ion mobility spectrum is several orders of magnitude longer than that of a time-of-flight mass spectrum, SID fragment ion spectra of all primary ions eluting the ion mobility cell from a single ionization event can be acquired simultaneously. The combination of MALDI, IMS, SID, and TOF therefore promises fast analysis of complicate solid samples, such as peptides: simultaneous peptide mapping and sequencing combined with fast sample throughput is possible (no special sample preparation is necessary and therefore in-solution species may be investigated). In figure 9.6, MALDI-IM-SID-TOF-spectra of a cytochrome c digestion solution are presented (Stone, Gillig et al. 2001). The analyte consisted of 2μL of digestion solution added to 20μL of a α-cyano/fructose matrix. 100 fmol of the analyte was deposited on the sample probe using the dried droplet method at a 1000:1 matrix-analyte ratio. MALDI ions were formed using a 337-nm nitrogen laser. Four prominent digest fragments in the two-dimensional mobility/mass map of cytochrome c were subjected to SID and analyzed by TOF-MS. Data were collected and signal averaged for 1-2- min. Data analysis was accomplished using protein identification programs.

In an earlier approach, secondary fragmentation by SID has been included in a tandem-MS experiment (Bier, Amy et al. 1987). Tandem-mass spectrometry (deHoffmann 1996) is a highly selective analytical method for basic research as well as application with a large variety of working modes. A very modern type of tandem-MS includes time-of-flight mass analysis. In particular reflectron-TOF instruments (Mamyrin, Karataev et al. 1973; Bergmann, Martin et al. 1990; Boesl, Weinkauf et al. 1992) are suited for tandem-MS experiments due to their special features concerning metastable ion decay. An early application for peptide sequencing involved MALDI, collisional excitation of primary molecular ions by residual gas molecules in the ion source and post-source decay (Kaufmann, Kirsch et al. 1994; Spengler 1997).

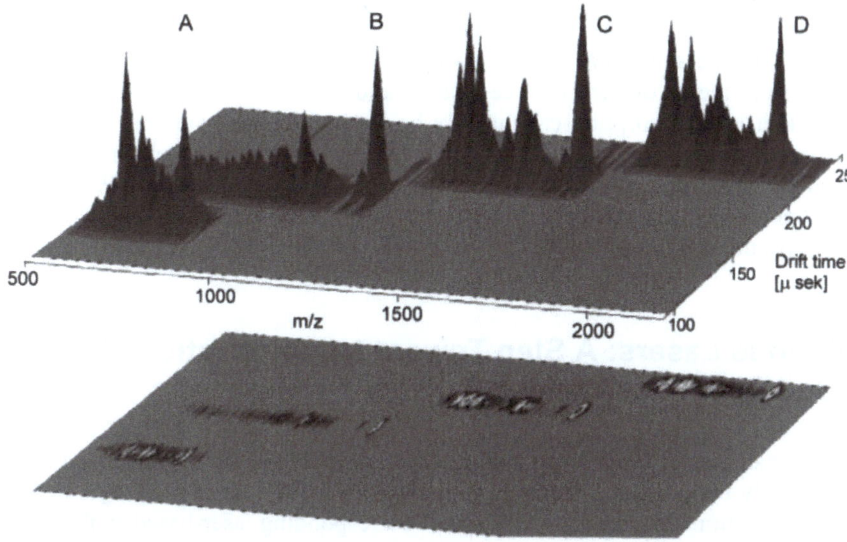

Fig. 9.6. MALDI-IM-SID-TOF-(MALDI ion mobility surface-induced dissociation time-of-flight) spectra of four species from an "in-solution" digest of cytrochrome c. Species A, B, C, and D are separated by ion mobility and then analyzed by time-of-flight mass spectrometry together with their secondary fragments formed by SID. Reprinted from (Stone, Gillig et al. 2001) with permission from the American Chemical Society.

Performing secondary fragmentation with photons or a collision gas at a special point in the TOF-spectrometer (the so-called space-focus) (Boesl, Neusser et al. 1982; Weinkauf, Walter et al. 1989) even promises a further progress for fast tandem-MS and application to fast peptide sequencing. Lasers can be included in this types of modern tandem-MS in a variety of ways, such as MALDI, laser desorption of neutral molecules, resonant ionization, or secondary excitation. An interesting method to analyze solid samples with two-dimensional selectivity and high speed also is the combination of neutral laser desorption, separation by gas chromatography and TOF-MS (Shahar, Dagan et al. 1998).

Of course, laser-assisted analysis of solid samples is not only based on laser desorption of neutral or ionized molecular systems. There exist numerous methods of laser spectroscopy of surfaces such as surface enhanced Raman and two-photon spectroscopy or electron energy loss and other types of photoelectron spectroscopy. A very fine method of studying particles on-line in an atmospheric environment and during its geometrical and chemical changes is levitation in a Pauls trap (Stockel et al. 2002). Lasers allow to observe optical properties such as refractive index and light scattering, to perform IR or Raman spectroscopy and even to induce evaporation followed by mass spectrometric analysis of trapped particles. Another way to study single particles is to introduce them via an aerodynamic gas inlet system into the mass spectrometer directly from atmosphere. The stray light of a first laser beam acts as a particle detector and triggers a second laser which induces laser desorption of positively and negatively charged ions. Both types of ions then are mass analyzed by two time-of-flight mass spectrometers which share the same ion source (Spengler, Hinz et al. 1996). For a review of mass spectrometry of aerosols see (Noble and Prather 2000). Another example shows that laser desorption cannot only be applied to solid surfaces or particles but also to fluids. The injection of liquid beams into the mass spectrometer followed by laser desorption and mass analysis has been achieved (Wattenberg, Sobott et al. 1999; Wattenberg, Sobott et al. 2000) with interesting applications to the study and analysis of solutions.

9.6 Diode Lasers: A Step Toward Miniaturization of Laser-Based Chemical Analysis?

All the examples of laser-based analysis described above, are characterized by the use of fairly bulky laser systems (e.g. excimer, Nd:YAG, argon-ion, dye, or OPO lasers). This limits the reduction of size of corresponding laser-based analytical instruments. Diode lasers may open new dimensions concerning miniaturization of laser based instruments for chemical analysis. A domain of diode lasers is vibrational IR-absorption spectroscopy with all its facets (e.g. cavity ring-down spectroscopy (Berden, Peters et al. 2000)). In chapter 11 the large variety of diode lasers, their stage of development and application is presented. Its title: (diode-laser sensors) indicates that diode laser based instrumentation has reached the stage of miniaturized, On-line capable analyzers already.

However, also diode lasers at the high-energy end (Nakamura and Fasol 1997) of the available wavelength range offer the potential for trace analysis. The first electronic transitions of larger conjugate hydrocarbons and of molecular systems with free electron pairs already can be excited by visible and near-UV light, which opens the possibility of detection by laser-induced fluorescence. The use of diode laser induced fluorescence as selective detector for capillary electrophoresis (Melanson, Boulet et al. 2001) has already been mentioned in section 9.3. Other types of trace compounds which can be excited by visible or near-UV light are atomic species (e.g. alkaline or earth-alkaline elements or excited atoms). Atomic species have advantageous absorption cross sections which are larger than molecular ones by many orders of magnitude; in addition, the narrow laser line widths (and thus high spectral density) available with diode lasers in conjunction with typically narrow atomic lines also are favorable for the low-concentration detection limit. Both effects compensate for the considerably lower peak power range of diode lasers in comparison to the above mentioned types of lasers. From the numerous examples of laser-diode based atomic trace analysis (e.g. (Quentmeier, Bolshov et al. 2001; Uhl, Franzke et al. 2001) two are discussed in the following.

Trace analysis of environmental pollutants often concentrates on classes of molecular systems with typical characteristics, e.g. the chlorine fluorine hydrocarbons CCl_2F_2, $CClF_3$, and $CHClF_2$. One way of monitoring these compounds is dissociation followed by chlorine and fluorine atomic detection. Dissociation may be achieved in plasma sources which can be operated on micro-chips. Combination with laser diodes therefore may allow a "lab-on-the-chip" analysis of this types of molecular classes. The approach of a microchip plasma on the basis of dielectric barrier discharge (which is used in flat panel plasma displays for color TV screens, for example) has been used to detect traces of CCl_2F_2 (Miclea, Kunze et al. 2001). In Figure 9.7, the plasma source and the experimental arrangement for detection by diode laser absorption spectroscopy is displayed. Helium and argon are used as plasma gases, metastable halogen atoms are produced during total molecular dissociation in the plasma, and diode laser atomic absorption at 837 nm and 685 nm has been involved for detection of chlorine and fluorine, respectively.

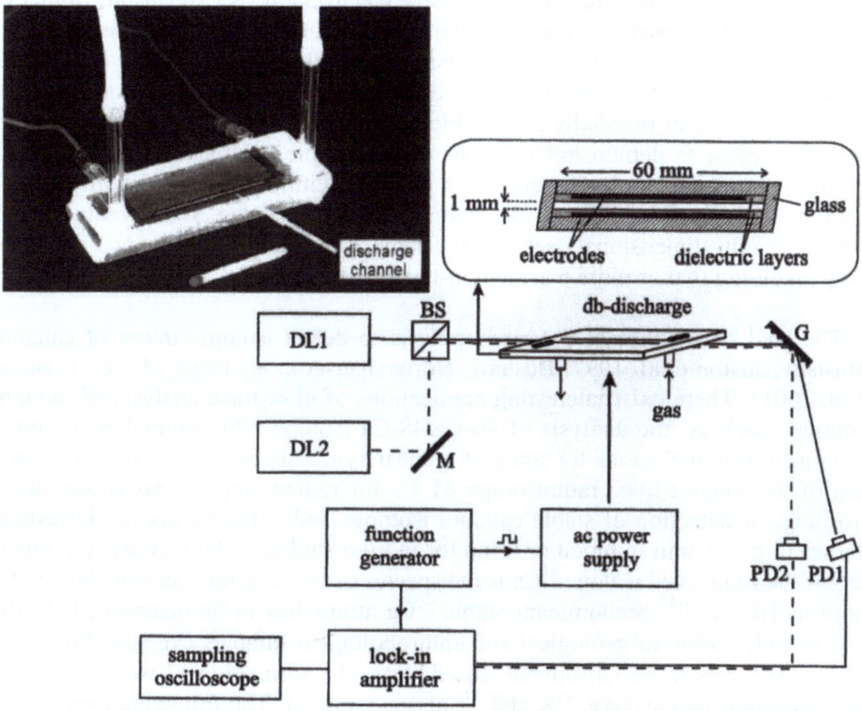

Fig. 9.7. Dielectric barrier discharge as a plasma dissociation source and experimental arrangement. DL1 and DL2: diode lasers, BS: beam splitter, M: mirror, G: grating, PD1and PD2: photodiodes. Reprinted from (Miclea, Kunze et al. 2001) with permission from Elsevier Science B.

The length of the discharge channel and therefore the absorption length was 60 mm. For the electric AC discharge a low power supply (<1W) was sufficient. To achieve simultaneous chlorine and fluorine detection two diode lasers have been coupled into the discharge channel. Modulated absorption spectroscopy has been applied by modulation of the plasma and phase-sensitive detection using a lock-in amplifier. Detection limits for CCl_2F_2 of 400 ppt and 2 ppb has been reached using the chlorine 837 nm and fluorine 685 nm line, respectively.

Problems with diode laser absorption spectroscopy arise when small traces of chemical compounds have to be detected and measured in complicate matrices since only a "one-dimensional" selectivity due to one absorbed laser wavelength is available. These matrices may comprise numerous different compounds of considerably larger concentrations, or complex isotopic mixtures with a rare isotope as the trace compound. Two-dimensional selectivity can be achieved by combination with other selective techniques. This is difficult for vibrational IR-diode laser spectroscopy, but has already been realized with visible diode lasers, e.g. using induced fluorescence as detector for capillary electrophoresis (Melanson, Boulet et al. 2001) as mentioned in section 9.3. The combination of laser spectroscopy with mass spectrometry is particularly promising for ultra-trace and on-line analysis as discussed in several of the above sections. The necessary ionization by multiphoton-absorption is usually achievable by pulsed lasers with high peak power, only. However, as demonstrated by the next example, resonance enhanced multiphoton-ionization of atomic species is possible with diode lasers supplying exceptionally high selectivity. This example also illustrates a different way to achieve "multi-dimensional" selectivity, namely by multiple resonance spectroscopy involving two or more wavelength-tunable diode lasers.

The goal of the project described now is to detect isotopic traces of calcium (Bushaw, Juston et al. 1997; Bushaw, Nörtershäuser et al. 1999; Müller, Bushaw et al. 2001). There exist interesting applications of ultra-trace analysis of calcium isotopes, such as: the analysis of 46-Ca/48-Ca isotope ratio anomalies in single grain meteorite inclusions for study of nuclear synthesis processes, the determination of the longest-lived radioisotope 41-Ca for radiodating of fossils and depth profiling, or detection of stable calcium isotopes and 41Ca tracers in biomedical studies (e.g. calcium in blood or urine for in vivo studies of human calcium kinetics). If the long-lived isotope ^{41}Ca is the species of interest, then an abundance of 1 atom in 10^9 to 10^{12} predominant stable ^{40}Ca atoms has to be determined. In the case of radiodating of geological and anthropological samples even an abundance of 1 in 10^{15} has to be considered. In addition, discrimination of isobaric atomic and molecular species (e.g. ^{41}K and ^{40}CaH) is required. The following experimental set-up promises to achieve these specifications. Samples are atomized with a graphite tube furnace. Three single-mode diode lasers with grating-tuned extended cavities supply three tunable wavelengths with linewidths in the 10MHz range. Fixed frequency lasers (Ar^+: 364nm or 515 nm, or CO_2: 10,6 μm) serve to ionize Rydberg states which have been excited by one, two or three diode-laser photons. The so-formed ions are detected in a quadrupole mass analyzer which is able to suppress neighbouring masses by 8 orders of magnitude. In Figure 9.8 three excitation schemes are shown adapted to the requirements of different applications.

For the medical application mentioned above a single-resonance excitation (scheme 1) is sufficient. With double and triple resonance ionization a 2 10^4 and 10^9 fold selectivity, respectively, of ^{41}Ca against ^{40}Ca has been achieved exploiting the isotope shifts and selected Rydberg states (Müller, Bushaw et al. 2000) of the calcium isotopes. An overall efficiency of 10^{-3} and an isotopic abundance sensitivity of 10^{13} has been reached in the meanwhile. Thus all goals, except the utmost difficult one (radiodating) have been reached by now. Figure 9.8 shows a double-resonance spectrum of minor calcium isotopes representing a two-dimensional selectivity with the parameters: wavelength 1 and wavelength 2 resembling the two-dimensionality of absorption/emission spectroscopy of section 9.4. The interesting feature here is that a narrow band, tunable three-wavelength diode laser system has been used which promises compactness, performance, reliability and low price, which possibly can be mounted in a container of the size of a shoe box.

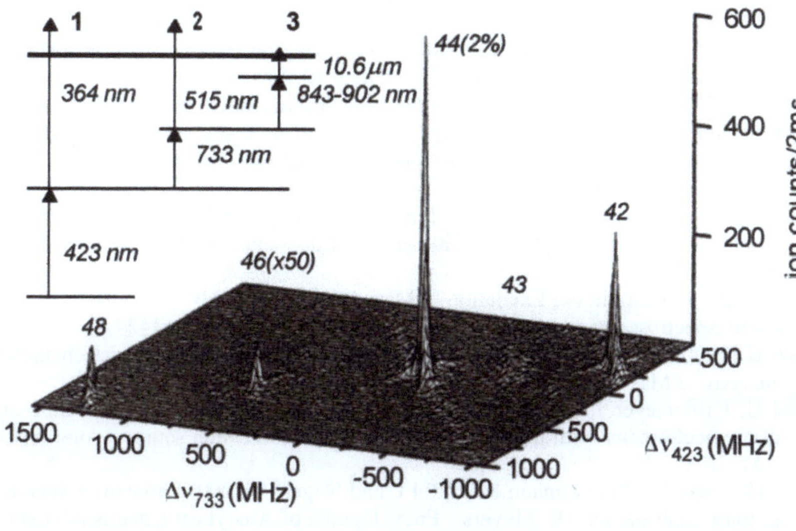

Fig. 9.8. Isotope shifts for the minor calcium isotopes observed in double-resonance three-photon ionization. Inset: Single-, double-, and triple-resonance multiphoton ionization schemes using diode lasers for the resonance steps. In the double- and triple-resonance schemes, the diode lasers as well as the fixed frequency lasers may induce ionization (364 and 515 nm Ar+ laser lines, 10.6 μm CO2-laser). The calcium 43 isotope (as the 41 isotope not shown here) exhibit hyperfine structure which is resolved in the spectrum. Reprinted from (Bushaw, Juston et al. 1997; Bushaw, Nörtershäuser et al. 1999; Müller, Bushaw et al. 2001) with permission from AIP.

References

Andrews D (1990) Lasers in chemistry. Springer, Berlin

Andrews D (1994) Applied laser spectroscopy: techniques, instrumentation and application. VCH Verlag Chemie, Weinheim

Antonov V, Knyazev I, Letokhov V, Matiuk V, Movshev V and Potapov V (1978) Stepwise laser photoionization of molecules in a mass spectrometer: a new method for probing and detection of polyatomic molecules. Optics Lett 3 :37 - 39

Antonov V, Letokhov V, Matveyets Y and Shibanov A (1982) Sputtering of neutral molecules and molecular ions from the adenine crystal surface induced by the UV picosecond laser pulse. Laser Chem 1 :37 - 43

Ban E, Nam H-S and Yoo Y (2001) Competitive immunoassay for recombinant hirudin using capillary electrophoresis with laser-induced fluorescence detection. J Chromatogr A 924 :337 - 344

Baumbach J and Eiceman G (1999) Ion mobility spectrometry. Appl Spectrosc 53 :338A - 353A

Berden G, Peters R and Meijer G (2000) Cavity ring-down spectroscopy: experimental schemes and applications. Int Rev Phys Chem 19 :565 607

Bergmann T, Martin TP and Schaber H (1990) High-resolution time-of-flight mass spectrometers: Part III, Reflectron Design. Rev Scientific Instrum 61 :2592 - 2600

Bernstein R (1982) Systematics of multiphoton ionization -fragmentation of polyatomic molecules. J Phys Chem 86 :1178 - 1184

Bier M, Amy J, Cooks R, Syka J, Ceja P and Stafford G (1987) Tandem quadrupole mass spectrometry for the study of surface-induced dissociation. Int J Mass Spectrom Ion Proc 77 :31 - 47

Boesl U (1999) Multiphoton Excitation in Mass Spectrometry. IN: C. McNeil; Encyclopedia of Spectroscopy and Spectrometry. Academic Press, 1411 - 1424

Boesl U (2000) Laser mass spectrometry for environmental and industrial chemical trace analysis. J Mass Spectrom 35 :289-304

Boesl U, Grotemeyer J, Walter K and Schlag E (1987) A high resolution time-of-flight mass spectrometer with laser desorption and laser ionization source. Anal Instrum 16 :151 - 171

Boesl U, Heger H, Zimmermann R, Püffel P and Nagel H (2000) Laser mass spectrometry in trace analysis. IN: R. Meyers; Encyclopedia of Analytical Chemistry. John Wiley & Sons, Chichester, 2087 - 2118

Boesl U, Neusser H and Schlag E (1982) Secondary excitation of ions in a multiphoton mass spectrometer. Chem Phys Lett 87 :1

Boesl U, Neusser HJ and Schlag EW (1978) Two-Photon Ionization of Polyatomic Molecules in a Mass Spectrometer. Z Naturforsch 33A :1546 - 1548

Boesl U and Rink J (2002) Organic trace compounds in soil samples: Investigation of laser desorption and resonant multi-photon ionization for fast on-site analysis. IN: *Laser Based Environmental and Process Measurement,* edited by R.Noll and W.Schade (Springer, proposed date 2003, ISBN 3-540-42945-X)

Boesl U, Weinkauf R and Schlag EW (1992) Reflectron time-of-flight mass spectrometry and laser excitation for the analysis of neutrals, ionized molecules and secondary fragments. Int J Mass Spectrom Ion Phys 112 :121 - 166

Bushaw B, Juston F, Nörtershäuser W, Trautmann N, Haan PV-d and Wendt K (1997) Multiple resonance RIMS measurements of calcium iostopes using diode lasers. IN: Resonance Ionization Spectroscopy 1997. AIP, 115 - 118

Bushaw B, Nörtershäuser W and Wendt K (1999) Lineshapes and optical selelctivity in high-resolution double-resonance ionization mass spectrometry. Spectrochim Acta B 54 :321 - 332

Clemett S, Chillier X, Gillette S, Zare R, Maurette M, Engrand C and Kurat G (1998) Observation of indigenous polycyclic aromatic hydrocarbons in 'giant' carbonaceous Antarctic micrometeorites. Origins-of-Life-and-Evolution-of-the-Biosphere 28 :425-448

Cotter R (1984) Laser and mass spectrometry. Anal Chem 56 :485A

deHoffmann E (1996) Tandem mass spectrometry: a primer. J Mass Spectrom 31 :129-137

Dreizler A, Sick V and Wolfrum J (1997) Applied laser spectrocopy in technical combustion systems. Ber Bunsenges Phys Chem 101 :771 - 782

Fassett J, Moore L, Travis J and DeVoe J (1985) Laser resonance ionization mass spectrometry. Science 230 :262 - 267

Gedanken A, Robin M and Kuebler N (1982) Nonlinear photochemistry in organic, inorganic, and organometallic systems. J Phys Chem 86 :4096 - 4107

Gobeli D, Yang J and El-Sayed M (1985) Laser multiphoton ionization-dissociation mass spectrometry. Chem Rev 85 :529 - 554

Haefliger O and Zenobi R (1998) Laser mass spectrometric analysis of polycyclic aromatic hydrocarbons with wide wavelength range laser multiphoton ionization spectroscopy. Anal Chem 70 :2660-2665

Hafner K, Zimmermann R, Rohwer E, Dorfner R and Kettrup A (2001) A capillary-based supersonic jet inlet system for resonance-enhanced laser ionization mass spectrometry: principle and first on-line process analytical applications. Anal Chem 73 :4171 - 4180

Hayes J and Small G (1982) Rotationally cooled laser-induced fluorescence/gas chromatography. Anal Chem:1202 - 1204

Hayes JM (1987) Analytical spectroscopy in supersonic expansions. Chemical Reviews 87 :745-760

Heger H, Boesl U, Zimmermann R, Dorfner R and Kettrup A (1999) On-line resonance-enhanced multiphoton ionization time-of-flight laser mass spectrometry for combined multi-component-pattern analysis and target-compound monitoring: non-chlorinated aromatics and chlorobenzene in flue gases of combustion processes. Eur Mass Spectrom 5 :51 - 57

Herrmann A, Leutwyler S, Schumacher E and Wöste L (1977) Multiphoton ionization: mass selective laser-spectroscopy of Na2 and K2 in molecular beams. Chem Phys Lett 52 :418 - 425

Illenseer C and Löhmannsröben H (2001) Investigation of ion-molecule collisions with laser-based ion mobility spectrometry. Phys Chem Chem Phys 3 :2388 - 2393

Johnson PM (1980) Molecular multiphoton ionization spectroscopy. Applied Optics 19 :3920 - 3925

Karas M, Glückmann M and Schäfer J (2000) Ionization in matrix-assisted laser desorption/ionization: singly charged molecular ions are the lucky survivors. J Mass Spectrom 35 :1 - 12

Kaufmann R, Kirsch D and Spengler B (1994) Sequencing of peptides in a time-of-flight mass spectrometer: evaluation of postsource decay following matrix-assisted laser desorption ionisation (MALDI). Int J Mass Spectr Ion Proc 131 :355

Ke C-B, Su K-D and Lin K-C (2001) Laser-enhanced ionization and laser-induced atomic fluorescence as element-specific detection methods for gas chromatography: Application to organotin analysis. J Chromatogr A 921 :247 - 253

Kleimeyer J, Rose P and Harris J (2001) Determination of ultratrace-level fluorescent tracer concentrations in environmental samples using a combination of HPLC separation and laser-excited fluorescence multiwavelength emission detection: Application to testing of geothermal well brines. Appl Spectrosc 55 :690 - 700

Kompa K, Sick V and Wolfrum V (1993) Laser diagnostics for industrial processes. Special Issue of Ber Bunsenges Phys Chem:97

Köster C, Grotemeyer J and Schlag E (1990) A high pressure pulsed valve for gases, liquids, and supercritical fluids. Z Naturforsch 45a :1285 - 1292

Kovalenko L, Maechling C, Clemett S, Philippoz J, Zare R and Alexander C (1992) Microscopic organic analysis using two-step laser mass spectrometry: application to meteoric acid residues. Anal Chem 64 :682 - 690

Letokhov V (1986) Laser analytical spectrochemistry. Adam Hilger, Bristol

Letokhov VS (1987) Laser photoionization spectroscopy. Academic Press, Orlando

Levy D (1981) The Spectroscopy of Very Cold Gases. Science 214 :263 - 269

Löhmannsröben H-G and Roch T (1996) Laserfluoreszenzspektroskopie als extraktionsfreies Nachweisverfahren für PAK und Mineralöle in Bodenproben. IN: H. Günzler; Analytiker Taschenbuch. Springer, Berlin, 217

Lubman D (1987) Optically Selective Molecular Spectrometry. Anal Chem 59 :31A - 40A

Lubman D and Kronick M (1982) Mass spectrometry of aromatic molecules with resonance-enhanced multiphoton ionization. Anal Chem 54 :660 - 665

Lubman DM (1990) Lasers and Mass Spectrometry. Oxford University Press, New York

Male K and Luong J (2001) Derivatization, stabilization and detection of biogenic amines by cyclodextrin-modified capillary electrophoresis-laser-induced fluorescence detection. J Cromatogr A 926 :309 - 317

Mamyrin BA, Karataev VI, Shmikk DV and Zagulin VA (1973) The mass-reflectron, a new nonmagnetic time-of-flight mass spectrometer with high resolution. Sov Phys - JETP 37 :45 - 48

Meijer G, deVries M, Hunziker H and Wendt H (1990) Laser desorption jet-cooling of organic molecules. Appl Phys B51 :395 - 403

Meijer G, deVries M, Hutziker H and Wendt H (1990) Laser desorption jet-cooling spectroscopy of para-amino benzoic acid monomer, dimer and clusters. J Chem Phys 92 :7625 - 7635

Melanson J, Boulet C and Lucy C (2001) Indirect laser-induced fluorescence detextion for capillary electrophoresis using a violet diode laser. Anal Chem 73 :1809 - 1813

Miclea M, Kunze K, Musa G, Franzke J and Niemax K (2001) The dielectric barrier discharge - a powerful microchip plasma for diode laser spectrometry. Spectrochim Acta B 56 :37 - 43

Morrical B, Fergenson D and Prather K (1998) Coupling two-step laser desorption/ionization with aerosol time-of-flight mass spetrometry for the analysis of individual organic particles. J Am Soc Mass Spectrom 9 :1068-1073

Müller P, Bushaw B, Blaum K, Diel S, Geppert C, Trautmann N and Wendt K (2001) Progress in 41Ca ultratrace determination by diode-laser-based RIMS. IN: Resonance Ionization Spectroscopy 2000. AIP, 155 - 160

Müller P, Bushaw B, Nörtershäuser W and Wendt K (2000) Iosotpe shifts and hyperfine structure in calcium 4snp 1P1 and 4snf F Rydberg states. Eur Phys J D 12 :33 - 44

Nakamura S and Fasol G (1997) The blue laser diode. Springer,

Noble C and Prather K (2000) Real-time single particle mass spectrometry: A historical review of a quarter century of the chemical analysis of aerosols. Mass Spectrom Rev 19 :248 - 274

Opsal R and Reilly J (1988) Ionization of alkylbenzenes studied by gas chromatography/laser ionization mass spectrometry. Anal Chem 60 :1060 - 1065

Oser H, Coggiola MJ, Faris GW, Young SE, Volquardsen B and Croseley DR (2001) Development of a jet-REMPI (resonantly enhanced multiphoton ionization) continuous monitor for environmental applications. Appl Opt 40 :859 - 865

Panne U, Dicke C, Duesing R, Niessner R and Bidoglio G (2000) Stimulated raman scattering as an excitation source for time-resolved excitation-emission fluorescence spectroscopy with fiber-optical sensors. Appl Spectrosc 54 :536 - 547

Quentmeier A, Bolshov M and Niemax K (2001) Measurement of uranium isotope ratios in solid samples using laser ablation and diode laser-atomic absorption specgtrometry. Spectrochim Acta B 56 :45 - 55

Schulz C, Sick V, Heinze J and Stricker W (1997) Laser-induced-fluorescence detection of nitric oxide in high-pressure flames with A-X(0,2) excitation. Appl Opt 36 :3227 - 3232

Schulz C, Sick V, Wolfrum J, Drewes V, Maly R and Zahn M (1996) Quantitative 2D singls-shot imaging and mathematical modeling of NO concnetrations and temperatures in a transparent SI engine. IN: Proceedings of the Twenty-Sixth International Symposium on Combustion. The Combustion Institute, Pittsburgh, Pa, 2597 - 2604

Schulz C, Yip B, Sick V and Wolfrum J (1995) A laser-induced fluorescence scheme for imaging nitric oxide in engines. Chem Phys Lett 242 :259 - 264

Shahar T, Dagan S and Amirav A (1998) Laser desorption fast gas chromatography-mass spectrometry in supersonic molecular beams. J Am Soc Mass Spectrom 9 :628 - 637

Smalley RE, Wharton L and Levy DH (1977) Molecular optical spectroscopy with supersonic beams and jets. Acc Chem Research 10 :139 - 145

Spengler B (1997) Post-source decay analysis in matrix-assisted laser desorption/ionization mass spectrometry of biomolecules. J Mass Spectrom 32 :1019-1036

Spengler B, Hinz K and Kaufmann R (1996) Airborne particle analysis. Science 274 :1993 - 1997

Stockel P, Vortisch H, Leisner T, Baumgärtel H (2002) Homogeneous nucleation of supercooled liquid water in levitated microdoplets. J Mol Liquids 96 :153 - 175

Stone E, Gillig K, Ruotolo B, Fuhrer K, Gonin M, Schultz A and Russell D (2001) Surface-induced dissociation on a MALDI-ion mobility-orthogonal time-of-flight mass spectrometer: sequencing peptides from an "in-solution" protein digest. Anal Chem 73 :2233 -2238

Tan W, Carnelly T, Murphy P, Wang H, Lee J, Barker S, Weinfeld M and Le X (2001) Detection of DNA adducts of benzo[a]pyrene using immunoelectrophoresis with laser-induced fluorescence: Analysis of A549 cells. J Chromatogr A 924 :377 - 386

Uhl R, Franzke J and Haas U (2001) Detection of argon and krypton traces in noble gases by diode laser absorption spectrometry. Appl Phys B 73 :71 - 74

v.Weyssenhoff H, Selzle HL and Schlag EW (1985) Laser-desorbed large molecules in a supersonic jet. Z Naturforsch 40a :674 - 676

Vertes A, Gijbels R and Adams F, (1993). Laser ionization mass analysis. John Wiley & Sons, New York

Wattenberg A, Sobott F, Barth H-D and Brutschy B (2000) Studying non-covalent protein complexes in aqueous solution with laser desorption mass spectrometry. Int J Mass Spectrom 203 :49 - 57

Wattenberg A, Sobott F, H-DBarth and Brutschy B (1999) Laserdesorption Mass spectrometry on liquid beams. Eur J Mass Spectrom 5 :71 - 76

Weickhardt C and Tönnies K (2002) Rapid analysis of complex mixtures by means of resonant laser mass spectrometry. IN: Laser assisted analytical methods in environmental sciences, atmosphere, soils and water. Springer, Berlin, to be published

Weickhardt C, Zimmermann R, Boesl U and Schlag EW (1993) Laser mass spectrometry of dibenzodioxin, dibenzofuran and two isomers of dichlorodibenzodioxins: selective ionization. Rap Commun Mass Spectrom 7 :183 - 185

Weickhardt C, Zimmermann R, Schramm K, Boesl U and Schlag EW (1994) Laser mass spectrometry of the di-, tri- and tetrachlorobenzenes: isomer selective ionization and detection. Rap Commun Mass Spectrom 8 :381 - 384

Weinkauf R, Walter K, Weickhardt C, Boesl U and Schlag E (1989) Laser tandem mass spectrometry in a time of flight instrument. Z Naturforsch 44a :1219

Zandee L and Bernstein R (1979) Resonance-enhanced multiphoton ionization and fragmentation of molecular beams: NO, I2, benzene, and butadiene. J Chem Phys 71 :1359 - 1371

Zenobi R, Philippoz J-M, Buseck P and Zare R (1989) Spatially resolved organic analysis of the Allende meteorite. Science 246 :1026

Zhan Q, Voumard P and Zenobi R (1995) Application of two-step laser mass spectrometry to the chemical analysis of aerosol particle surfaces. Rap Commun Mass Spectrom 9 :119 -127

Zimmermann R, Heger H, Blumenstock M, Dorfner R, Schramm K, Boesl U and Kettrup A (1999) On-line Measurement of Chlorobenzene in Waste Incineration Flue Gas as a Surrogate for the Emission of Polychlorinated Dibenzo-p-Dioxins/Furans (I-TEQ) Using Mobile Resonance Laser Ionization Time-of-Flight Mass Spectrometry. Rap Comm Mass Spectrom 13 :307-314

Zimmermann R, Lenoir D, Kettrup A, Grebner T, Neusser H and Boesl U (1995) The Ionization Energies of Polychlorinated Dibenzo-p-dioxins: New Experimental Results and Theoretical Studies. Int JMass Spectrom Ion Proc 145 :97 - 1008

Zimmermann R, Lermer C, Schramm KW, Kettrup A and Boesl U (1995) Three-dimensional trace analysis: Combination of gas chromatography, supersonic beam UV spectroscopy and time-of-flight mass spectrometry. Eur Mass Spectrom 1 :341-351

Zimmermann R, Rohwer E and Heger H (1999) In-line catalytic derivatization method for selective detection of chlorinated aromatics with a hyphenated gaschromatography/laser mass spectrometry technique: A concept for comprehensive detection of isomeric ensembles. Anal Chem 71 :4148 - 4153

10 Rapid Analysis of Complex Mixtures by Means of Resonant Laser Ionization Mass Spectrometry

Christian Weickhardt and Karen Tönnies

10.1 Introduction

Unlike any other field of instrumental analysis, mass spectrometry is in a phase of highly dynamic development, in which new areas of applications and technical or methodical innovations stimulate one another. Its popularity arises from its universal applicability, its high sensitivity, the quickness of its measurements and their high information content. On the instrumental side, besides the continuous improvement of the conventional types of mass spectrometers, the development of novel ones on the basis of magnetic and electrodynamic ion traps (March and Todd 1995) has to be mentioned. However, the most important improvements in the field of mass spectrometry within the last decade are probably due to the development of new ionization techniques. These are, on the one hand side methods which allow the intact ionization of large, in particular biologically and medically relevant, molecules (e.g. Electrospray Ionization (Gaskell 1997) and Matrix Assisted Laser Desorption/Ionization (Karas et al. 1991) and on the other side techniques which involve a certain degree of selectivity in the ionization step. Selectivity is imperative for the analysis of complex mixtures and is usually added to a mass spectrometric measurement in form of chromatographic preseparation. However, chromatographic techniques eliminate one of the major advantages of mass spectrometry: the quickness of the measurement. Furthermore, they require a considerable effort for the sample preparation, which as far as time consumption is concerned in many cases exceeds that for the instrumental analysis step by far. Finally, sample preparation and clean-up are a major source of quantitative errors.

Selective ionization methods in mass spectrometry are an attempt to obtain selectivity in a way which does not affect the rapidity of a mass spectrometric measurement or the sample treatment required. While being of general interest this is an indispensable prerequisite in cases when the analytical problem involves the real-time measurement of trace components with high temporal resolution or high sample throughput respectively. The two most important principles for selective ionization are chemical ionization (CI) (Harrison 1992) based on the charge transfer reaction between preformed ions and the analyte molecules and photo ionization. Chemical ionization and photo ionization resulting from absorption of a single vacuum ultraviolet photon allow the discrimination of substances due to their ionization energy and can therefore, be called partially selective. Substances with an ionization energy below a certain given value are ionized while those with a higher are not. Compared to them another form of photo ionization termed resonant multiphoton ionization (Letokhov 1987; Powis 1995; Weickhardt et al. 1996) which requires intense light in the UV/VIS range and involves excited states of the substance to be ionized offers an exceptionally high degree of selectivity which even allows to distinguish between isomers (Tembreull et al. 1985; Lubman 1987; Weickhardt et al. 1994). Because of the comparatively small transition cross sections for such photo processes resonant multiphoton ionization they depend on the availability of intense light which can only be provided by pulsed lasers up to now.

Within the last decade laser based techniques have successfully entered the field of mass spectrometry (Lubman 1994). Matrix assisted laser desorption / ionization was established as a powerful tool for the analysis of large non-volatile compounds and has become a routine technique in pharmaceutical and biochemical analysis. For elemental analysis laser desorption has proven an ideal method for evaporation of solid material prior to its ICP-MS analysis (Becker and Dietze 1992) and is widely used meanwhile. Besides their capabilities an important factor for the broad acceptance of laser techniques was the development of reliable and economic pulsed laser systems for analytical applications. Compared to other analytical instruments high intensity lasers are still sensitive and expensive devices but the highly dynamic progress in this area has opened their way into routine applications and gives rise to the expectation that soon also more complex systems offering e.g. wavelength tunability and/or ultrashort light pulses will be available for analytical purposes.

In this chapter the principle and features of resonant multiphoton ionization and its coupling to mass spectrometry will be discussed first before two typical examples are presented which demonstrate the capabilities and specific advantages of the resulting analytical technique.

10.2 Principles of Laser Ionization Mass Spectrometry

10.2.1 Resonant Multiphoton Ionization

A way to achieve very high selectivity in the ionization step is the involvement of the analyte molecule´s UV/VIS absorption spectrum. As the energetic position of excited states is characteristic for a certain compound the molecules of interest can be selectively photo excited by choice of a proper wavelength. If the lifetime of the excited state is sufficiently long and the light intensity is sufficiently high a photo excited molecule may absorb a further photon which takes it over the ionization energy and thus, leads to the formation of an ion. Processes of this kind are shown schematically in figure 10.1.

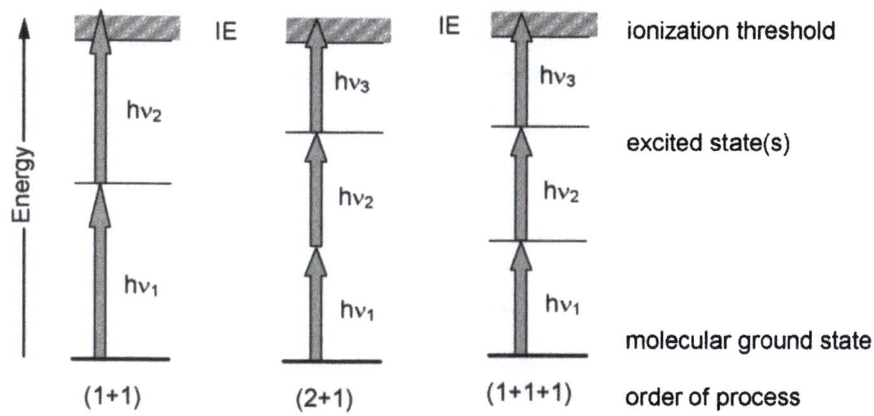

Fig. 10.1. Multiphoton processes illustrated.

In this way the UV/VIS absorption spectrum of the substance of interest is transferred into a spectrum of ion current which is as characteristic.

Photo ionization processes of atoms or molecules via real intermediate states are called resonance enhanced or resonant multiphoton ionization (REMPI). They were predicted theoretically by Göppert-Meyer (Göppert-Mayer 1931) as early as 1931 but it took until the development of the first lasers that they could also be observed experimentally (Vernov and Delone 1965; Berezhetskaya 1970; Chin 1971). In order to obtain a useful ionization probability REMPI is usually carried out by pulsed lasers at light intensities ranging typically from 10^6 W/cm^2 to some 10^8 W/cm^2. Besides the resonant multiphoton processes also non-resonant transitions may occur which can be viewed as the absorption of two or more photons via virtual intermediate states. As their lifetime which can be estimated by the uncertainty relation is very short the probability for them to occur is greatly reduced compared to the resonant case and they only become important at very high laser intensities. From the analytical point of view non-resonant processes are usually undesirable as they lead to unselective ionization as long as they do not involve any resonant level.

Obviously the degree of selectivity achievable by resonant multiphoton ionization depends on the separation or overlap of absorption bands of the substances in the mixture to be analyzed. As the spectral width of vibronic bands depends on the temperature the selectivity of REMPI can be influenced by either cooling or heating the sample gas. While an effusive inlet system at room temperature may permit sufficient selectivity for a routine analytical task it might be necessary to efficiently cool the sample if special requirements as i.e. isomer specific analysis are demanded. Cooling of a gaseous sample can readily be achieved by its adiabatic expansion into the vacuum system of the mass spectrometer through a small nozzle either pulsed or continuous (Smalley et al. 1977; Levy 1981).

A further advantage of resonant multiphoton ionization for analytical mass spectrometry is the fact that the ions are formed with very little excess energy. As a result the degree of fragmentation is usually small but may be increased in a controlled way (e.g. in order to extract structural information) by raising the laser intensity leading to the absorption of further photon(s) by the ion and the subsequent formation of fragments (Boesl et al. 1982). In figure 10.2 the mass spectra of p-xylene obtained by REMPI at different laser intensities demonstrate this effect.

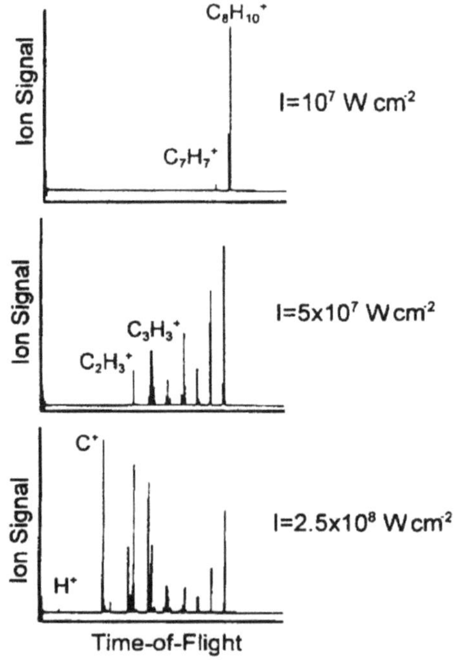

Fig. 10.2. resonant multiphoton ionization mass spectra of p-xylene at three different laser intensities in the ionization volume demonstrating the tunable degree of fragmentation. Whereas at low intensities almost no fragmentation of the molecular ion is observable, the dominating signals at high intensities emerge from small fragments and atomic ions. Laser wavelength: 272.2 nm. (By kind permission of IM Publications)

For a number of analytically important groups of substances, such as the aromatic hydrocarbons, REMPI can reach high ionization probabilities and thus, high detection sensitivity (Boesl 2000; Oser et al. 2001). However, fast intramolecular relaxation processes which may occur in the excited state may compete with the absorption of a further photon and reduce the ionization probability. Examples for such processes are Internal Conversion (IC) and Inter System Crossing (ISC).

While IC is the major relaxation process in molecules possessing a large number of internal degrees of freedom ISC is the dominating process in molecules containing heavy atoms (e.g. polychlorinated and polybrominated compounds) as well as in certain aldehydes and ketones. Both processes lead to a vibrationally highly excited molecule which shows vanishing Franck-Condon factors for a photo transition to a higher level.

The rate constants governing these processes may reach values up to 10^{13} sec^{-1}. Furthermore, a photo excited molecule may dissociate before it is ionized as is frequently observed in metal organic compounds. If this happens at a high rate compared to the inverse duration of the ionizing laser pulse, mostly fragment ions can be detected which complicates the interpretation of the mass spectra and often makes the identification of the substance impossible.

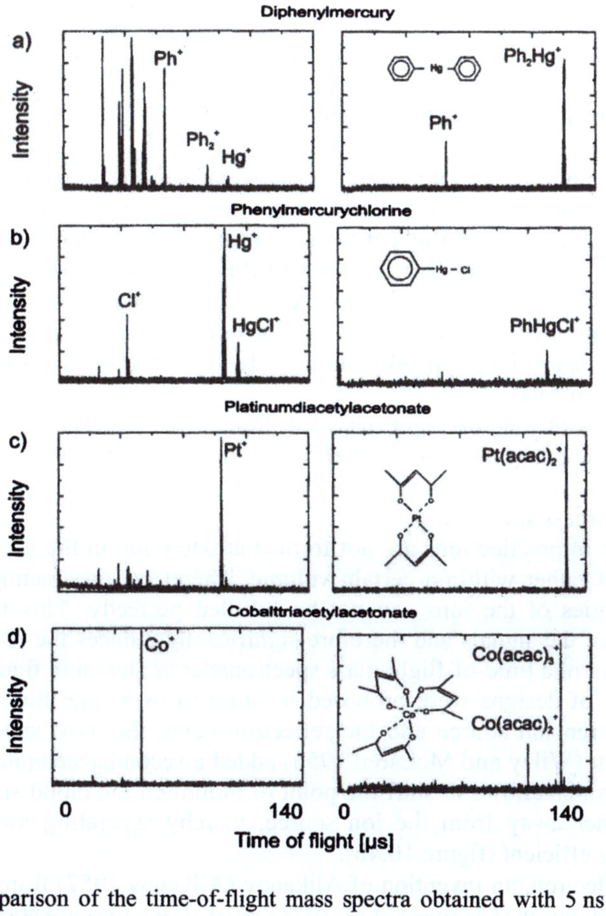

Fig. 10.3. Comparison of the time-of-flight mass spectra obtained with 5 ns laser pulses (left column) and 0.5 ps laser pulses (right column) of a) diphenylmercury, b) phenylmercurychlorine, c) platinumacetylacetonate, d) cobaltacetylacetonate. (By kind permission of Elsevier Science B.V.)

A straight forward way to overcome the problem related to these fast intramolecular relaxation processes which affect many analytically important groups of substances is the use of ultrashort laser pulses with durations below the picosecond range in order to finish the ionization before the intermediate state(s) can be significantly depleted (Ledingham 1997). Figure 10.3 compares the mass spectra of several molecules obtained by multiphoton ionization using pulses with a duration of about 5 ns to such recorded after ionization by 0.5 ps pulses with the same center wavelength.

In all cases the ultra short pulses reduce the degree of fragmentation significantly and lead to the formation of well interpretable mass spectra with an intense molecular ion signal. It could be shown in this context that the high intensities of the ultrashort laser pulses which was around 10^{10} W/cm^2 did not lead to a reduction of the selectivity as a result of non-resonant multiphoton ionization (Weickhardt et al. 1998).

10.2.2 Time-of-flight mass spectrometry

As resonant multiphoton ionization is usually carried out by intense pulsed lasers it is advantageous to combine it with discontinuously working mass spectrometers. In particular time-of-flight mass spectrometry (Weickhardt et al. 1996) has proven to be an ideal combination for REMPI. Based on a very simple principle this type of instruments is characterized by its high transmission, its unlimited mass range and the fact that it allows the registration of a complete mass spectrum within a very short time (typically on the order of 100 µs) without the necessity to scan any parameters.

The basic idea of time-of-flight mass spectrometry is to accelerate all ions to the same kinetic energy and measure their flight time to a detector. As this is proportional to the square root of the ion mass the flight time spectrum can easily be converted into a mass spectrum.

However, as in practice ions are not formed at one point in the accelerating electric field but rather within a certain volume. Therefore, the assumption of equal kinetic energies of the ions can not be fulfilled perfectly. This fact leads to a broadening of the signals and therefore significantly reduces the resolution of the simplest form of a time-of-flight mass spectrometer as shown in figure 10.4a.

A couple of designs were proposed in order to overcome this limitation, the Wiley-McLaren ion source and the reflectron being the most successful. Wiley and McLaren (Wiley and McLaren 1950) added a second acceleration field to the ion source which allows to shift the point of minimum ion cloud size in flight direction further away from the ion source, thereby separating ions of different masses more efficient (figure 10.4b).

In the reflectron, an invention of Alikanov (Alikanov 1957) from 1957 and optimized by Mamyrin (Mamyrin et al. 1973) in 1973, the ions are stopped in a static electric field and reaccelerated before they reach the detector (see figure 10.4c). Ions with a higher kinetic energy penetrate deeper into this reflector field and spend more time in it than those with less kinetic energy. If the dimensions and the field strength in the reflector are chosen properly the dispersive effect of the reflector just cancels the flight time differences in the field free drift regions and all ions of the same mass hit the detector surface at about the same time.

With the original design the mass resolution remained far beyond what was required for applications in chemical analysis. However, for the Wiley-McLaren instrument resolving powers exceeding 4600 were reported (Opsal et al. 1985) under optimized conditions and with the reflectron time-of-flight mass spectrometer resolutions on the order of several ten thousands are achieved routinely with commercial instruments.

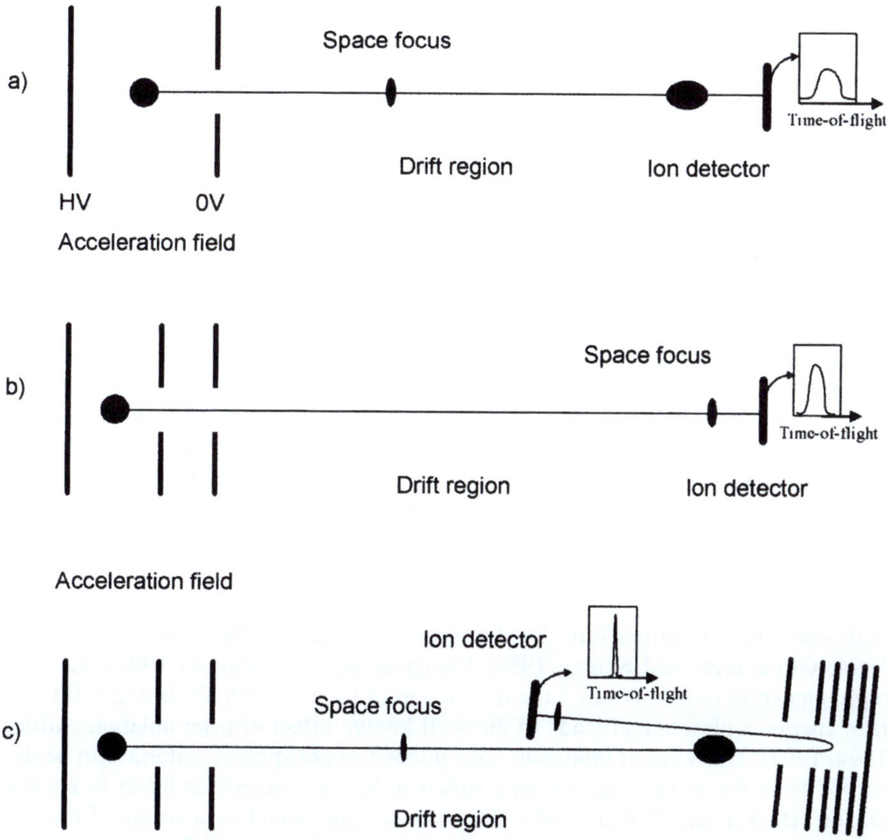

Fig. 10.4. Three types of time-of-flight mass spectrometers: a) linear instrument with single stage acceleration field b) linear instrument with Wiley & McLaren (double stage) ion source c) Reflectron-TOF-MS. The black ellipses illustrate the size of the ion cloud at various points along its flight path.

Besides the development of such highly resolving instruments the recent availability of inexpensive fast detectors and signal processing electronics contributed significantly to the increasing popularity of time-of-flight mass spectrometers.

The combination of resonant multiphoton ionization and time-of-flight mass spectrometry is commonly referred to as laser ionization mass spectrometry. The short and well collimated laser pulse with a typical duration of a few nanoseconds or below provides a perfect starting signal for the flight time measurement and forms the ions within a small volume from where efficient transfer into the mass spectrometer can be achieved. Besides these instrumental advantages the employment of resonant multiphoton ionization in time-of-flight mass spectrometry offers further desirable features for chemical analysis. While the quickness of the measurements by the time-of-flight mass spectrometer is completely maintained and only limited by the repetition rate of the laser (which can be up in the kilohertz range) REMPI adds a high degree of selectivity to the analysis. In many cases this makes chromatographic preseparation steps unnecessary and laser ionization mass spectrometry the method of choice when selectivity has to be combined with high temporal resolution and/or high sample throughput.

10.2.3 Laser desorption

The application of laser ionization mass spectrometry as described above requires the analyte molecules to be in the gas phase prior to multiphoton ionization. This necessitates an additional preceding desorption step in cases where the substances of interest do not posses a sufficiently high vapor pressure. The easiest way to transfer analyte molecules into the gas phase would be to continuously heat the sample. However, if thermally labile substances are to be analyzed this simple method is of limited use. Furthermore, being a continuous method it leads to a small duty cycle when combined with a pulsed technique such as laser ionization mass spectrometry resulting in a reduced sensitivity. In order to increase the duty cycle attempts were made to apply the heat contact in a pulsed way or to use pulsed particle beams for desorption (Demirev 1995). However, it turns out that the most promising way to combine efficient and mild sample evaporation with a high duty cycle is to perform the desorption by means of laser pulses (Lubman 1994; Grotemeyer and Schlag 1989; Meijer et al. 1990; Zenobi 1994). At very high intensities ($>10^{10}$ W/cm^2) the interaction of light with an absorbing solid surface creates a plasma and leads to the well-known effect of laser ablation (Miller 1994, 1997). However, if moderate laser intensities are applied material can be desorbed from the sample surface in a soft way leaving molecules intact to a great extend (Becker and Gillen 1984). This can be understood as a result of the extremely high heating rate achievable by laser irradiation which causes a rapid evaporation of the molecule before energy may flow into its internal degrees of freedom and initiate dissociation. The individual mechanisms involved in the laser desorption process and their interaction are still discussed controversially and may vary according to the excitation mechanism and the specific composition and structure of the substrate (Träger 1989). However, a large variety of laser types with different wavelengths and pulse durations were tested in this context and almost all proved to be useful with astonishingly little difference in their performance. In any case material is desorbed for typically some tens of microseconds after the absorption of the laser pulse. The large part of the molecules is emitted normal to the substrate surface. Therefore, laser desorption is able to offer a high density of neutrals at the point where the ionization is to take place during a short period of time.

10.3 Application Examples

In order to demonstrate the capabilities of laser ionization mass spectrometric techniques in analytical applications two very different examples will be discussed. The on-line trace analysis of the exhaust gas of a combustion engine demonstrates the ability of laser ionization mass spectrometry to combine very high temporal resolution with selective multicomponent analysis, resulting in a unique method for combustion analysis. The second example, the ultra fast trace analysis in soil samples shows how high sample throughput and easy sample preparation compared to standard methods can be achieved by an efficient combination of laser techniques and mass spectrometry.

10.3.1 On-line exhaust gas analysis

The global changes of the climate and the air pollution particularly in the centers of population and industry obviously become two of the most important factors affecting our quality of life in the centuries to come. This fact has drawn great attention towards the sources of atmospheric pollutants and the development of less harmful technologies.

In the centers of population the exhaust gases of the road traffic are a major and rapidly growing source of air pollution. Besides their toxicity certain components are the starting point of photochemical reactions leading to the formation of ozone at ground level (Campbell 1977). Therefore, the further development and the optimization of combustion engines is of great importance in order to improve the air quality in and around our cities.

As a matter of fact modern engines turn out to work sufficiently clean when operated under static conditions particularly when equipped with a well adjusted catalyst. However, it is well known that many atmospherically important substances are emitted by combustion engines during dynamic operation phases in high concentrations within very short periods of time. As this short processes contribute significantly to the average production of pollutants they have to be investigated and understood in order to develop cleaner engines and more sophisticated motor management systems. This problem is further complicated by the fact that substances of different toxicity and importance for air chemistry are emitted with very different temporal characteristics and relation to the motor operation conditions. Obviously this situation calls for analytical techniques which allow high enough temporal resolution and at the same time offer the ability to selectively detect individual substances down to the trace level.

While time resolving analytical instruments exist only for a few exhaust gas components such as CO, NO, O_2 and the sum of all hydrocarbons (Clement et al. 1993), off line techniques based on chromatography allow the quantification of a large number of individual components with high sensitivity. Attempts were made to introduce modern spectroscopic techniques, in particular FTIR and diode laser spectroscopy as well as chemical ionization mass spectrometry to the field of on-line exhaust gas analysis (Villinger et al. 1996). However, all of them are either only applicable to a limited number of substances or suffer from selectivity restrictions.

As discussed above laser ionization mass spectrometry appears to be an advantageous alternative when highly selective trace analysis has to be carried out with high temporal resolution. It therefore seems to match the requirements of modern exhaust gas analysis perfectly. The required technology and an appropriate instrument were developed within the frame of a research project of the "Forschungsvereinigung Verbrennungskraftmaschinen" in the early 1990s. In the following the design and performance of this instrument for on-line exhaust gas analysis will be discussed.

Instrumental Setup

For a set of over 20 exhaust gas components which were selected according to their contribution to the ozone formation (California 1991) the detection parameters were determined by multiphoton spectroscopy and time-of-flight mass spectrometry (Weickhardt et al. 1994). This group of substances comprised aromatic and unsaturated hydrocarbons, aldehydes and ketones, nitric oxides, SO_2, oxygen and ammonia. The list of detection parameters contained the information about the wavelength at which a certain substance can be selectively ionized or simultaneously together with certain other trace components and about the mass number at which it can be recorded in the laser ionization mass spectrum.

In order to provide laser light with the appropriate wavelengths (between 260 nm and 320 nm) available the frequency doubled output of a dye laser was used for ionization. While at that time this was the only possibility to produce intense wavelength tunable laser light nowadays all solid state systems (optical parametric oscillators) are available which combine higher output energies with faster and easier tuning. The laser system delivered pulses with a duration of about 5 ns and energies up to 200 μJ. Its repetition rate was variable and could be synchronized with the number of rotations at which the motor was running.

For the operation at a motor test bed a special reflectron time-of-flight mass spectrometer was developed with particular emphasis to compact and robust design (Frey et al. 1995). All ion optical components and the detector fitted into a stainless steel cylinder of diameter of 25 cm and length 1 m. The cylinder and all components contained in it were heatable in order to avoid condensation. The mass spectrometer including vacuum pumps and power supplies were placed in a mobile housing with dimensions 118 cm (height) x 100 cm (length) x 58 cm (width). The whole instrument as well as the laser system was controlled by a remote computer system. Figure 10.5 presents a scheme of the complete setup (Boesl et al. 1993).

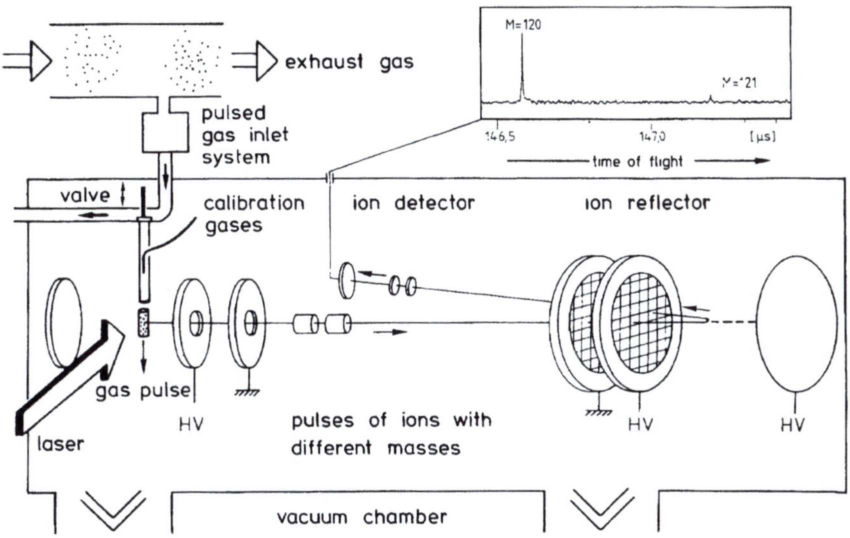

Fig. 10.5. Apparatus for on-line trace analysis of automobile exhaust gas with high temporal resolution. (By kind permission of the Society of Automotive Engineers)

If the high temporal resolution offered by laser ionization mass spectrometry is to be maintained in a specific application the development of a suitable sample transfer system is a crucial point. Not only has the stoichiometry of the sample be left unchanged but also its temporal structure should be transferred from the point where it is taken to the inlet of the mass spectrometer without significant broadening. Additional requirements which have to be met include a high density of neutrals within the ionization volume while at the same time maintaining a sufficient vacuum inside the drift region of the mass spectrometer ($< 10^{-7}$ mbar). These conditions can only be fulfilled by special pressure reduction devices in which dead volumes and flow obstacles have to be carefully avoided. For real time exhaust gas analysis a sample could be taken at different positions along the exhaust tube. From there it was pumped through a heated stainless steel tubing (inner diameter 3 mm) to a mechanical vacuum pump. A small part of the sample gas was branched off into a thin metal tubing (0.5 mm ID, length 3 cm) delivering it right into the center of the ion source of the mass spectrometer. The sample gas could either be introduced into the MS continuously or in a pulsed mode for which the bypass was opened to the thin tubing by means of a magnetic valve.

Results

In figure 10.6 the time-of-flight mass spectrum of an exhaust gas obtained after electron ionization is compared to those observed using resonant multiphoton ionization at different wavelengths (Weickhardt and Boesl 1993).

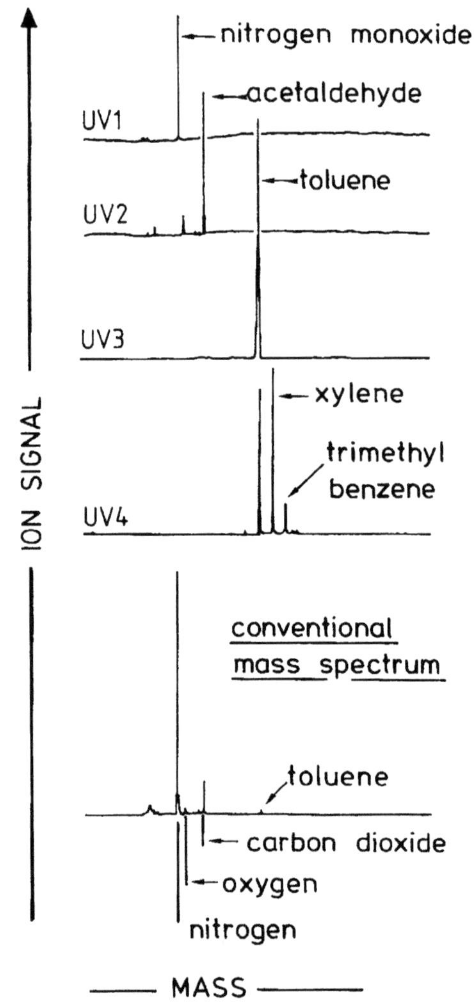

Fig. 10.6. Mass spectra of an exhaust gas. At the bottom: non-selective electron ionization. On top (UV1-UV4): selective laser ionization at four different laser wavelengths. (By kind permission of the Society of Automotive Engineers)

While the EI spectrum exhibits only the main constituents of the exhaust gas which are not very relevant from the point of view of toxicology and atmospheric chemistry trace components can be selectively detected with high sensitivity under REMPI conditions using appropriate wavelengths. Using well-defined mixtures of the substances of interest in nitrogen detection limits of 1 ppm could be reached for all compounds in the list. For some, particularly the aromatic hydrocarbons, laser ionization mass spectrometry turned out to be even more sensitive because of their favorable ionization wavelengths.

The method for quantifying the measurements has to take into account short term fluctuations, particularly caused by variations of the laser pulse energy and beam profile as well as long term drifts of the mass spectrometer's transmission and detector sensitivity. Therefore, a relative measurement in respect to an internal standard seems to be the most promising way. Suitable substances must not interfere with analyte substances in the mass spectrum, have a sufficient vapor pressure and be ionized by the same multiphoton process and at the same wavelength as the substance to be quantified. Using a set of three different calibration substances it was possible to quantify all substances in the target list mentioned above. The internal standards were mixed to the exhaust gas in the thin tubing leading into the ion source. Using this technique quantitative results were obtained within an accuracy of ± 10 % using only one single laser shot. If laser intensities are applied which are able to saturate individual optical transitions in the analyte or internal standard substances a simple relative measurement does not yield useful results. For such conditions a quantification procedure involving two internal standards with different dependencies of their ion yield on laser intensity was developed (Boesl et al. 1993; Weickhardt et al. 1994). In this case the relative ion signal of the two internal standards is used as a sensor for the laser intensity in the ionization volume.

In order to check the temporal resolution of the instrument the concentration of toluene in the exhaust gas sampled right at the outlet of one cylinder was measured while turning off the ignition manually for single ignition strokes (Weickhardt et al. 1994). As toluene is added to modern fuels as a component to reduce knocking, its concentration in the exhaust gas is supposed to increase significantly when the ignition is turned off and it is ejected unburned. This is exactly what can be seen in figure 10.7 where the measured toluene concentration in the exhaust gas is displayed for every single emission stroke of one of the cylinders of a four cylinder engine running at 2160 rpm.

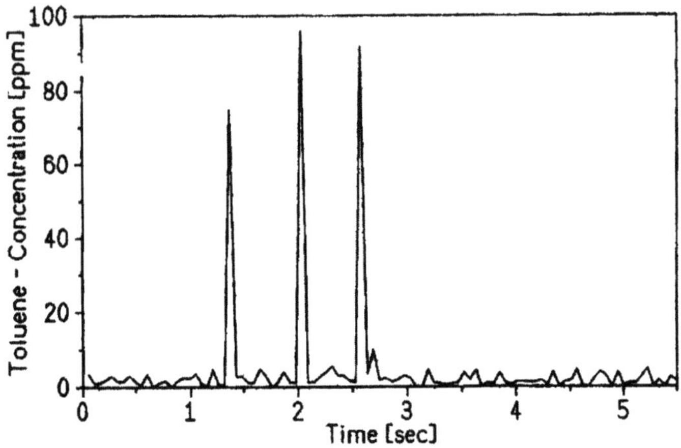

Fig. 10.7. Toluene concentration in the exhaust gas of a four-stroke engine running at 2160 rpm (36 s^{-1}). A measurement is taken for each individual emission stroke. The peaks correspond to manually induced ignition failures. (By kind permission of the American Chemical Society)

Three times the ignition was turned off for one ignition stroke resulting in an immediate increase of the toluene concentration in the exhaust gas. When for the following stroke the ignition was turned on again the toluene signal was instantly back to normal values. The time between two emission strokes is 55 ms in this case. The result of the experiment demonstrates that no signal broadening by the sample transfer or inlet system is observed on this time scale and that the whole system is able to analyze the composition of the exhaust gas for every single emission stroke.

Following these fundamental investigation the capabilities of the instrument were demonstrated in motor test bed measurements under realistic dynamic conditions (Boesl et al. 1998). First the emission of aromatic hydrocarbons was followed during a cold start for about seven minutes. The exhaust gas samples were taken behind the catalytic converter. Benzene, toluene and xylene were measured simultaneously using the fourth harmonic of a Nd:YAG laser (wavelength 266 nm) with a repetition rate of 50 Hz for ionization. The concentrations of these compounds are displayed in figure 10.8 as a function of time.

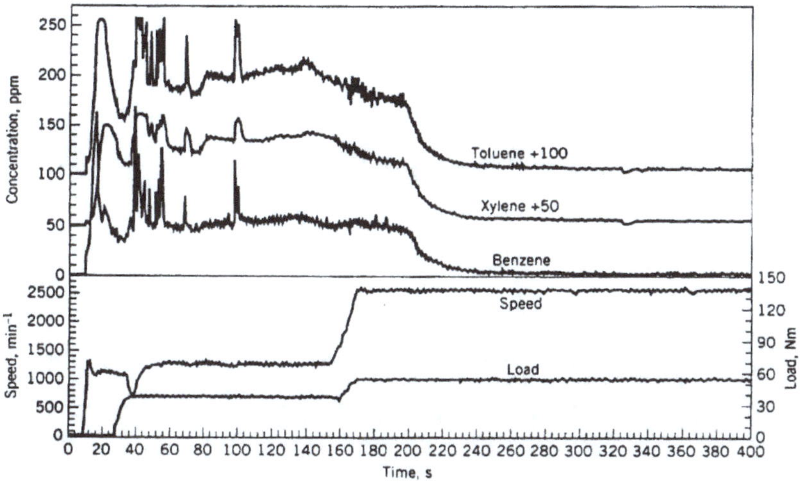

Fig. 10.8. Cold start emission of an engine measured behind the catalytic converter with a measurement rate of 5 Hz (summation over 10 points at a repetition rate of 50 Hz). Xylene (toluene) displayed with an offset of 50 ppm (100 ppm). (By kind permission of John Wiley & Sons, Inc.)

Seven seconds after the beginning of the measurement the engine is ignited and operated at low speed and load. Due to the emission of unburned fuel and the incomplete combustion in the cold engine aromatic hydrocarbons are emitted at high concentrations during this first phase. A detailed investigation shows that benzene, toluene and xylene are detected with certain delays at the end of the exhaust pipe which is due to a storage effect of the catalyst. Semi volatile hydrocarbons are adsorbed on its cold surface and only transmitted after saturation is reached. Next, the concentrations of all three compounds drop quickly but reach high values again as the load on the engine is increased. They stabilize at values between 50 and 150 ppm until about 200 seconds after the start of the motor. At this point the catalyst reaches its working temperature and the concentrations of the aromatic compounds drop to values below 20 ppm. The high values of the concentrations and their complex temporal behavior already demonstrate the necessity of fast online analytical techniques for the development of cleaner engines and optimized catalytic converters.

Another important group of dynamic processes regarding the formation of toxic trace compounds are sudden changes of the speed and/or load of an engine. They occur frequently in stop-and-go traffic or during fast passing maneuvers.

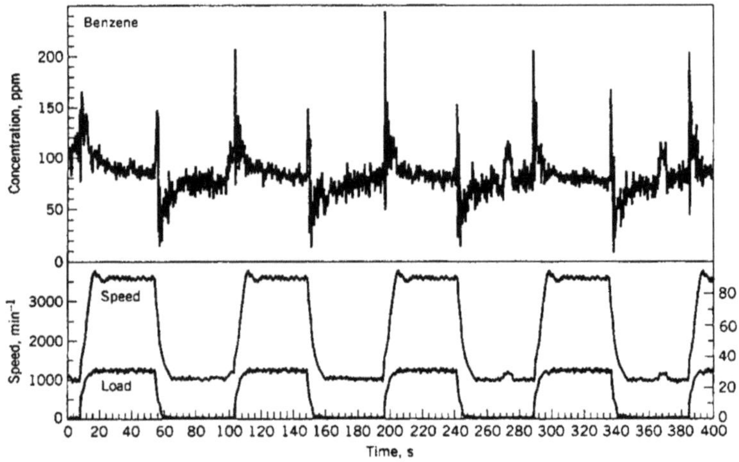

Fig. 10.9. Benzene concentration in an exhaust gas measured during a cycle including sudden speed and load changes. The sample was taken behind the exhaust valve and the benzene concentration determined with a rate of 5 Hz. (By kind permission of John Wiley & Sons, Inc.)

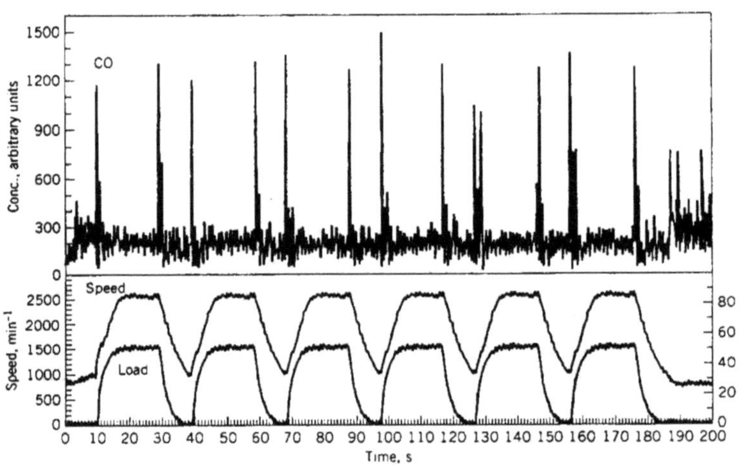

Fig. 10.10. Carbon monoxide concentration in an exhaust gas measured during a cycle including sudden speed and load changes. The sample was taken behind the exhaust valve and the CO concentration determined at a rate of 5 Hz. Laser wavelength: 230.01 nm. (By kind permission of John Wiley & Sons, Inc.)

Figure 10.9 displays the benzene concentration in the exhaust gas right after the outlet valve of an engine measured by laser ionization mass spectrometry during a cycle of several acceleration and deceleration procedures. These results can be compared to figure 10 where a similar measurement was carried out for carbon monoxide. While benzene is emitted with an almost constant concentration of about 80 ppm during static operation conditions, its concentration rises instantaneously by a factor of four to five at the moment when the acceleration starts and returns back to normal within a few seconds. The beginning of the deceleration phase correlates to a brief reduction of the benzene emission. The changes in the carbon monoxide emission measured with an ionization laser wavelength of 230.01 nm during such a cycle are even more dramatic. The CO concentration increases by almost an order of magnitude for a single emission stroke whenever the operating conditions are changed, e.g. at the very beginning of every acceleration and deceleration phase.

It becomes obvious from these measurements that not only high temporal resolution is required to provide the motor engineer with the information he needs to optimize an engine but that multicomponent capability is imperative too.

10.3.2 Soil Analysis

There exist about 198.000 residual waste areas in the Federal Republic of Germany (Umweltbundesamt 1998). Besides former military areas industrial sites are among them. The characterization of a military or industrial residual waste site is an important precondition for its successful remediation. The work performed in this field up to know shows that in particular the problem of the analytical investigation of sites contaminated by explosives and the danger emerging from them has not been successfully solved yet (Kühnel 1991; Wöstmann and Zentgraf 1992; Rapsch 1991; Mies and Bauer 1991; Spyra 1991; Kayed 1990).

Besides the evaluation of residual waste sites of the world wars I and II, the analysis of areas recently left by the german army or its allies becomes increasingly important because these areas are aimed for public use. In this context sound analytical results are necessary in order to obtain an unequivocal grading of the risk.

Due to the inhomogeneous distribution of the contaminants typical for residual waste sites usually a large number of soil samples have to be analyzed in order to obtain a clear picture of the situation. Such a vast number of individual samples can hardly be dealt with by conventional soil analytical techniques because most of them require time and money consuming sample preparation steps. Among them are the highly developed chromatographic separation techniques, such as high resolution gas chromatography (Yinon 1996, 1981) using electron capture detectors (ECD) (Weinberg and Hsu 1983) and High Performance Liquid Chromatography (HPLC) with UV detection (Yinon and Zitrin 1993). These highly precise methods are capable of detecting explosives in the ultra trace concentration region and even of distinguishing between individual isomers (Tarradellas et al. 1997). Furthermore, thin layer chromatography, mass spectrometry after chromatographic preseparation, tandem MS, UV/VIS photometry, ion chromatography and supercritical fluid chromatography are currently employed for the detection of explosives and other contaminants in soil samples.

The problematic point about all these techniques is the expensive sample preparation they require. Thus, the trace analysis of organic contaminants as e.g. 2,4,6,-TNT in soil may take several days (Hessische Landesanstalt für Umwelt 1998).

The clear identification and quantification of nitro aromatic compounds in soil samples still causes severe analytical problems. In Germany a standardized analytical procedure does not exist and the regulations developed for water analysis are applied. But in contrast to water samples the analysis of soils is complicated by the heterogeneity of the matrix. Furthermore, typical concentrations of nitrobenzenes and nitrotoluenes in water samples are in the ultra trace region between some ng/l up to several mg/l while their concentration in soil samples may reach values up to 360 g/kg (Dornberger and Welsch 1995).

In order to develop standardized guidelines a round robin test was conducted in Germany using conventional analytical techniques. At higher concentrations of the contaminants good results were obtained and the danger value of 300 mg/kg could be well handled in general. However the action value for living areas (20 mg/kg) could only be analyzed unsatisfactory by most of the laboratories (Hessische Landesanstalt für Umwelt 1998). These results clearly demonstrate the need for new soil analytical techniques in particular screening techniques which allow high sample throughput and can avoid lengthy preparation procedures. A further goal is the reduction of the number of sample manipulations and the corresponding sources of quantitative error. A direct analysis of the contaminants within a short timeframe would be desirable.

The demands on a suitable technique for soil analysis, e.g. selective trace detection in a complex matrix, high sample throughput and simple sample preparation, point towards laser ionization mass spectrometry as a powerful alternative. If REMPI's selectivity was sufficiently high a chromatographic preseparation would become unnecessary. In this case sample clean-up and extraction procedures could be avoided. Only a laser desorption step of the solid material has to precede the laser ionization mass spectrometric analysis, thereby consequently minimizing the sample handling required.

Of course it has be kept in mind that soil is not at all a well defined matrix but may vary drastically in composition, structure and homogeneity. Therefore, the analytical methods developed in this context could only be investigated and tested using a few typical soil types up to now. Although these examples were very different it cannot be ruled out that the results presented in the following may not hold for certain very special soils.

Instrumental Setup

Soil samples were collected from several sites in and around Cottbus which resemble different types of soil ranging from sandy via humus and loamy soil to clay. For quantitative measurements weighed out amounts of chemicals (anthracene, pyrene, fluorene, 9-chloroanthracene, phenazine) which were obtained in technical quality and used without further purification were added to the soil samples. Afterwards the mixtures were homogenized in a ball mill for ten minutes.

The soil samples were pressed to small rods with a diameter of 3 mm, introduced into the mass spectrometer through a vacuum lock and placed in position for laser desorption by means of a manipulator.

Laser desorption could be carried out directly in the ion source of the mass spectrometer or in front of a pulsed nozzle with subsequent acceleration of the desorbed material towards the ionization region by means of the supersonic jet (see figure 10.11).

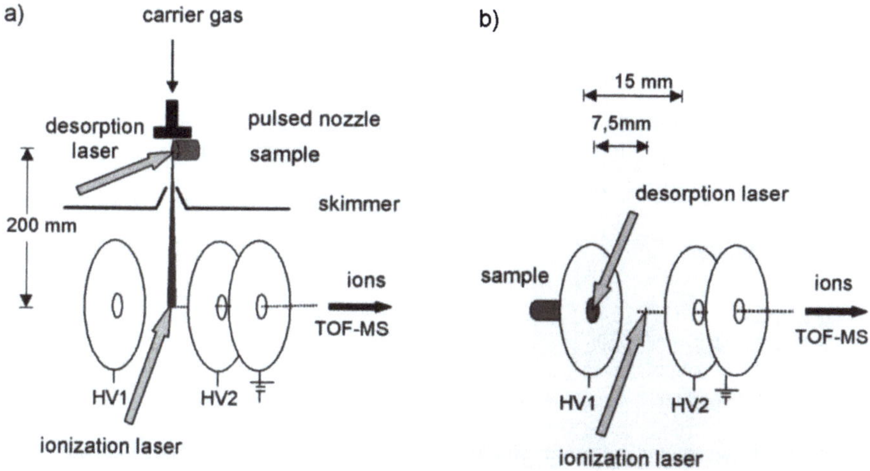

Fig. 10.11. Different laser desorption setups for REMPI: a) laser desorption coupled to the supersonic expansion in front of a pulsed nozzle and b) laser desorption direct within the ion source.

While the first possibility results in a high density of neutrals at the point of ionization the latter is advisable when cooling of the molecular degrees of freedom is required in order to increase the selectivity of the ionization. All measurements discussed in this chapter were carried out using the setup without supersonic beam (fig. 10.11 b). Desorption was performed by the frequency quadrupled output of a Nd:YAG laser with an energy of 3 mJ per pulse. The laser beam irradiated the whole cross section of the sample.

For ionization either the frequency doubled output of a nanosecond dye laser system or the forth harmonic of a Ti:Sapphire laser system generating pulses with a temporal width of about 150 fs could be used. Typical pulse energies used were about 0.5 mJ for the nanosecond and 20 μJ for the femtosecond laser. The beam of both lasers was focussed into the ionization region by means of a quartz lens with a focal width of 200 mm. After photo ionization the ions were transferred into a homebuilt gridless reflectron time-of-flight mass spectrometer with a total drift length of about 1.5 m and equipped with a Wiley-McLaren type ion source. The flight time spectrum was recorded by a microsphere plate detector whose signal is directly transmitted to a digital oscilloscope coupled to a PC for data storage and processing. The complete setup for soil analysis by LD-LI-MS can be seen schematically in figure 10.12.

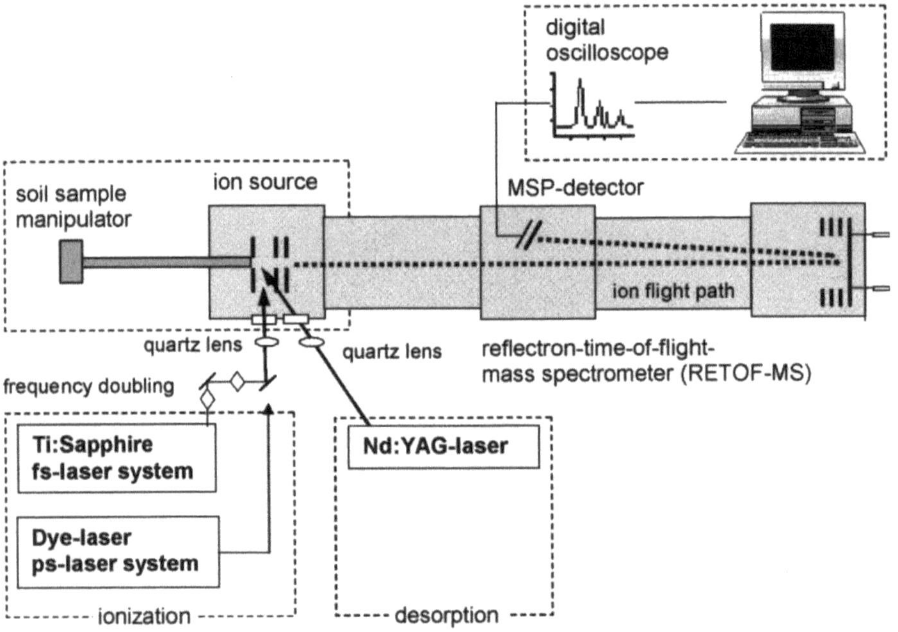

Fig. 10.12. Experimental setup for soil analysis by laser desorption – laser ionization mass spectrometry (LD-LI-MS)

Results

In the first step a possible influence of the soil type on the measurements was investigated. Samples from three different soils, sandy soil, humus and clay, were prepared. They were studied in respect to desorption efficiency, possible mass spectrometric interferences and their influence on the quantification. In order to make sure that the laser desorption and all other parts of the instrument were working properly, pyrene was added to each sample in a concentration of 10 μg/kg and each mass spectrum was checked for the presence of the pyrene signal. In figure 10.13 the time-of-flight spectra obtained using nanosecond laser pulses with a wavelength of 266 nm which is well suited for the simultaneous ionization of polycyclic aromatic hydrocarbons and are shown for the different soil types.

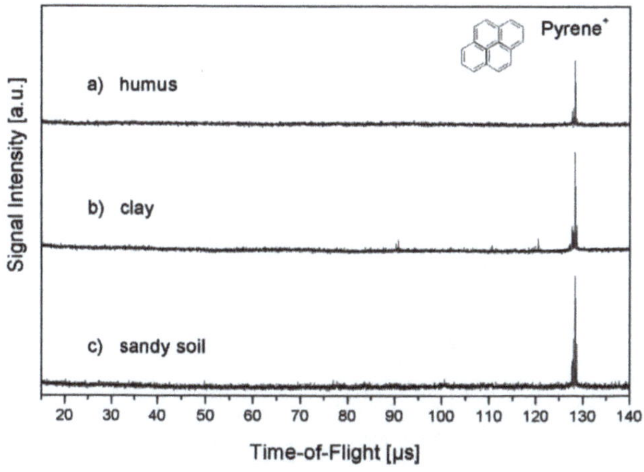

Fig. 10.13. Laser ionization mass spectra of pyrene (concentration: 10 µg/kg) detected in different types of soil by laser desorption laser ionization mass spectrometry. Note the absence of possibly disturbing signals.

They clearly demonstrate that a certain influence of the soil on the desorption efficiency is present but that no interfering signals complicate the detection of aromatic compounds at least down to the concentration values investigated here.

In connection with the analysis and remediation of military residual waste sites the substances of greatest interest are explosives and chemical warfare agents. From the point of view of laser ionization mass spectrometry such chemicals are somehow problematic because their REMPI spectra have not been measured yet and they tend to quickly dissociate after photo excitation. For nitroaromatic or metal organic compounds these dissociation or other relaxation processes which may occur in the molecule after the absorption of a photon may take place with rate constants significantly larger than $10^9\ s^{-1}$ thus giving rise to the detection of only small and uncharacteristic fragment ions in the mass spectrum. As pointed out above, several groups demonstrated within the last years that by the use of ultrashort light pulses with durations in the sub picosecond region it is possible to overcome the problem related to fast intramolecular relaxation processes which appears to be a rather general one for a broad variety of analytically important substances. These short and intense pulses are aimed to finish the ionization process before the majority of the excited molecules can undergo relaxation processes.

Figure 10.14 shows the mass spectra of 3-nitrotoluene and 2,4,6-TNT in technical quality and introduced into the mass spectrometer by means of a Knudsen cell. Those on the left hand side were obtained by multiphoton ionization using conventional nanosecond laser pulses. The mass spectra recorded after ionization by light pulses with a duration of ca. 150 fs are shown on the right hand side. Each spectrum represents an average over 100 laser shots (Tönnies et al. 2001).

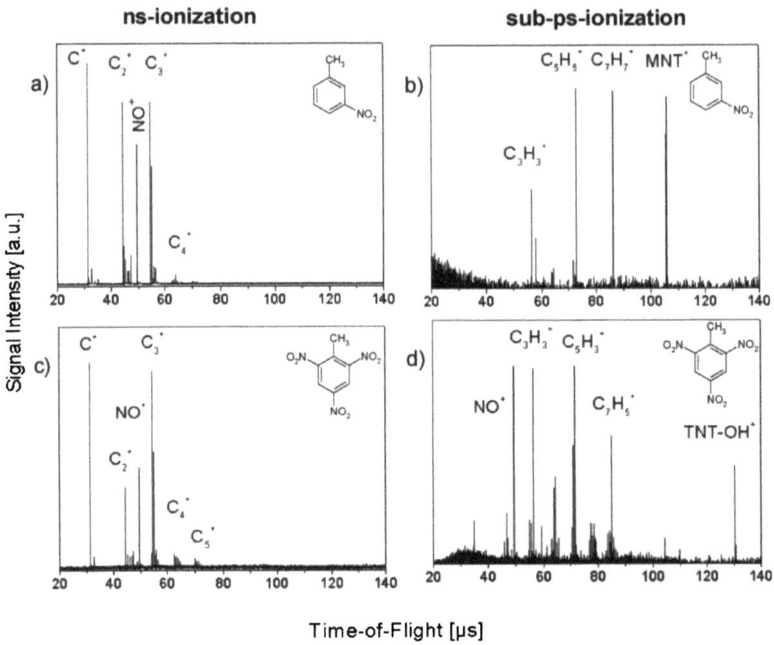

Fig. 10.14. Laser ionization mass spectra of 3-nitrotoluene and 2,4,6-trinitrotoluene using a) + c) nanosecond laser pulses, b) + d) 150 sub-picosecond laser pulses. Wavelength: 206 nm.

In both cases the wavelength was set to 206 nm. While for the monosubstituted toluene the molecular ion signal is still the base peak of the mass spectrum, the dissociation of TNT is so fast that even under the conditions of ultrafast multiphoton ionization no molecular ions are observed. However, TNT can readily be identified by its OH-loss signal. In the nanosecond case the high degree of fragmentation makes a clear identification of any of the compounds impossible.

Another problem related to fast intramolecular relaxation processes is the fact that they broaden the bands in the REMPI spectrum and thus, reduce the degree of selectivity achievable. In the case of TNT no wavelength dependence of the ionization efficiency could be observed within the range from 200 to 210 nm.

Quantification of the measurements was obtained by performing relative measurements in respect to an internal standard added to the soil sample as early as possible, i.e. prior homogenization in the ball mill. It should be kept in mind that the method applied here is basically a surface sensitive technique because the material desorbed by laser ablation comes from a thin surface layer of the sample. Furthermore, the interaction of the desorption laser with the sample surface may change its chemical composition and stoichiometry. Against this background it is obvious that reliable quantitative data can only be obtained from the first few laser shots onto the sample. In figure 10.15 the temporal course of the signal of laser desorbed and laser ionized decacyclene desorbed from clay and ionized by 266 nm nanosecond laser pulses is plotted for a series of 6000 laser shots applied within 10 minutes. The signal first decreases with a time constant of about 150 s (= 1500 laser shots) and afterwards remains almost constant at about 20 % of the maximum level. In order to check whether part of the signal is due to thermal desorption induced by laser heating of the surface the desorption laser was blocked from time to time for one or two shots and the resulting ion signal was compared to the average signal level at that point in time. These events show up in the lower trace in figure 10.15 as sudden signal setbacks.

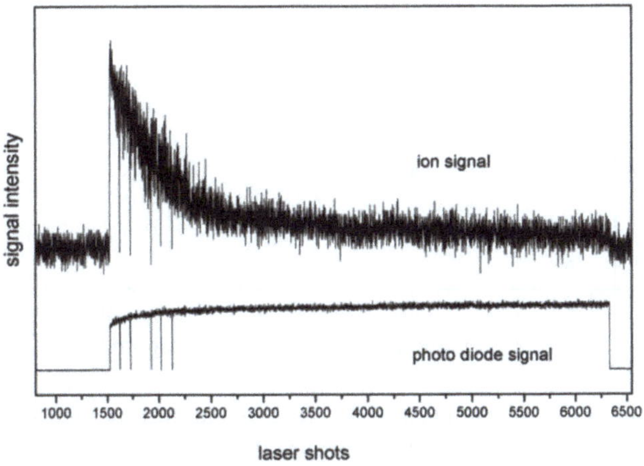

Fig. 10.15. Temporal course of the decacyclene ion signal in a laser desorption laser-MS measurement on clay. When the desorption laser is blocked for single shots the signal vanishes completely and returns immediately at the next shot hitting the sample.

As the blocking of the desorption laser reduces the signal down to baseline at all times after the start of the measurement thermal desorption can be ruled out as a significant contribution. We therefore assume that the intense signal at the beginning of the measurement can be attributed to analyte molecules located within a surface laser with a thickness of about the penetration depth of the laser from where they can easily escape into the vacuum. This surface layer is depleted within several hundreds of laser shots. Afterwards the analyte molecules can only be transferred into the gas phase at the rate with which the laser digs a hole into the soil material and sets free lower layers. This process leads to a reduced signal but can continue for a long time.

A comparison of the temporal behavior of the ion signal of two different substances shows that their surface concentration is decreased with different rates. Consequently, the quantification has to be carried out within a series of laser shots short enough to assure that the ratio of the two signals is constant to a good approximation. At the desorption laser parameters used in these experiments this was usually the case for up to 30 to 50 pulses. Therefore, the way to obtain good quantitative results is a tradeoff between getting rid of shot-to-shot fluctuations by averaging over many individual measurements on the one hand side and finishing the measurement within a time short enough to ensure that the relative signal intensity of analyte and internal standard does not change significantly.

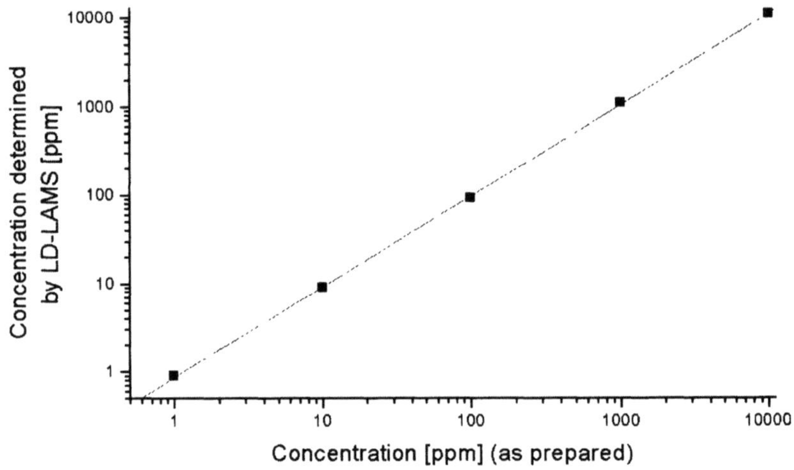

Fig. 10.16. Comparison of the phenanzine concentration in clay samples determined by LD-LI-MS with the values prepared by weighing (all in ppm by mass).

We determined the concentration of phenazine in clay using 9-chloroanthracene at a concentration of 10 mg/kg as internal standard and nanosecond laser pulses with a wavelength of 266 nm for desorption as well as for ionization. Phenazine concentrations ranged from 1 mg/kg to 10 g/kg. Figure 10.16 compares the values determined by laser desorption laser ionization mass spectrometry with the concentrations as they were prepared by weighing. The value obtained for the highest phenazine concentration was used for calibration. All data points are located within a ± 20 % range around the solid line indicating the one-to-one correspondence. This is of course far from being a precision measurement but acceptable for a rapid scanning technique which requires only about 1 minute of time for each sample. In this way it is possible to quickly locate the "hot spots" of a residual waste site which then may be investigated in detail by standard methods.

10.4 Conclusion

The examples of applications for laser ionization mass spectrometry given in this chapter were chosen in order to demonstrate that this new analytical technique offers highly interesting features when rapid detection of trace substances in complex mixtures is demanded. For certain problems it is even the only available solution up to now. However, it should be clearly pointed out that laser ionization mass spectrometry is not a well established technique in analytical chemistry yet. This is mainly due to the fact that it requires tunable laser systems which still are expensive and sensitive devices. Within the last decade an important progress was achieved in this field by the development of commercial tunable all solid state laser systems based on the optic parametric effect (Tang and Cheng 1995). Compared to dye lasers these instruments are characterized by their broad wavelength tuning range, high output power and long term stability. As far as reliability and handling are concerned it can be expected that they will soon reach a state in which they can be described as stand alone instruments. Hopefully, their price trend too will pave their way into routine applications.

As pointed out above fast intramolecular relaxation processes following photo excitation can set severe restrictions to "conventional" laser ionization mass spectrometry, i.e. when nanosecond laser pulses as produced by usual dye or solid-state laser systems, are used for ionization. This problem is not at all limited to a small number of compounds but is frequently observed in analytically important groups such as the polychlorinated and polybrominated aromatic hydrocarbons, the aldehydes and ketones, metal organic compounds and explosives. It was successfully demonstrated how this limitation can be overcome by the use of ultra short light pulses with durations in the picosecond range or below and that in this way the advantages of resonant multiphoton ionization can be extended to quickly relaxing molecules (Grun et al. 1999). In respect to analytical applications what is true for conventional laser systems applies all the more for lasers delivering such ultra short pulses. However, the development in this field is even more dynamic than for conventional laser systems although a higher degree of specialization can be recognized. While instruments offering full flexibility are still far from being user friendly and economic systems are available meanwhile which i.e. are adjusted to a fixed wavelength but are placed in a single housing without need for further adjustments. These developments are driven by the advantageous applications ultra short laser pulses offer for material processing and laser ionization mass spectrometry can benefit from them. Also from the economic point of view the developments on the market of short pulse lasers can be called very promising.

On the mass spectrometry side LI-MS profits from the decreasing prices for high frequency electronics necessary for time-of-flight instruments. However, resonant multiphoton ionization is not limited to this type of mass spectrometers. In particular the use of quadrupole ion traps has become an interesting alternative. Besides their reduced vacuum requirements they offer tandem or even multiple stage MS capability and can be built very robust and compact. Probably they will be the mass spectrometers of choice for LI-MS when not ultimate time resolution is demanded. Furthermore, as they can be operated as ion storage devices too, they allow the ideal coupling of high repetition rate lasers with mass spectrometry. The ionized molecules can be accumulated for an almost arbitrary number of laser shots before the mass analysis. As the data system sets a limit to the maximum repetition rate in time-of-flight mass spectrometers the duty cycle as well as the signal-to-noise ratio can be enhanced by the use of an ion trap when high repeating lasers are employed.

It can be summarized that laser ionization mass spectrometry offers unique features for analytical chemistry making it the preferred or even sole solution to specific problems. However, in the current situation it always has to be checked thoroughly, if these advantages justify the additional financial effort so that laser ionization mass spectrometry can compete with the established techniques.

Acknowledgement

Financial Support by the German Federal Foundation for the Environment (Deutsche Bundesstiftung Umwelt) and the German Federal Ministry for Research and Technology is gratefully acknowledged. The authors are indebted to Prof. Grotemeyer, University of Kiel, for loan of instrumentation and continuing discussion and Prof. Reif, Brandenburg, Technical University Cottbus, for loan of the femtosecond laser system.

References

Alikanov SG (1957) A new impulse technique for ion mass measurement. Sov Phys JETP 4: 452

Becker CH, Gillen KT (1984) Surface analysis by nonresonant multiphoton ionization of desorbed or sputtered species. Anal Chem 56: 1671

Becker JS, Dietze H-J (1992) Laser ionization mass spectrometry in inorganic trace analysis. Fresenius J Anal Chem 344: 69

Berezhetskaya NK, Varanov GV, Delone GA, Delone NB, Piskova GK (1970) Effect of a strong optical-frequency electromagnetic field on the hydrogen molecule. Sov Phys JETP 31: 403

Boesl U, Neusser HJ, Schlag EW (1982) Secondary excitation of ions in a multiphoton mass spectrometer. Chem Phys Lett 87: 1

Boesl U, Weickhardt C, Schmidt S, Nagel H, Schlag EW (1993) Calibration method for the quantitative analysis of gas mixtures by means of multiphoton ionization mass spectrometry. Rev Sci Instrum 64: 3482

Boesl U, Weickhardt C, Zimmermann R, Schmidt S, Nagel H (1993) Fast exhaust gas probe for multicomponent analysis: Scientific / techniqual principle. SAE Technical Papers Series, No. 960083, SAE, Warrendale, PA

Boesl U, Nagel H, Weickhardt C, Frey R, Schlag EW, Meyers RA (eds) (1998) Vehicle exhaust emission; analysis by laser mass spectrometry. The Encyclopedia of Environmental Analysis and Remediation, John Wiley & Sons: 5001

Boesl U (2000) Laser mass spectrometry for environmental and industrial chemical trace analysis. J Mass Spectrom 35: 289

California Air Resource Board (1991) Proposed Reactivity Adjustment Factors for Transitional Low-Emission Vehicles. Staff Report and Technical Support Document, State of California Air Resources Board, Sacramento, Calif, Sept 27

Campbell IM (1977) Energy and the Atmosphere. John Wiley & Sons Ltd, Chichester, U.K.

Chin SL (1971) Multiphoton ionization of molecules. Phys Rev A 4: 992

Clement RE, Koester CJ, Eiceman G (1993) Environmental Analysis. Anal Chem 65: 85R

For a review see: Cotter RJ (ed) (1994) Time-of-flight mass spectrometry. American Chemical Society, Washington

Demirev PA (1995) Particle-induced desorption in mass spectrometry .1. Mechanisms and processes. Mass Spectrom Rev 14: 279

Dornberger U, Welsch T (1995) Explosivstoffe in Altlasten der Rüstungsproduktion. Z Umweltchem Ökotox 7: 302

Frey R, Nagel H, Franzen J, Rikeit H-E (1995) Time-resolved measurement of individual aromatic hydrocarbons in automotive exhaust gas at transient engine operation. SAE Technical Papers Series No. 951053, Warrendale, PA

Gaskell SJ (1997) Electrospray: principles and practice. J Mass Spectrom 32: 677

Göppert-Mayer M (1931) Über Elementarakte mit zwei Quantensprüngen. Ann Phys 401: 273

Grotemeyer J, Schlag EW (1989) Biomolecules in the Gas Phase: Multiphoton Ionization-Mass Spectrometry. Acc Chem Res 22: 399

Grun C, Heinicke R, Weickhardt C, Grotemeyer J (1999) The application of ultra-short light pulses for the analysis of quickly relaxing organic molecules by means of laser mass spectrometry. Int J Mass Spectrom 185/186/187: 307

Harrison AG (1992) Chemical ionization mass spectrometry. CRC Press, Boca Raton
Hessische Landesanstalt für Umwelt (1998) Abschlußbericht Methodenvergleich Rüstungsaltlasten. Umweltplanung. Arbeits-und Umweltschutz: 251
Karas M, Bahr U, Gießmann U (1991) Matrix-assisted desorption ionization mass spectrometry. Mass Spectrom Rev 10: 335
Kayed A (1990) Abfallwirtschaft in Forschung und Praxis 39: 75
Kühnel G (1991) Die Rüstungsaltlastenproblematik in der Bundesrepublik Deutschland. Müll und Abfall 3: 155
For a recent review see: Ledingham KWD Singhal RP (1997) High intensity laser mass spectrometry - A review. Int J Mass Spectrom Ion Proc 163: 149
Letokhov VS (1987) Laser Photoionization Spectroscopy. Academic Press, Orlando, FL
Levy DH (1981) The Spectroscopy of very cold gases. Science 214: 263
Lubman DM (1987) Optically selective molecular mass spectrometry. Anal Chem 59: 31A
Lubman DM (1994) Lasers and Mass Spectrometry. Oxford University Press, New York
Mamyrin BA, Karataev DV, Shmikk DV, Zagulin VA (1973) The mass reflectron, a new non-magnetic time-of-flight mass spectrometer with high resolution. Sov Phys JETP 37: 45
March RE, Todd JFJ (eds) (1995) Practical aspects of ion trap mass spectrometry. CRC Press, Boca Raton
Meijer G, deVries MS, Hunziker HE, Wendt HR (1990) Laser desorption jet-cooling of organic molecules – cooling characteristics and detection sensitivity. Appl Phys B 51: 395; (1990) Laser desorption jet-cooling spectroscopy of para-amino benzoic acid monomer, dimer, and clusters. J Phys Chem 94: 4394
Mies B-M, Bauer H-J (1991) Rüstungsaltlasten in Nordrhein-Westfalen. Müll und Abfall 7: 442
Miller JC (ed) (1994) Laser Ablation. Springer-Verlag, Berlin
Miller JC, Haglund RF (eds) (1997) Laser Ablation and Desorption. Academic Press, Orlando, FL
Opsal RB, Owens KG, Reilly JP (1985) Resolution in the linear time-of-flight mass spectrometer. Anal Chem 57: 1884
Oser H, Coggiola MJ, Faris GW, Young SE, Volquardsen B, Crosley DR (2001) Development of a jet-REMPI (resonantly enhanced multiphoton ionization) continuous monitor for environmental applications. Appl Optics 40: 859
Powis I, Baer T, Ng C-Y (1995) High resolution laser photoionization and photoelectron studies. John Wiley & Sons, Chichester
Rapsch H-J (1991) Rüstungsaltlasten in Niedersachsen. Müll und Abfall 4: 221
Smalley RE, Wharton L, Levy DH (1977) Molecular Optical Spectroscopy with Supersonic Beams and Jets. Acc Chem Res 10: 139
Spyra W (1991) Untersuchungen von Rüstungsaltlasten. EF-Verlag für Energie und Umwelttechnik, Berlin
Tang CL, Cheng LK (1995) Fundamentals of Optical Parametric Processes and Oscillators. In: Laser Science and Technology. Vol 20, Harwood Academic Pub, Newark, NJ
Tarradellas J, Bitton G, Rossel D (1997) Soil ecotoxicology. CRC Lewis, New York
Tembreull R, Sin CH, Li P, Pang HM, Lubman DM (1985) Applicability of resonant two-photon ionization in supersonic beam mass spectrometry to halogenated aromatic hydrocarbons. Anal Chem 57: 1186
Tönnies K, Schmid RP, Weickhardt C, Reif J, Grotemeyer J (2001) Multiphoton ionization of nitrotoluenes by means of ultrashort laser pulses. Int J Mass Spectrom 206: 245
Träger F (1989) Photoacoustic, photothermal and photochemical processes at surfaces and in thin films. In: Hess P (ed) Springer-Verlag, Berlin

Umweltbundesamt (1998) UBA-Informationen zur Altlastenerfassung. http://www.umweltdaten.de

Vernov GS, Delone NB (1965) JETP Lett 1: 66 (Atoms)

Villinger J, Federer W, Dornauer A, Weissnicht A, Hönig M, Mayr T (1996) Dynamic Differentiated Hydrocarbon Monitoring in Direct Engine Exhaust: A Versatile Tool in Engine Development. SAE Technical Papers Series, No. 960063, SAE, Warrendale, PA

Weickhardt C, Boesl U (1993) Time resolved trace analysis of exhaust gas by means of laser mass spectrometry. Ber Bunsenges Phys Chem 97: 1716

Weickhardt C, Boesl U, Schlag EW (1994) Laser mass spectrometry for time-resolved multicomponent analysis of exhaust gas. Anal Chem 66: 1062

Weickhardt C, Boesl U, Schlag EW (1994) Method and apparatus for calibrating strongly fluctuating measuring signals for quantitative analysis of gas mixtures by resonant laser mass spectrometry. German Patent DE 4,305,981, 1. Sept

Weickhardt C, Zimmermann R, Boesl U, Schlag EW (1994) Laser mass spectrometry of the di-, tri- and tetrachlorbenzenes: Isomer-selective ionization and detection. Rapid Commun Mass Spectrom 8: 381

Weickhardt C, Moritz F, Grotemeyer J (1996) Time-of-flight mass spectrometry: State-of-the-art in chemical analysis and molecular science. Mass Spectrom Rev 15: 139

Weickhardt C, Moritz F, Grotemeyer J (1996) Multiphoton ionization mass spectrometry: principles and fields of application. Eur Mass Spectrom 2; 151

Weickhardt C, Grun C, Grotemeyer J (1998) Fundamentals and features of analytical laser mass spectrometry with ultrashort laser pulses. Eur Mass Spectrom 4: 239

Weinberg DS, Hsu JP (1983) J High Resolut Chromatogr Commun 6: 404

Wiley WC, McLaren IH (1950) Time-of-Flight Mass Spectrometer with Improved Resolution. Rev Sci Instrum 26: 1150

Wöstmann U, Zentgraf C (1992) Altlastensituation im Bereich eines militärischen Zwischenlagers für Sonderabfälle. Müll und Abfall 10: 719

Yinon J, Zitrin S (1981) The analysis of explosives. John Wiley, New York

Yinon J; Zitrin S (1993) Modern methods and applications in analysis of explosives. John Wiley, Chichester

Yinon J, Zitrin S (1996) Modern methods and applications in analysis of explosives. John Wiley, New York

Zenobi R (1994) Advances in surface-analysis and mass-spectrometry using laser-desorption methods. Chimica 48: 64

11 Diode-Laser Sensors for In-Situ Gas Analysis

Peter W. Werle

11.1 Absorption Spectroscopy

Optical sensors based on semiconductor lasers are at the threshold of routine applications in gas analysis and increasingly these sensors are used for industrial and environmental monitoring applications whenever sensitive, selective and fast in-situ analysis in the near- and mid-infrared spectral region is required. With the increasing complexity of processes, online gas analysis is becoming an issue in automated control of various industrial applications such as combustion and plasma diagnostics, investigations of engines and automobile exhaust measurements. Other challenges are online analysis of high purity process gases, medical diagnostics and monitoring of agricultural and industrial emissions (VDI 2002). The need to meet increasingly stringent environmental and legislative requirements has led to the development of analyzers to measure concentrations of a variety of gases based on near- and mid-infrared absorption spectroscopy.

Absorption spectrometers generally contain a radiation source and an appropriate detector together with the species under investigation in an absorption cell for concentration measurements based on Beer´s law. As a prerequisite to obtain the required selectivity a dispersive element has to be inserted in to the optical path. Modern gas analyzers use semiconductor lasers, where the selective element is the radiation source itself. Various techniques and designs have been developed to meet specific requirements of different measurement challenges and for high sensitivity in-situ applications several techniques are available. In photo-acoustic spectroscopy (see chap. 16) intensity modulated light is absorbed by a target gas at a specific wavelength. The absorbed photon energy is transformed into translation energy by collisions, resulting in a modulation of gas temperature and pressure respectively. Using a sensitive microphone to measure this signal, very low concentrations can be detected. Photo-acoustic trace detectors have shown their value in the fields of medical sciences (e.g. breath tests, see chap. 12) and environmental studies. Cavity ring down (see chap. 7 and 14) spectroscopy is an other sensitive absorption technique in which the rate of absorption rather than the magnitude of the absorption of a light pulse confined in an optical cavity is measured.

Tunable diode-laser absorption spectroscopy (TDLAS) is increasingly used as an attractive technique for analytical instrumentation. In such instruments a single narrow laser line is tuned by injection current changes over an isolated absorption line from ν_1 to ν_2 of the species under investigation (Fig. 11.1a). To achieve the highest selectivity, analysis is made at low pressure, where absorption lines are not substantially pressure broadened. This type of measurements has developed into a very sensitive and general technique for monitoring atmospheric trace species (Schiff et al. 1994). The main requirement is that the molecule should have an infrared line-spectrum which is resolvable at the Doppler limit, which in practice includes most molecules with up to five atoms (as for example CO, CO_2, NO, N_2O, NO_2, HNO_3, NH_3, CH_4, CH_2O, H_2O, H_2O_2) together with some larger molecules. Because TDLAS operates at reduced pressure it is not restricted in wavelength to the atmospheric windows at 3.4-5 μm and 8-13 μm. Direct absorption measurements have to resolve small changes ΔI in a large signal offset I_0. Therefore, most applications of TDLs in atmospheric research required long-path absorption cells to provide high sensitivity local measurements (Brassington 1995).

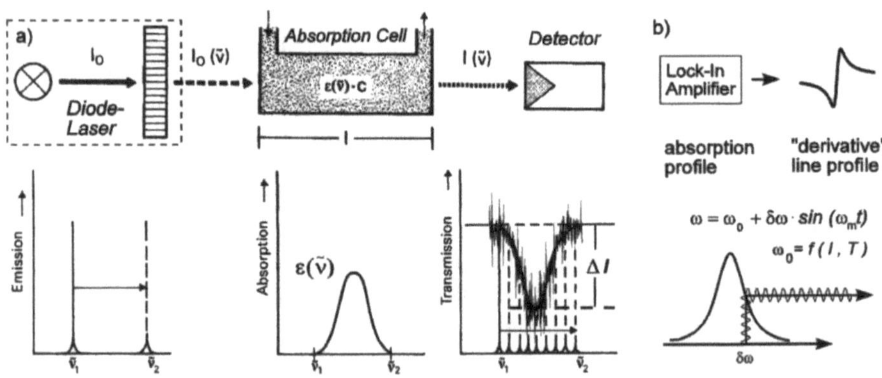

Fig. 11.1a. Diode-laser absorption spectroscopy **b.** wavelength modulation spectroscopy

Signal averaging increases the signal-to-noise ratio (SNR) and for signal levels $\Delta I/I_0$ below 10^{-3} additional noise suppression can be achieved by the application of modulation techniques. In modulation spectroscopy, the laser injection current is modulated at ω_m while the laser wavelength is tuned repeatedly over the selected absorption line to accumulate the signal from the lock-in amplifier with a digital signal averager (Fig. 11.1b). This produces a derivative line profile with an amplitude proportional to the species concentration. Scanning over the line gives increased confidence in the measurement, because the characteristic spectrum of the measured species is clearly seen and unwanted spectral features due to interfering species or étalon fringes can easily be identified. The benefits of modulation spectroscopy are twofold : Firstly, offsets are eliminated (zero baseline technique) as it produces a derivative signal, directly proportional to the species concentration and, secondly, it allows narrowband detection of the signal at a frequency at which the laser noise is reduced.

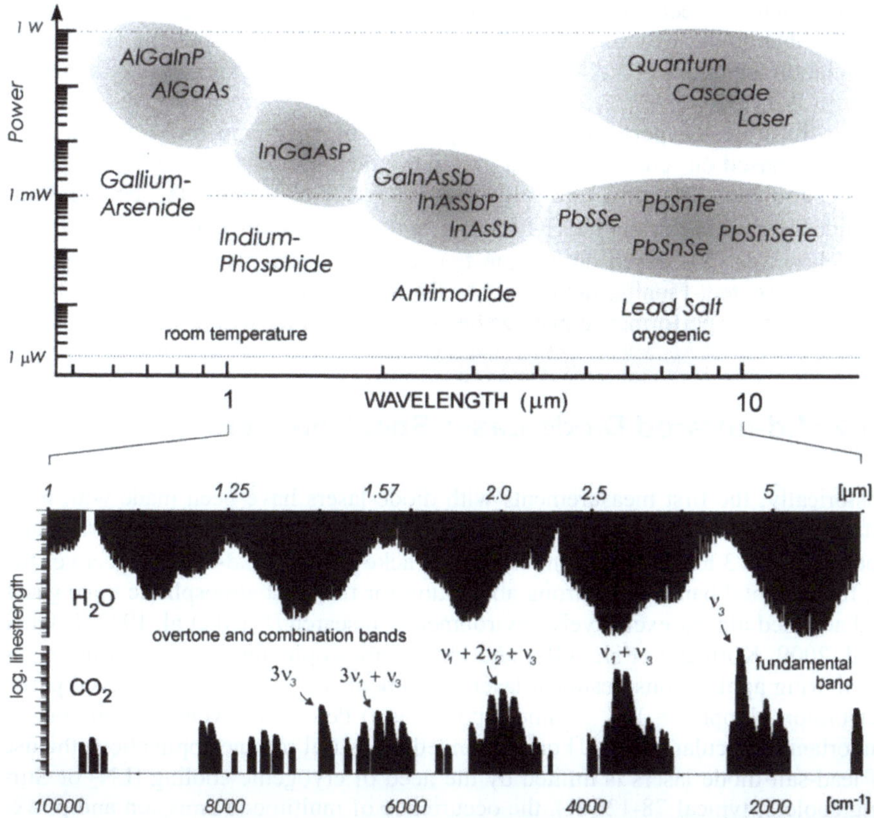

Fig. 11.2. Typical available diode-laser continuous wave output power versus spectral coverage of the visible to infrared region by different semiconductor material systems together with an example of absorption cross sections for CO_2 and interfering water vapor (Rothman et al. 1992)

Different molecules show absorption of light at different wavelengths based on their fingerprint-like absorption spectrum (Fig. 11.2). In the spectral range from the visible to the infrared combination and overtone bands as well as the fundamental bands can be covered by diode-lasers based on Gallium-Arsenide, Indium-Phosphide, Antimonides and Lead-salts (Werle et al. 2002) and the recently developed quantum cascade lasers (QCL) (Faist et al 1994, Beck et al. 2002). While in the past only mid-infrared diode-lasers operated at cryogenic temperatures covered the fundamental absorption bands required for high sensitive gas analysis, near-infrared room temperature diode lasers gave access only to the significantly weaker overtone and combination bands. Therefore, the selection of the operating regime for a gas analyzer is always a trade off between the required sensitivity, system complexity and operational cost. While predicted sensitivities are based on known line strengths and system performance, it is in the nature of field measurements that optimum performance is not always achieved due to instrumental drifts, interferometric effects and turbulent refractive index fluctuations. To cope with these problems, for example approaches based upon signal-processing and double modulation techniques (Werle and Lechner 1999) have been successfully applied.

To illustrate the performance and operation of near- and mid-infrared spectrometers based on tunable diode-lasers, in the next sections selected applications of spectrometers applying lead-salt diode-lasers for NO_2 and CH_4 sensing, antimonide lasers for CH_4 and HCHO sensing in the 3-4 μm range and a near-infrared gas sensor for CO_2 based on a room temperature 2 μm Indium-Phosphide laser will be presented. Finally, the impact of the mid-infrared quantum cascade lasers on spectrometer performance that has been obtained so far will be discussed.

11.2 Mid-Infrared Diode-Laser Spectrometers

Historically, the first measurements with diode-lasers have been made with mid-infrared lead-salt devices. They are based on IV-VI semiconductor materials and operate in the 3 to 30 μm spectral region (Tacke 1995). Lead-salt lasers cover the IR fundamental bands with strong absorption for the most atmospheric trace gases and are used almost exclusively environmental research (Fried et al. 1997, Fischer et al. 2000, Kormann et al. 2001) and for spectroscopic applications. In trace gas monitoring applications, lead-salt laser instruments have routinely achieved parts-per-billion (1 ppbv = 10^{-9} volume mixing ratio) detection levels of a number of important molecular species. For unattended industrial routine applications the use of lead-salt diode lasers is limited by the need of cryogenic cooling (LN_2 or Stirling coolers, typical 78-120 K), the occurrence of multimode emission and power levels, which are typically several hundred microwatts. Compared to GaAs lasers, lead-salt diode-lasers are at a relatively early stage of their development due to a much smaller market. In order to improve TDLAS detection speed and detection limits high frequency modulation (FM) techniques have been introduced. These techniques determine the absorption or dispersion of a narrow spectral feature by detecting the heterodyne beat signal that appears when the optical spectrum of the probe wave is distorted by the spectral feature of interest. Advances in laser-optical gas analyzers based on these techniques have been reviewed (Werle 1998) and therefore only the essentials will be summarized here. The major difference to conventional modulation spectroscopy is the application of radio frequency modulation (rf) instead of conventionally used kHz frequencies. This allows faster scanning and signal detection at MHz to GHz frequencies, where laser excess noise does not dominate detection and therefore, in principle, a detection limit close to the quantum limit can be obtained (Werle et al. 1989). In a FM spectrometer a rf-current of typically about 100 MHz is used to modulate a DC current with a superimposed ramp via a bias-T to decouple the different current sources. The modulated current, $i_L(t)$, generates a frequency modulated electromagnetic field, $E_l(t)$, which interacts resonantly with the rotational-vibrational absorption of the molecules in the sample cell. The number of photons emitted from the laser depends upon the number of electrons in the conduction band and, depends upon the number of electrons in the conduction band and, therefore, from the current through the pn-junction of the diode laser. The higher the current, the higher the number of photons available and the amplitude of the electromagnetic field depends on photon density, i.e. changes in the laser current will lead to an amplitude modulation of the laser. The index of refraction in the pn-junction of the laser depends on the carrier density. Therefore, there is a coupling between amplitude and frequency modulation for the electric field. The phase modulated electrical field, $E_1(t)$, with residual amplitude (AM) modulation is

$$E_1(t) = E_0(t)\cdot [1+M \sin(\omega t+\psi)] \cdot \exp\{ i (\Omega t + \beta \sin(\omega t))\}, \tag{11.1}$$

where $\omega \equiv$ modulation frequency, $\Omega \equiv$ laser carrier frequency, $\beta \equiv$ FM-index, $M \equiv$ AM-index and $\Psi \equiv$FM-AM-phaseshift. For low modulation indices we obtain in the frequency domain an upper and a lower sideband, which are displaced $\pm\omega$ from the laser carrier Ω. The principle setup of a FM-TDLAS system is shown in Fig. 11.3a. A fraction of the original laser beam is required for active line locking using the reference channel, while about 90% of the laser intensity is used for the sample gas detection in a multipass absorption cell. The electrical field, $E_2(t)$, after interaction with the sample can be described by

$$E_2(t) = E_1(t) \exp\{- \delta(\omega)-i\phi(\omega)\} \tag{11.2}$$

where $\delta(\omega)$ is the absorption and $\phi(\omega)$ is the dispersion of the sample gas. The electrical field after the probe induces a detector current, $i_{\mathrm{rf}}(t)$, in a photovoltaic Mercury Cadmium Telluride (MCT) detector.

$$i_{\mathrm{rf}}(t) = / E_2(t /^2 \tag{11.3}$$

The amplified and filtered current is fed into the rf-input of a double balanced mixer for phase sensitive detection at the modulation frequency. For a selected phase shift between the local rf-oscillator i_{LO} and the detector signal i_{rf} we record at the intermediate frequency IF mixer output port the lowpass (τ) filtered product

$$i_{\mathrm{IF}}(t) = \langle i_{\mathrm{LO}}(t) \cdot i_{\mathrm{rf}}(t) \rangle_\tau \tag{11.4}$$

After this phase sensitive detection at the modulation frequency, the demodulated signal, $i_{\mathrm{IF}}(t)$, is proportional to the concentration of the trace gas in the absorption cell and by adjusting the detection phase either the absorption or the dispersion signal can be selected (Werle 1998). The reference beam passes through a reference cell, which provides at high signal-to-noise ratio a signal from the spectral feature under investigation. This channel is used for line-locking and online drift correction. A line locking procedure monitors the deviation of the signal position from a given set-point and compensates for drifts. The sample and the reference signals are then digitized and further processed by digital filters, line locking algorithms, calibration procedures and an intensity normalization to cope with laser power fluctuations (Werle et al. 1994). The corrected signals are then further stored in a computer for digital signal processing and referenced to a previously recorded calibration spectrum to provide final concentrations in different units (ppbv, molec/cm^3, μg/cm^3) together with the calculated measurement precision.

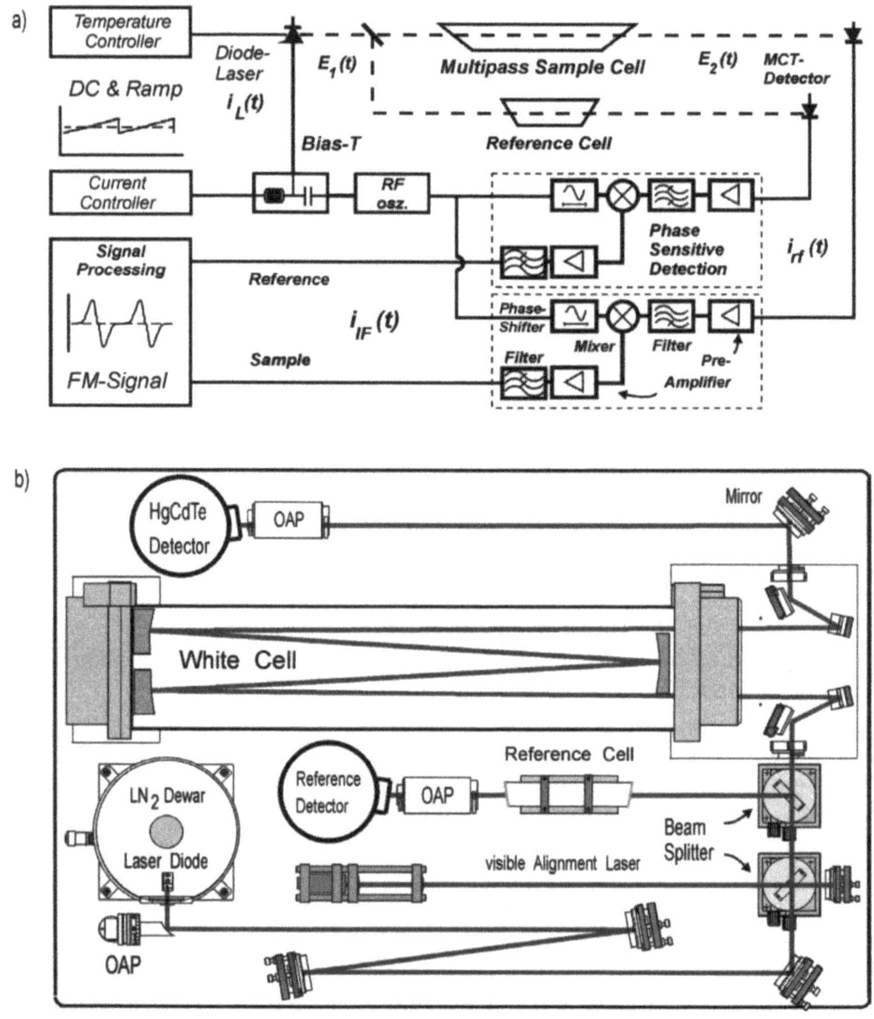

Fig. 11.3a. FM-detection scheme **b.** Mid-infrared lead-salt diode-laser spectrometer

The mid infrared TDLAS system shown in Fig. 11.3b is based on lead-salt lasers and has been used for spectroscopic in-situ detection of NO_2. For the experiments a lead-salt diode-laser was mounted in a liquid-nitrogen (LN_2) cooled dewar, which has been used for a spectral characterization in a laser test setup prior to the spectroscopic measurements. To accommodate for a possible deviation angle between the cone of laser emission and the laser mount axis, the LN_2 - dewar is mounted on a xyz-stage alignable within ± 30°. The beam from the TDL is first collimated by an off-axis parabola (OAP) and then directed by a sequence of mirrors through the sample cell and onto a LN_2-cooled HgCdTe photovoltaic detector.

A visible alignment laser beam can be combined via a pellicle beam splitter with the invisible infrared beam to assist during the system alignment phase. For typical line-strengths an ambient concentration of 1 ppbv produces an absorption of only 1 part in 10^7 over a 10 cm path-length. Conventional absorption spectroscopy would not be able to measure such small absorption. TDLAS overcomes this problem by using a multi-pass cell with folded optical paths of 100 m or even more (White 1976, Herriott and Schulte 1965). The White cell used in this system has a base length 62.5 cm and an adjustable path length, L, of up to 100 m. For optimum SNR, the absorption path length is adjusted to 27.5 m. The system operates at a gas flow of 10 l/min and the pressure inside the cell is actively regulated using a MKS Baratron to maintain a pressure of 26.7 hPa. The optical setup is mounted on a 100 x 60 cm optical breadboard and is enclosed in a box flushed with dry nitrogen to improve the thermal stability. The frequency of the laser was tuned over the selected NO_2 absorption. For NO_2 measurements an absorption line at 1600.413 cm^{-1} was chosen since its background was free of disturbance from the pressure broadened H_2O lines nearby. The NO_2 line consists of two unresolved lines of equal line strength of $1.17 \cdot 10^{-19}$ cm/molecule.

Trace gas measurements near to the detection limit are usually performed by measuring the ambient air spectrum and the spectrum of zero air, i.e. air devoid of the target substance, which is referred to as the background spectrum. The background spectrum still contains the disturbing spectral signatures from interfering fringes and therefore can be subtracted from an ambient spectrum to obtain a clean spectrum. Another prerequisite for quantitative measurements is a calibration spectrum, which can for example be obtained by measuring gas from a commercial certified gas cylinder after dilution to the required concentration level. For calibration purposes higher concentrations are usually used with corresponding signals that are much larger than the fringes in the spectrum. Provided that the laser frequency is kept constant by line locking, the acquisition of the calibration spectrum can then be omitted from the measurement sequence. This is advantageous since a substantial part of the time is needed to exchange the gas in the White cell after switching from ambient air to zero air for background recording.

The instrument performance in terms of the detection limit and detectable optical density has been determined from NO_2 measurements in ambient air. A calibration, background and ambient spectrum as well as the background corrected spectrum is shown in Fig. 11.4a, where 256 spectra have been averaged within 740 ms. The electronics bandwidth of 1.5 kHz leads to an effective bandwidth of 5.86 Hz. The mixing ratio of NO_2 was calculated by least square fitting to the calibration spectrum taken at 12 ppbv (1ppbv = 10^{-9} volume mixing ratio). From a least squares fit a mixing ratio of 1.17 ppbv with a 1 σ precision of 31.5 pptv has been obtained. For quality assurance additional quantitative information on system stability and the maximum signal averaging time has been derived from an Allan variance analysis, which has been discussed in detail together with the aspects of background stability by Werle et al. 1993. An Allan plot has been generated from a continuous measurement of zero air spiked with 12 ppbv of NO_2 from a calibration source for a period of 600 s with a time resolution of 1.5 s. As the linearly decreasing part of the Allan variance is dominated from white noise, it is in this part equivalent to the statistical variance and, consequently, the square root of the Allan variance gives a prediction of the detection limit. From the recorded time series data in Fig. 11.4b we obtain for an integration time of 25 s a detection limit of 10 pptv from the Allan variance, corresponding to a detectable change in optical density of $5\cdot10^{-7}$. At longer integration times the Allan variance, and with it the instrument detection limit, will start to deteriorate as a consequence of instrumental drifts. In practical terms this means that the complete measurement sequence consisting of the acquisition of the ambient, background and calibration gas spectra has to be completed within 60 s.

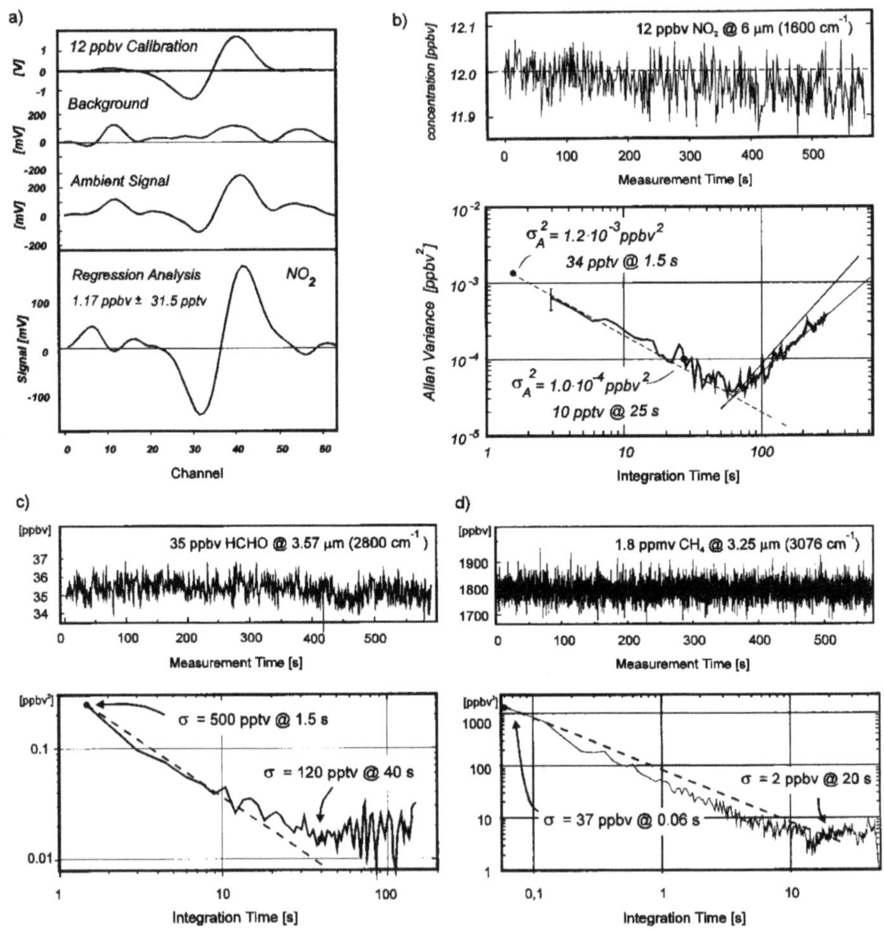

Fig. 11.4a. calibration, background and ambient spectra for NO_2 and **b-d.** time series data and corresponding Allan Plot (Werle et al. 1993) for NO_2, H_2CO, CH_4

The number of manufacturers for lead-salt diode-lasers is limited world-wide to one or two, which changes slightly from time to time. Therefore, alternatives in the infrared spectral region are desperately asked for. For wavelengths below 4 μm down to 1.8 μm Antimonide lasers, based on III-V compounds such as AlGaAsSb, InGaAsSb, and InAsSbP, can be used (Nicolas et al. 1998). Room temperature lasing from 2 to 2.4 μm has been reported from simple double heterostructure antimonide diode-lasers. As wavelength increases up to 3.7 μm, the maximum operating temperature decreases as a result of increasing optical and electrical losses. Laser devices used in the experiments described here are based on InAsSb/ InAsSbP double heterostructure devices and were grown by liquid phase epitaxy on InAs substrate at the Ioffe Physico Technical Institute in St. Petersburg, Russia and cover the spectral range from 3-4 μm at LN_2 temperatures. Such devices are well suited for the detection of HCHO at 3.6 μm and CH_4 at 3.26 μm (Werle and Popov 1999). For gas sensing applications lasers have been selected for formaldehyde emitting at 3.57 μm (2800.2 cm^{-1}) and for methane operating at 3.25 μm (3076.5 cm^{-1}). For the formaldehyde measurements the previously described NO_2 instrument with the 6 l White cell now at L=30 m total pathlength and a pressure of 30 hPa has been used. From experiments we determined a detection limit for HCHO of 120 pptv with 40 s integration time (Fig. 11.4c) or in terms of minimal detectable optical density $(\alpha L)_{min} = 10^{-6}$ at $\Delta f = 1$ Hz. The methane measurements aimed at a higher time resolution for flux measurements and the White cell was replaced by a 5 l Herriott cell with a total pathlength of 100 m. For CH_4 a precision of 37 ppb has been obtained with 0.06 s integration time corresponding to $(\alpha L)_{min} = 2.7 \cdot 10^{-4}$ at $\Delta f = 1$ Hz (Fig. 11.4d). While the results for formaldehyde were quite satisfying, the performance of the methane measurements was worse due to the fact that the spectral response of the HgCdTe detectors is degrading near 3 μm. Furthermore the relative low power of 200 μW and the 100 m optical pathlength with the corresponding strong power attenuation due to multiple reflection (Werle and Slemr 1991) led to a low power level at the detector. With an optimized system with respect to optical power transmission and an antimonide laser that emits at higher injection currents, providing higher power, the potential of the increased line strength in the ν_3 band of CH_4 according to the Hitran database (Rothman et al. 1992) should be feasible. Antimonide lasers might offer operational benefits compared to lead-salt lasers, while still maintaining high sensitivity by probing fundamental ro-vibrational absorption transitions.

Modern atmospheric research on gas exchange between the biosphere and the atmosphere requires sensitive, reliable and fast-response chemical sensors. Therefore, techniques for fast and simultaneously sensitive trace gas measurements based on tunable diode-laser absorption spectroscopy have been successful applied to micrometeorological trace gas flux measurement techniques as the eddy covariance technique (Zahniser et al. 1995, Kormann et al. 2001). The availability of such sensors allows for example a validation of closed chamber measurements and also can provide information about CH_4 emissions on a larger scale, which is the basis for any up-scaling effort from a regional to a global scale.

The eddy correlation technique directly determines the flux of an atmospheric constituent through a plane that is parallel to the surface. Ideally, the meteorological conditions controlling the state of the turbulence should not vary over the course of the measurements and the surface viewed by the sensors should be horizontally uniform, both in its physical and chemical-biological aspects. Because the eddy correlation method may be considered as defining the instantaneous upward or downward transport of the constituent and then averaging contributions to give the net flux, it must take into account the frequency range of the turbulence for vertically transporting the constituents in the atmosphere. The technique requires simultaneous fast and accurate measurements of the vertical wind velocity and the concentration of the trace species in question.

The key element of such a field instrument is the diode-laser. When starting to select a laser, the first task is to select from mode maps a combination of base temperature and drive current at which the laser produces a strong, preferably single mode emission, tuned to the absorption line being monitored. Due to the limited sensitivity obtained with the antimonide laser described before, a 7.8 μm (ν_4-band) lead-salt diode-laser was the optimum choice for CH_4 flux measurements. The optomechanical components of the spectrometer are mounted on an 50 x 90 cm optical breadboard (Fig. 11.5a). The lead-salt diode-laser is mounted on a cold-head within a LN_2-dewar. For injection currents between 400 and 600 mA at temperatures ranging from 85 to 95 K single mode operation (Fig 11.5b) with an average power level of 200 μW was ensured and isolated CH_4 absorption lines could be reproducibly selected for the measurements even after repetitive thermal cycling, which was an important criterion for the planned field measurements.

The experimental setup of the eddy correlation system has been described in detail by Werle and Kormann 2001 and is similar to the one shown in Fig. 11.3. The White cell has been replaced by a Herriott cell with a very small internal volume of 0.3 l designed for applications requiring fast gas flow and exchange to allow high time resolution. A rotary vacuum pump provides the gas flow of about 18 slm through the Herriott cell at a pressure of about 50 hPa. A dust filter is at the inlet of the measurement head to protect the gas system and the mirrors of the Herriott cell from pollution. A calibration system allowed programmed sequences of measurements of background signals, calibration gas and ambient air. The calibration system is based on a dynamic gas dilution system, where calibration gas from steel cylinders is diluted with N_2 down to ambient concentration levels. With this spectrometer ambient methane concentrations around 2 ppmv can be detected with a precision better than 1 % at a 10 Hz repetition rate and a typical 30 min data set contains 18000 individual concentration values (Fig. 11.5c). Each concentration value has been obtained by averaging individual spectra followed by a background correction as described previously. The "noisy" structure in the high resolution time series data reflects the turbulent nature of transport in the atmosphere and has frequency contributions from 0.01 Hz up to 10 Hz. For each concentration measurement the corresponding vertical wind speed has been measured.

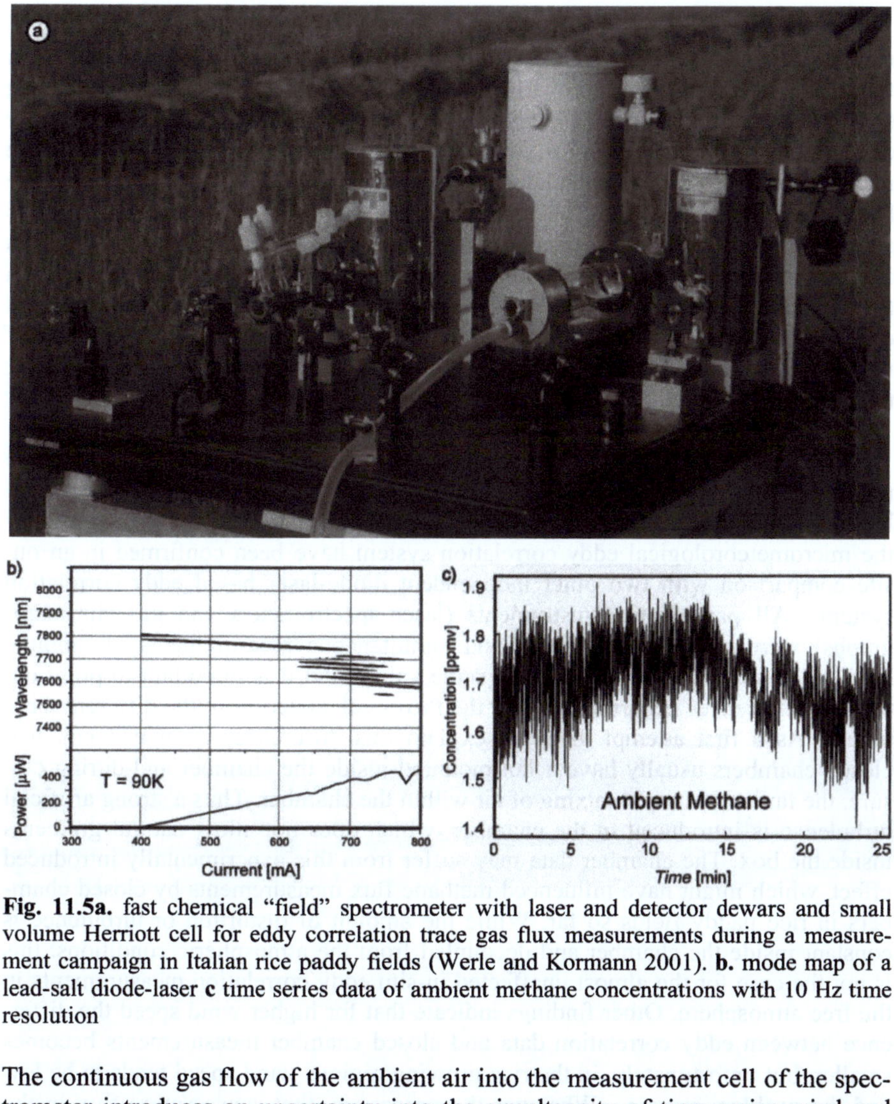

Fig. 11.5a. fast chemical "field" spectrometer with laser and detector dewars and small volume Herriott cell for eddy correlation trace gas flux measurements during a measurement campaign in Italian rice paddy fields (Werle and Kormann 2001). **b.** mode map of a lead-salt diode-laser **c** time series data of ambient methane concentrations with 10 Hz time resolution

The continuous gas flow of the ambient air into the measurement cell of the spectrometer introduces an uncertainty into the simultaneity of time series wind and concentration data. Therefore, a correlation analysis was used to find the time lag and the fluxes. The first step in the eddy correlation process is to calculate the perturbation values of the data points. For the measured time series of concentration values we subtract the mean from each data point to yield a time series of perturbations c′. We can similar find a time series of vertical wind velocity perturbations w′. Multiplying the respective values together yields a time series w′c′. The average of this series <w′c′> gives the turbulent vertical flux. An advantage of this method is that it is direct and simple, and fluxes can be calculated at whatever height or location the original time series was measured.

In the frame of an interdisciplinary research project eddy correlation measurements of methane emissions from rice paddy fields have been performed during a field campaign to allow a comparison with data from a set of on-site monitoring systems based on the closed chamber technique. A typical gas collection chamber covers a surface of about 0.4 m^2 and is fitted with a removable plexiglass. The methane emission rate is calculated from a temporal increase of CH_4 inside the box during a 30 min closure time using a gas chromatograph. Spatial variability is a great problem in using chambers to measure fluxes from a field or ecosystem. In addition, chambers disturb the natural air turbulence, decouple the rice plant from the ambient turbulent atmosphere and alter the temperature, solar radiation and gas concentration in the measurement environment. Therefore, the extrapolation of methane emissions, based on flux rates obtained by use of small closed chamber measurements, to field, landscape and regional levels is not so well established.

The measurements based upon the 'state-of-the-art' closed chamber technique report about 60-90% higher methane emissions than the simultaneous eddy correlation measurements (Werle and Kormann 2001). The lower fluxes measured by the micrometeorological eddy correlation system have been confirmed in an on-site comparison with two other independent diode-laser based eddy correlation systems. All participating instruments (laser spectrometers and gas chromatographs) were calibrated routinely and simultaneous measurements of ambient methane *concentrations* reported the same values and it is important to point out that the differences occurred only for the *fluxes* calculated from the different techniques. As a first attempt to try to explain this difference, we may recall that closed chambers usually have a fan mounted inside the chamber and during closure, the fan causes rapid mixing of air within the chamber. Thus a strong artificial turbulence is introduced in the chamber, which does not allow natural gradients inside the box. The chamber data may suffer from this experimentally introduced effect, which might have influenced methane flux measurements by closed chambers in rice paddy fields so far. While the amount of distortion or turbulence is constant inside the chamber and decoupled from the atmospheric conditions, this is not the case for the almost unaffected in-situ eddy correlation measurements in the free atmosphere. Other findings indicate that for higher wind speed the difference between eddy correlation data and closed chamber measurements becomes smaller, but unfortunately, in the rice growing regions wind speed tends to be low and the problem remains. Whatever the process is, that causes more flux in the closed chamber with fans on, so far the consensus is that it only accounts for a fraction of the difference between chambers and micrometeorological measurements. The discrepancy between micrometeorological measurements and the closed chamber technique has not been resolved completely yet, but this finding is important for atmospheric research in the context of greenhouse gases. Such fast and highly sensitive measurements as described here would not be possible with near-infrared systems due to the lack of sensitivity and the results shown here demonstrate, that tunable diode-laser absorption spectroscopy can be a valuable tool for quality assurance and quality control.

11.3 Near-Infrared Overtone Spectrometer

For many industrial applications or field measurements the use of liquid nitrogen must be avoided, closed cycle coolers are too expensive and only thermoelectrical elements are acceptable (D´Amato and De Rosa 2002). Several molecular species have absorption features in the near infrared spectral region. Near-IR absorptions are overtone or combination bands that are typically one to several orders of magnitude weaker than the IR-fundamental band. Nevertheless, many molecules of interest have near-IR absorption bands that are strong enough for detection at parts-per-million (1 ppmv = 10^{-6} volume mixing ratio) and even parts-per-billion (ppbv) levels.

The overtone or combination band transitions can be accessed by Gallium-Arsenide and Indium-Phosphide lasers, which are commercially made from the III-V group of semiconductor materials. These diode lasers emit from the visible to near-infrared wavelengths from 0.63 μm to above 2 μm including the InGaAsP/InP lasers. The technology of the 1.3 μm and 1.55 μm InGaAsP/InP diode-lasers developed for fiber-optic telecommunication has been extended to fabricate lasers that emit up to more than 2 μm. These near- infrared multiple-quantum-well distributed-feedback (DFB) lasers have the advantages of single-mode outputs at power levels up to several milliwatts and additionally room-temperature operation.

InP-DFB-lasers developed at the Sarnoff Research Center (Princeton, NJ) with room temperature single-mode emission at $\lambda \approx 2$ μm have been used for the design of a fast carbon dioxide sensor. The DFB-laser is held inside a Peltier-cooled mount, which is fixed on a xyz-stage (Fig. 11.6a) and the laser beam is collimated by an off-axis parabola (OAP) with 10 mm diameter and 12 mm focal length. The beam is focused by a spherical mirror (f= 1m) into the center of a commercial 5 l Herriott cell. After 181 reflections, corresponding to an optical pathlength of 100 m, the beam exits the cell and is focused onto a temperature-stabilized extended InGaAs detector by another OAP. About 8% of the laser beam is coupled off by a beam-splitter and directed through a 28 cm reference cell. The optical system is prealigned with a visible diode-laser, coupled into the setup by a pellicle beam-splitter, which has to be removed during the measurements to optimize power throughput. In order to provide static as well as flux measurements at defined cell pressures, the measurement cell is equipped with a pressure sensor (MKS Baratron) and on/off-valves (at the inlet and outlet) as well as with a needle-valve at the inlet and a throttle valve at the outlet, which is part of an active pressure stabilization loop during flow measurements. The reference cell is filled with a high concentration CO_2 mixture and sealed off and is connected to the measurement cell by a temperature bridge and a differential pressure sensor. With appropriate laser power and gas concentration in the reference cell the signals from both detectors can according to Beer´s law be adjusted to have identical amplitude and shape and after system calibration using certified gas mixtures, the reference signal can be used as a secondary calibration standard.

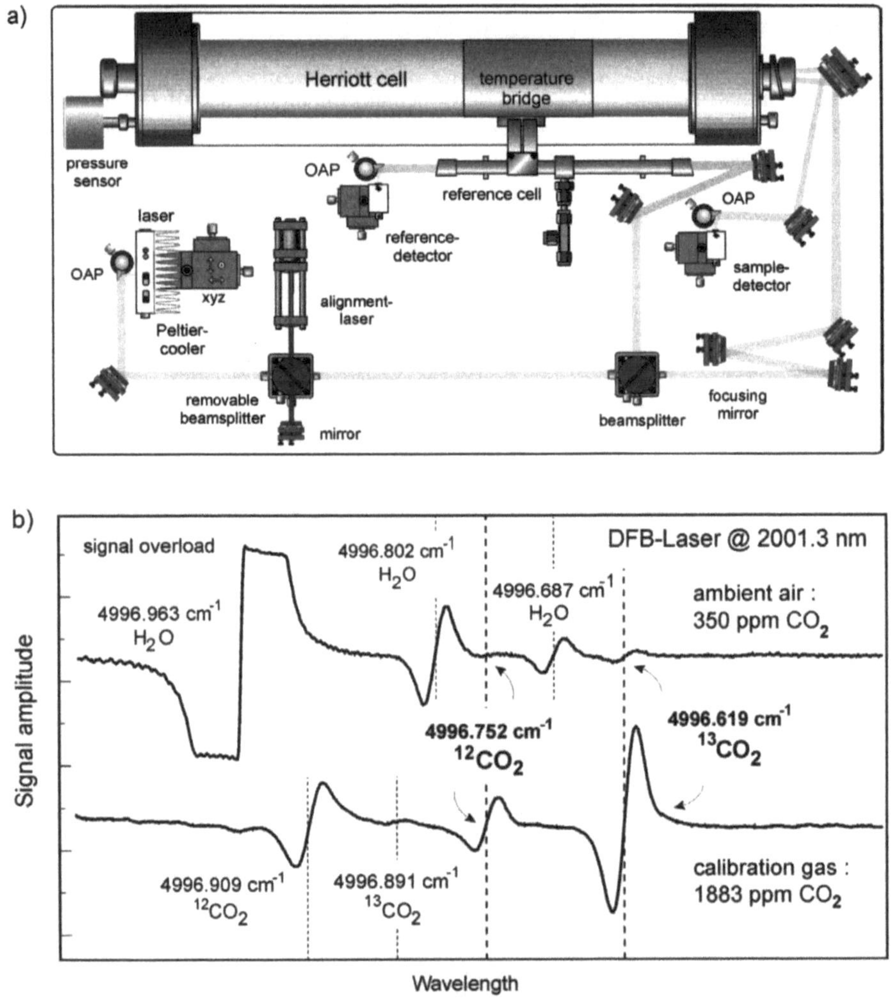

Fig. 11.6a. Optical layout of a *near-infrared* FM-spectrometer with a optical multipass cell **b.** Identification of $^{12}CO_2$ and $^{13}CO_2$ absorption near 2 μm

With this instrument a series of ambient air measurements have been performed. Time series data obtained from a 358 ppmv carbon dioxide calibration gas cylinder have been recorded and from an Allan Variance analysis, as discussed in the previous sections on the infrared measurements, a precision of about 300 ppbv has been obtained for an integration time of 1 sec. This corresponds to a minimum detectable optical density $(\alpha L)_{min}$ of 10^{-4}. The major limitation during these measurements were high transmission losses after 181 reflections in the 100 m fixed pathlength multipass Herriott cell, leading to relative low optical power levels at the detector.

The calculated detection limits in the near- and mid-infrared spectral regions are listed in Tab. 11.1 for a minimum detectable optical density of 10^{-6} for a pressure of 150 hPa and 25 m optical pathlength. The corresponding carbon dioxide spectrum is shown in Fig. 11.2.

Table 11.1 Calculated near - and mid-infrared detection limits for carbon dioxide

CO_2 Band	Wavelength λ [cm^{-1}]	[μm]	Linestrength [cm/molec]	Detection limit [ppbv]	[$\mu g/m^3$]
$3\nu_3$	6983.01	1.432	$6.043 \cdot 10^{-23}$	73	144
$2\nu_1 + 2\nu_2 + \nu_3$	6359.96	1.572	$1.846 \cdot 10^{-23}$	220	430
$\nu_1 + 4\nu_2 + \nu_3$	6240.10	1.603	$1.838 \cdot 10^{-23}$	235	461
$2\nu_1 + \nu_3$	5109.31	1.957	$4.003 \cdot 10^{-23}$	107	210
$\nu_1 + 2\nu_2 + \nu_3$	4989.97	2.004	$1.332 \cdot 10^{-21}$	3.1	6.1
$\nu_1 + \nu_3$	3597.96	2.779	$3.525 \cdot 10^{-20}$	0.11	0.22
ν_3	2361.46	4.235	$3.524 \cdot 10^{-18}$	0.002	0.004

The NIR system described above has been applied to investigate the feasibility of carbon dioxide isotopic ratio measurements and Fig. 11.6b shows an example of $^{13}CO_2/\,^{12}CO_2$ line pairs in the 2 μm region from non-linear oscilloscope traces recorded during an investigation of line pairs (Werle et al. 1998). It is obvious from Tab. 1 that this spectral region has a significant advantage versus the 1. 57 μm absorption band in the NIR, where the line strength is about 2 orders of magnitude weaker. At 2 μm the line strength is still weaker than in the fundamental band, but room temperature operation of diode-lasers is possible for continuous wave (cw) applications (Webber 2001). Future developments of antimonide lasers might give access to the $\nu_1+\nu_3$ band near 2.78 μm, where again a significant increase in detection sensitivity can be expected. Besides atmospheric measurements, this type of instrument can be used for isotopic ratio measurements in medical diagnosis.

11.4 Quantum Cascade Lasers

Until recently all semiconductor lasers, regardless of their operating wavelength, relied upon direct band-to-band transitions in bulk material as shown in Fig. 11.7a. In such semiconductor lasers electrons recombine at the pn-junction with positively charged "holes" to release single photons with a wavelength that is determined by the bandgap, E_g, and thus the chemical composition of the semiconductor sandwich. The interband transitions between the conduction and the valence bands provide the laser radiation.

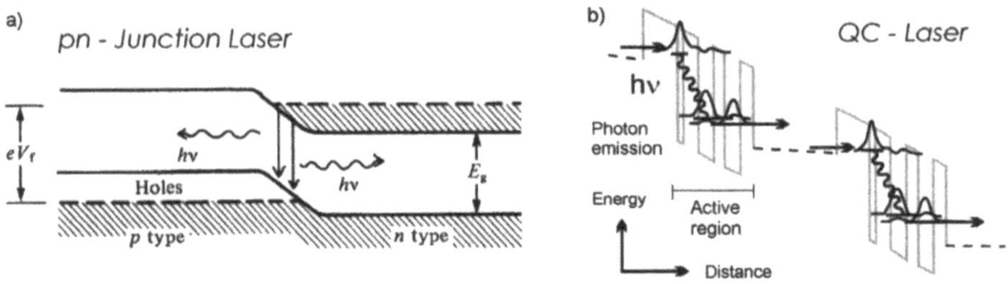

Fig. 11.7a. pn-semiconductor laser and **b.** quantum cascade laser (Faist et al. 1994)

Quantum cascade lasers are based on a completely different approach than the lasers described so far. Their operation is based on intersubband transitions, i.e. transitions within the conduction band (Fig. 11.7b) of a cascaded multiple quantum well structure. Although the basic concept was proposed as early as 1971, it took more than 20 years until an actual device was demonstrated in 1994 (Faist et al. 1994). In a pictorial way, this laser is freed from bandgap-slavery as the emission wavelength depends only on the layer thickness and not on the bandgap of the constituent materials. The quantum well structures are grown using molecular beam epitaxy as alternating layers with a thickness of a few atoms are grown of alloy materials (e.g. InGaAs and InAlAs).

The operation of the quantum cascade laser can be understood us follows. The different materials of the semiconductor in the active region have different band gaps, which leads to the creation of quantum wells. These quantum wells have discrete energy levels due to the thinness of the layers comparable to the electrons de Broglie wavelength. The electrons motion is restricted in the direction perpendicular to the plane of the layers but can move freely in the plane of the layers. An electron in the upper level of the active region will first in a slow process scatter to an intermediate sub-band producing a photon and then fast into the lowest subband. The energy levels are determined by the thickness of the layers in the active region. The stages of the QC laser consist of an area with closely spaced layers (the injection region) followed by more widely spaced layers (active region). The stack of active regions is clad with two thick semiconductor layers of low refractive index, that serve as a wave-guide to direct the produced light along the active regions. In a QCL typically 30 to 75 alternating structures of active regions and injector/relaxation regions are stacked. Once an electron is injected from the contact regions, it is forced to pass through all the periods of active regions and injectors sequentially (cascading). Once the device exceeds lasing threshold, it will emit one photon per period. Adding stages to QC lasers thus increases their output power. In lasers developed in 1999 electrons cascade down 75 steps, instead of 20 or 30 as in earlier QC lasers (i.e. producing up to 75 photons for one electron). In this way QC lasers can provide more than a thousand times the output power of any commercial semiconductor laser operating in the mid-infrared region. Such a QC laser can operate in a large number of modes at wavelengths around the one determined by the energy difference between the upper and intermediate levels. To produce stable, single-mode emission from these QC lasers, as is needed for spectroscopic applications, a grating is integrated into the laser wave-guide producing a distributed feedback device. The grating selects a single mode that satisfies the Bragg condition. Thus, continuous, single mode emission is produced with tuning ranges of about 100-150 nm (at 3-15 μm). The tuning here takes place by changing the temperature of the laser, which changes the refractive index of the wave-guide material, and thus the wavelength at which the Bragg condition holds.

Quantum Cascade-Distributed Feed Back (QC-DFB) lasers can operate either in pulsed mode up to room temperature or in cw mode, operating from cryogenic to above liquid nitrogen temperature (Köhler et al. 2000) and even room temperature cw emission of up to 17 mW at 9.1 μm has already been reported (Beck et al. 2002). In pulsed mode, heating occurs during the current pulse. This changes the emission wavelength slightly, resulting in a dynamic line-width of the laser of a few hundred MHz. Therefore, for application of high resolution spectroscopy in trace gas detection, the laser is preferably used in cw mode, in which case line-widths of a few kHz are attainable (Williams et al. 1999).

QC-DFB lasers have been reported for various wavelengths between 5.2 and 16 μm and have already been used to study gases as NO (Sharpe et al. 1998), N_20 (Namjou et al. 1998), NH_3 (Sharpe et al. 1998), CH_4 (Kosterev et al. 1999), and C_2H_4 (Hvozdara er al. 2000). A QCL-system has been flown on NASA's ER-2 high altitude aircraft to measure stratospheric N_2O and CH_4 (Webster et al. 2001) and the number of applications is rapidly increasing (Kosterev and Tittel 2002). With the development of a QC-DFB laser operating at 4.6-4.7 μm medically important gases like CO and CO_2 and their isotopes, that have their strongest rotational-vibrational bands between 4 and 5 μm have come within range as well (Köhler et al. 2000). Quantum cascade lasers based on InGaAs/InAlAs are already commercially available and have been demonstrated in the wavelength range from 3.4 μm to 13 μm, with room temperature operation from 5 μm to 11.5 μm. Using super-lattice active regions also operation at 17 μm was demonstrated. An advantage of this super-lattice type of laser is that they can carry higher electrical currents than conventional QC lasers, which potentially provides higher output powers (0.5 W at room temperature). Using a novel design where surface plasmon modes are exploited instead of conventional dielectric wave-guides, lasers operating up to 19 μm have recently been achieved (Tredicucci et al. 2000). Other materials are also being used, e.g. GaAs/AlGaAs Systems have been demonstrated for 9.6 μm and 13 μm, and in DFB mode for 10 μm (Schrenk et al. 2000). Output of the QC lasers so far is limited on the short wavelength side of the mid-infrared spectrum by the band-offset between the quantum-well and the barrier materials. For shorter wavelengths deeper quantum wells are needed, which requires different materials. New developments are directed towards developing, lasers, which can produce shorter wavelengths by identifying and implementing new material systems, e.g. based on group III nitrides (Hofstetter et al. 2000).

11.5 Quantum Limited Spectroscopy

Diode laser spectroscopy is a valuable technique for gas analysis. The ability to provide unambiguous measurements qualifies TDLAS as a reference technique against which other methods are often compared. The technique is universally applicable to smaller infrared active molecules and the same instrument can easily be converted from one species to another by changing the laser and calibration gases. The time resolution of TDLAS measurements can be traded off against sensitivity and this allows fast measurements with millisecond time resolution. In order to improve sensitivity various types of modulation spectroscopy have been employed in which the diode laser wavelength is modulated while being scanned across an absorption line. These modulation techniques allow absorption as low as 1 part in 10^6 to be measured within a 1 Hz bandwidth. In combination with optical multipass cells this is equivalent to detection limits of around 20 pptv for the most strongly absorbing species and better than 1 ppbv for almost all species of interest.

The ultimate detection capability is, in principle, only limited by quantum noise (Ye et al. 1998). The signal-to-noise ratio (SNR) is a figure of merit for a detection system. Usually absorption spectrometers are designed in a way that the detected signal is proportional to the laser power arriving at the detector. The total noise is given by the sum of contributions from excess noise, photon induced shot noise and thermal noise, which is independent from power. If an appropriate detection scheme is selected and sufficient power is available, shot noise dominates over thermal noise. The SNR under such "quantum limited" conditions is proportional to the square root of the power impinging on the detector (Werle 1998). Such quantum limited conditions have been obtained with single optical paths (Werle et al. 1989).

In order to discuss problems that are connected with the application of multipass cells with different optical pathlength L, detection limits and other characteristic data from instruments based on White and Herriott cell designs are summarized in Table 11.2. As a figure of merit the observed minimum detectable optical density $(\alpha L)_{min}$ normalized to a $\Delta f = 1$ Hz bandwidth is included, ranging from $1 \cdot 10^{-4}$ to $5 \cdot 10^{-7}$ for different multipass setups. The highest sensitivities have been obtained with a White cell instrument, where the optical pathlength has been reduced from 100 m down to about 30 m and with a fast eddy correlation system, where the pathlength of the Herriott cell has been set to 18 m instead of the possible 36 m. In order to understand the advantage of the reduced pathlength, we have to recall that in the mid-infrared a minimum power at the detector of about 100 μW is required to make shot noise the dominating contribution and, therefore, too many reflections in the optical multipass cells deteriorate system performance significantly (Werle and Slemr 1991). With respect to the discussion of quantum limited performance, it can be seen from Table 11.2 that the best performance has been obtained for high laser power and if pathlength is reduced below maximum, as a trade-off between absorption pathlength and power throughput.

Table 11.2. Summary of characteristics and performance data of optical multipass systems

Instrument	Mid Infrared High sensitivity		High Speed		Near Infrared
Target Gas	NO_2	H_2CO	CH_4	CH_4	CO_2
Wavenumber	1600 cm^{-1}	2800 cm^{-1}	3076 cm^{-1}	1290 cm^{-1}	4990 cm^{-1}
Wavelength	6.25 μm	3.57 μm	3.25 μm (ν_4)	7.8 μm (ν_3)	2.004μm
S [cm/molec]	~2x10^{-19}	~6x10^{-20}	~2x10^{-19}	~5x10^{-20}	~1x10^{-21}
Cell type	White	White	Herriott	Herriott	Herriott
Volume	6 l	6 l	5 l	0.3 l	5 l
Pressure	26.7 hPa	30 hPa	30 hPa	50 hPa	100 hPa
Path Length	27.5 m	30 m	100 m	18 m	100 m
Laser Power	1000 μW	400 μW	< 200 μW	200 μW	1700 μW
Cooling	LN_2	LN_2	LN_2	LN_2	Peltier
Detector	HgCdTe	HgCdTe	HgCdTe	HgCdTe	InGaAs
Calibration	12 ppbv	35 ppbv	1.8 ppmv	2 ppmv	358 ppmv
Type	Permeation	Permeation	Gas Cylinder	Gas Cylinder	Gas Cylinder
Precision Integr. Time	10 pptv @ 25 sec (0.08%)	120 pptv @ 40 sec (0.3%)	37 ppbv @ 0.06 sec (2%)	9 ppbv @ 0.1 sec (0.5%)	300 ppbv @ 1 sec (0.08%)
$(\alpha L)_{min}$ @ Δf=1 Hz	$5 \cdot 10^{-7}$	$1 \cdot 10^{-6}$	$2.7 \cdot 10^{-4}$	$1.5 \cdot 10^{-5}$	$1 \cdot 10^{-4}$

Rapid progress has been reported in quantum cascade lasers and these lasers appear to offer the prospect of significantly higher cw-power required for quantum limited multipass systems. With a laser power of a few hundred mW a quantum limited performance is feasible together with the improvements in SNR according to the square root relationship mentioned before. Additionally, the pathlength could easily be extended and the reported detection limits would scale accordingly. For applications, where shot noise limited sensitivities are not required, an increase in signal-to-noise ratio can be used to simplify signal processing, allow less maintenance and, therefore, help to reduce operational cost. An increasing number of spectroscopic measurements with quantum cascade lasers have been reported and the commercial availability of these lasers will promote the development of new operational systems that allow new sensitive measurements based on the strong fundamental IR transitions.

TDLAS has made the transition from a technique mainly of interest to instrument developers into one which produces results of real value to industrial gas analysis and atmospheric research. The near- and mid-infrared spectral regions will provide complementary systems. For a limited number of species, where ultra-high sensitivity is not required, the near-infrared systems will provide advantages of size, simplicity and cost. For other species, requiring a more universal and sensitive system, mid-infrared lasers will continue to provide a highly specific device to meet the requirements of current and future measurement challenges.

References

Beck M, Hofstetter D, Aellen T, Faist J, Oesterle U, Ilegems M, Gini E, Melchior H (2002) Continuous wave operation of a mid-infrared semiconductor laser at room temperature. Science 295:301-305

Brassington DJ (1995) Tunable diode laser absorption spectroscopy for the measurement of atmospheric species. In : Clark RJH, Hester RE (eds) Spectroscopy in environmental science, Wiley, New York

D'Amato F, De Rosa M (2002) Tunable diode-lasers and two-tone frequency modulation spectroscopy applied to atmospheric gas analysis. Optics & Lasers in Eng 37:533-551

Fischer H, Wienhold FG, Hoor P, Bujok O, Schiller C, Siegmund P, Ambaum M, Scheeren HA, Lelieveld J (2000) Tracer correlations in the northern high latitude lowermost stratosphere : Influence of cross-tropopause mass exchange. Geophys Res Lett 27:97-100

Fried A, Sewell S, Henry B, Wert B, Gilpin T, Drummond JR (1997) Tunable diode laser absorption spectrometer for ground-based measurements of formaldehyde. J Geophys Res 102:6253-6266

Faist J, Capasso F, Sivco DL, Sirtori C, Hutchinson AL, Cho AY (1994) Quantum cascade laser. Science 264:553-555

Herriott DR, Schulte HJ (1965) Folded optical delay lines. Appl Opt 4:883-889

Hofstetter D, Faist J, Bour DP (2000) Mid-infrared emission from InGaN/GaN-based light emitting diodes. Appl Phys Lett 76:1495-1497

Hvozdara L, Gianordoli S, Strasser G, Schrenk W, Unterrainer K, Gornik E, Murthy Ch, Kraft M, Pustogov V, Mizaikoff B (2000) GaAs/AlGaAs quantum cascade laser – a source for gas absorption spectroscopy. Physica E 7:37-39

Köhler R, Gmachl K, Tredicucci A,Capasso F, Sivco DL, Chu SNG, Cho AY (2000) Single mode tunable, pulsed and continuous wave quantum cascade distributed feedback lasers at $\lambda \cong 4.6$-$4.7\ \mu m$. Appl Phys Lett 76:1092-1094

Kormann R, Müller H, Werle P (2001) Eddy flux measurements of methane over the fen Murnauer Moos, 11°11'E, 47°39'N, using a fast tunable diode laser spectrometer. Atmos Env 35:2533-2544

Kosterev AA, Curl RF, Tittel FK, Gmachl C, Capasso F, Sivco DL, Baillargeon JN, Hutchinson AL, Cho AY (1999) Methane concentration and isotopic composition measurement with a mid-infrared quantum-cascade laser. Opt Lett 24:1762-1764

Kosterev AA, Tittel FK (2002) Chemical sensors based on quantum cascade lasers, IEEE J of Quant Elect 38:582-591

Namjou K, Cai S, Whittaker EA, Faist J, Gmachl C, Capasso F, Sivco DL, Cho AY (1998) Sensitive absorption spectroscopy with a room-temperature distributed-feedback quantum-cascade laser. Opt Lett 23:219-221

Nicolas JC, Baranov AN, Cuminal Y, Roillard Y, Alibert JC (1998) Tunable diode laser absorptions spectroscopy of carbon monoxide around 2.35μm, Appl Opt 37:7906-7911

Rothman LS, Gamache RR, Tipping RH, Rinsland CP, Smith MAH, Brenner DC, Malathy Devi V, Flaud JM, Camy-Peyret C, Perrin A, Goldman A, Massie ST, Brown LR, Toth RA (1992) Hitran molecular database. J Quant Spectrosc Radiat Transfer 48:469-507

Schiff HI, Mackay GI, Bechara J (1994) The use of tunable diode laser absorption spectroscopy for atmospheric measurements. In : Sigrist MW (ed) Air monitoring by spectroscopic techniques, Wiley, New York

Schrenk W, Finger N, Gianordoli S, Hvozdara L, Strasser G, Gornik E (2000) GaAs / Al-GaAs distributed feedback quantum cascade lasers, Appl Phys Lett 76:253-255

Sharpe SW, Kelly JF, Hartman JS, Gmachl C, Capasso F, Sivco DL, Baillargeon JN, Cho AY (1998) High-resolution (Doppler-limited) spectroscopy using quantum cascade distributed-feedback lasers. Opt Lett 23:1397-1398

Tacke M (1995) New developments and applications of tunable IR lead-salt lasers. Infrared Physics and Technology 36:447-463

Tredicucci A, Gmachl C, Wanke MC, Capasso F, Hutchinson AL, Sivco DL, Chu SNG, Cho AY (2000) Surface plasmon quantum cascade lasers at $\lambda \cong 19$ μm. Appl Phys Lett 77:2286-2288

VDI Kompetenzfeld Optische Technologien (2002) Applications and trends in optical analysis technology - VDI Berichte 1667. VDI Verlag, Düsseldorf

Webber ME, Claps R, Englich FV, Tittel FK, Jeffries JB, Hanson RK (2001) Measurements of NH_3 and CO_2 with DFB diode lasers near 2.0 μm in Bioreactor vent gases. App Opt 40:4395-4403

Webster CR, Flesh GJ, Scott DC, Swanson JE, May RD, Woodward WS, Gmachl C, Capasso F, Sivco DL, Baillargeon JN, Hutchinson AL, Cho AY (2001) Quantum cascade laser measurements of stratospheric methane and nitrous oxide. Appl Opt 40:321-326

Werle P (1998) A review of recent advances in semiconductor laser based gas monitors. Spectrochimica Acta A54:197-236 with 197 references

Werle P, Slemr F (1991) Signal-to-noise ratio analysis in laser absorption spectrometers using optical multipass cells. Appl Opt 30:430-434

Werle P, Lechner S (1999) Stark-modulation-enhanced FM-Spectroscopy. Spectrochimica Acta A55:1941-1955

Werle P, Popov A (1999) Application of Antimonide lasers for gas sensing in the 3-4 μm range. Appl Opt 38:1494-1501

Werle P, Kormann R (2001) Fast chemical sensor for eddy correlation measurements of methane emissions from rice paddy fields. Appl Opt 40:846-858

Werle P, Mücke R, Slemr F (1993) Limits of signal averaging in atmospheric trace gas monitoring by tunable diode laser absorption spectroscopy. Appl Phys B57:131-139

Werle P, Scheumann B, Schandl J (1994) Real time signal-processing concepts for trace-gas analysis by diode-laser spectroscopy. Opt Eng 33:3093-3105

Werle P, Slemr F, Gehrtz M, Bräuchle Chr (1989) Quantum-limited FM-spectroscopy with a lead-salt diode-laser. Appl Phys B49:99-108

Werle P, Mücke R, D'Amato F, Lancia T (1998) Near-infrared trace-gas sensors based on room-temperature diode lasers. Appl Phys B67:307-315

Werle P, Slemr F, Maurer K, Kormann R, Mücke R, Jänker B (2002) Near- and mid-infrared laser-optical sensors for gas analysis. Optics & Lasers in Eng 37:101-114

White JU (1976) Very long optical paths in air. J. Opt.Soc. Am. 66:411-416

Williams RM, Kelly JF, Hartmann JS, Sharpe SW, Taubmann MS, Hall JL, Capasso F, Gmachl C, Sivco DL, Baillargeon JN, Hutchinson AL, Cho AY (1999) Kilohertz linewidth from frequency stabilized mid-infrared quantum cascade lasers. Opt Lett 24:1844-1846

Ye J, Ma LS, Hall JL (1997) Ultrasensitive detections in atomic and molecular physics : demonstration in molecular overtone spectroscopy. J Opt Soc Am B 15:6-14

Zahniser MS, Nelson DD, McManus JB, Kebabian PL (1995) Measurement of trace gas fluxes using tunable diode laser spectroscopy. Phil. Trans. R. Soc. Lond. 351:371-382

Part IV

Applications in Life Science

12 Laser Analytics of Gas Samples in Life Science

Manfred Mürtz and Peter Hering

12.1 Introduction

Lasers have found a wide-spread application in life sciences, in particular in the field of biomedical research and clinical diagnostics. To date, these applications mainly involve laser-based instruments for imaging purposes or for therapeutic use, the latter exploiting the thermal or ablative effect of laser radiation interacting with biological tissue. However, modern laser systems get more and more useful also for analytical purposes in biomedical research. This contribution is intended to introduce the particular demands, advantages and problems of laser-based analytical techniques in life sciences, and to discuss in particular the medical and clinical aspects.

In the past decade, an increasing number of publications dealt with the investigation of biogenic trace gases. In particular, the role of volatile diseasemarkers released by the human body have gained growing interest. The quantitative analysis of exhaled breath can provide important information about the health status of a living subject. Sampling and analysis of breath is preferable to a direct measurement of the metabolites from blood samples because it is non-invasive, and the measurements are much simpler in the gas phase than in a complex biologic fluid, like blood or urine. Current breath tests involve, e. g., the analysis of carbon dioxide (13C breath test) or hydrogen (Lactose intolerance test).

Besides the major components, like carbon dioxide and water, exhaled human breath contains several hundred volatile species of endogenous origin. Most of them are present in volume fractions on the order of one part per billion (ppb) or lower. A few years ago, it has been found that trace gases like nitric oxide, carbon monoxide and various hydrocarbons are formed in the human organism. Some of these exhaled volatile species are considered to be diseasemarkers. For example, ethane and pentane are potential markers of lipid peroxidation; nitric oxide is considered to be an important marker for airway inflammation. There is strong evidence that the analysis of these and other trace constituents in exhaled breath could provide a new way of non-invasive monitoring of inflammation, oxidative stress and other processes in the airways and lungs. Also, various exhaled volatiles which are detectable in exhaled breath due to inhalation of polluted air are interesting as markers for exposition to toxic compounds. The non-invasive nature of the measurement of exhaled markers makes breath tests ideally suited for the serial monitoring of patients.

The development of rapid and sensitive analysis techniques for measurements of relevant volatile compounds released, e. g., in exhaled breath or from the skin, is still a challenge. The performance of laser spectroscopic methods for analytical purposes has made significant progress in the past ten years. The excellent properties of laser radiation – like spectrally narrow emission and low intensity fluctuations – enabled the development of analytical techniques with high sensitivity and specificity. Laser spectroscopic methods have already proven to be powerful tools for ultra-sensitive analysis of trace compounds in gaseous samples.

Milestones in laser technology progress during the past decade include the development of compact and high-performance laser sources for the infrared spectral region and the development of novel ultra-sensitive and precise laser spectroscopy methods. Recently, novel non-linear optical materials such as periodically poled lithium niobate (PPLN), and frequency-stable single-mode diode lasers in the near-infrared spectral region, became commercially available. Modern laser-based analytics allows measurements with unprecedented sensitivity and specificity; trace gas fractions down to sub-ppb levels can be quantitatively analysed in few seconds. The advances in laser technology and laser-based analytical methods open up a number of interesting new applications in biomedical research and diagnostics which will be in the focus of this contribution.

This contribution is organized as follows: In Sect. 2, an introduction to the various origins of biogenic gas samples is given. The particular properties and problems of laser-assisted analytical instrumentation in life sciences are discussed in Sect. 3. The advantages and limitations of laser analytics in biomedical research and diagnostics are discussed in comparison to alternative techniques. This Sect. focusses on laser-spectroscopic analytical methods; laser-assisted non-spectroscopic techniques like laser-assisted mass spectrometry or laser-assisted gas chromatography are not discussed here. In the final Sect. 4, an overview over current and potential applications of human breath tests is given.

12.2 Sources of Biological Gas Samples

A most interesting biogenic source of gas samples is exhaled human breath. Many volatile metabolites carried by the blood pass the alveolar-capillary membrane and can thus be found in exhaled breath. The various origins of volatile compounds in breath are described in the following. Apart from exhaled breath, there exist many other sources of volatile biogenic samples. Of medical interest are, for example, emissions from the skin or from the gastro-intestinal system. Moreover, bacteria and plants do exhibit volatile emissions which carry important information. These topics are briefly outlined here.

12.2.1 Composition of exhaled breath

Besides the main constituents nitrogen, oxygen, and water, exhaled human breath contains a number of volatile metabolites formed in the organism. The major metabolite in breath is carbon dioxide (CO_2) which is exhaled in volume fractions of about 4 percent. Though CO_2 is not a diseasemarker itself it plays an important role for breath testing when ^{13}C-labelled pharmaceuticals are applied. The role of such tracers for medical diagnostics is described in more detail below. Also, the contribution by Paldus *et al.* (Chap. 15) deals with a prominent example of isotopic breath testing, the non-invasive verification of a *Helicobacter-pylori* infection in the gastro-intestinal tract by means of a $^{13}CO_2$ breath test.

Additionally, a number of components can be found in breath that are present in very low fractions, typically in the low ppb region. Many of these trace gases are formed endogenously, for example nitric oxide (NO) and carbon monoxide (CO). Such volatile compounds originate in the organism from various metabolism processes. The quantitative measurement of these exhaled trace gases potentially provides important information about the physiological status and metabolic disorders of the organism. It is well-known for long that certain diseases are accompanied with a specific odour of exhaled breath; for example, the odour of acetone is related to the severe stage of diabetic metabolism (keto-acidosis) and the smell of sulfur compounds indicates liver impairment.

Table 12.1. Important endogenous trace gases found in exhaled breath and their average fraction in the breath of healthy humans (Phillips 1992; Risby and Sehnert 1999; Knutson et al. 1999; Hyspler et al.2000; Kharitonov and Barnes 2001).

Trace gas	Average fraction
Methane (CH_4)	0 – 10 ppm
Ethane (C_2H_6)	0 – 10 ppb
Pentane (C_5H_{12})	0 – 10 ppb
Nitric Oxide (NO)	1 – 20 ppb
Carbon Monoxide (CO)	0.5 – 5 ppm
Carbonyl Sulfide (OCS)	50 – 100 ppb
Nitrous Oxide (N_2O)	1 – 20 ppb
Isoprene (C_5H_8)	50 – 200 ppb
Ammonia (NH_3)	0 – 1 ppm
Acetone ($(CH_3)_2CO$)	0 – 1 ppm

Table 12.1 summarizes a number of important endogenous trace gases found in exhaled breath. Many other compounds, in particular many volatile organic compounds (VOCs) have been found up to now, however many of these VOCs which are detectable in exhaled air are of exogenous origin and have just been inhaled with polluted ambient air. The presence of such compounds in breath does not indicate a disease but may act as an indicator of recent exposure to these gases.

In the following the most important trace constituents found in exhaled breath are briefly characterized.

Nitric oxide

The most prominent diseasemarker in exhaled human breath is nitric oxide (NO). Little more than a decade ago, nitric oxide was mainly regarded as a noxious gaseous component of air pollution. Since that time, intense basic and clinical investigation has revealed that nitric oxide is produced by a variety of human tissues. Nitric oxide is now known to be a central mediator in biological systems (Arnal et al. 1999; Ignarro et al. 1999). Endogenous nitric oxide is derived from L-arginine by the enzyme NO synthase (NOS), of which at least three distinct isoforms exist (c-NOS, e-NOS, i-NOS) (Moncada et al. 1991; Culotta and Koshland Jr. 1992; Dillon et al. 1996).

The presence of endogenous nitric oxide in exhaled breath of animals and humans was first described in 1991 (Gustafson et al. 1991). Nitric oxide is formed in both, upper and lower respiratory tract, and diffuses into the lumen, thus conditioning exhaled gas with nitric oxide (Barnes 1993). The average fraction of exhaled nitric oxide (eNO) is generally in the low ppb region. The levels of nitric oxide originating from the upper respiratory tract are a hundred-fold higher than exhaled nitric oxide measured in the lower respiratory tract (1 to 9 ppb) (Borland et al. 1993; Dillon et al. 1996). This fact is mostly due to its high production in human paranasal sinuses.

Breath nitric oxide production has been attributed variably to the pulmonary vascular endothelium, the epithelium of small airways, the nose, and the paranasal sinuses (Dillon et al. 1996). Currently, it is believed that eNO is likely to be of epithelial rather than of endothelial origin, and most eNO is derived from airways rather than from alveoli (Kharitonov and Barnes 2001). The absence of a correlation between oral breath NO levels and plasma and urine NO_2^- and NO_3^- concentrations suggests that eNO reflects local rather than systemic nitric oxide production (Dillon et al. 1996).

Nitric oxide is the most extensively studied exhaled marker and abnormalities in eNO have been documented in several lung diseases, particularly asthma (Kharitonov 1999).

Volatile organic compounds. It is known for many years that exhaled human breath contains a mixture of several hundred volatile organic compounds (VOCs) (Phillips 1992). VOCs found in breath are predominantly blood-borne and are exhaled via the blood/breath interface in the lungs. Among the various hydrocarbons found in breath, alkanes (ethane, pentane, etc.) have been extensively studied since they are end-products of the oxidative degradation of polyunsaturated fatty acids (lipid peroxidation). Lipid peroxidation is related to oxidative stress which is the imbalance of oxidative chemical reactions and anti-oxidative defense in the organism in favour for the oxidative processes. It has been reported that the fraction of ethane and pentane in exhaled breath allows estimation of the magnitude of in-vivo lipid peroxidation. The first report of breath ethane as a marker of in-vivo lipid peroxidation was published in 1974 by Riely et al. (Riely et al. 1974). Since then, a number of studies have shown increased exhalation of breath alkanes in response to various oxidant stresses (Andreoni et al. 1999; Aghdassi et al. 2000). Increased fractions of exhaled ethane (C_2H_6) and/or pentane (C_5H_{12}) have been observed subsequent to hyperbaric oxygen, cigarette smoking, total body irradiation, and acute aerobic exercise (Knutson et al. 2000).

Another major breath hydrocarbon is isoprene (C_5H_8). Isoprene is a by-product of cholesterol synthesis (Hyspler et al. 2000). Its value as a diseasemarker is uncertain yet. Other hydrocarbons such as methane, propane and are mainly derived from the fecal flora.

Carbon monoxide (CO)

Carbon monoxide is generally considered just to be a noxious gaseous component of air pollution, mainly originating from traffic and industry. However, it is well-known for many years that CO is endogenously formed (Coburn 1970). Exhaled carbon monoxide (eCO) originates from three major sources: enzymatic degradation of heme, non–heme-related release (lipid peroxidation, xenobiotics, bacteria) and exogenous CO (Kharitonov and Barnes 2001). The predominant endogenous source of carbon monoxide (85%) in the body is from the degradation of hemoglobin by the enzyme heme oxygenase (HO). CO is a by-product of oxidative cleavage of hemoglobin by heme oxygenase (HO), which exists in three isoforms (HO-1, HO-2, HO-3). Heme is converted by HO to biliverdin and then to bilirubin, with the formation of CO and ferritin. The HO-1 isotype is widely distributed and rapidly induced by various stress associated stimuli (cytokines, bacterial toxins, hyperoxia, etc.). HO-1 is considered to protect against oxidative stress; however, the precise mechanisms for this protection are not fully understood (Kharitonov and Barnes 2001).

It is currently assumed that the alveoli are the predominant site of eCO in normal subjects. There is also a small contribution of carbon monoxide derived from the airways, in particular from the nose and and the paranasal sinuses.

The exhaled breath of smokers contains an increased fraction of carbon monoxide as compared to non-smokers. So, the measurements of eCO allows conclusions about the smoking status of the tested person (cf. Paldus *et al.* in this book). Furthermore, exhaled CO measurements are discussed as a means to monitor bilirubin production, including hyperbilirubinemia in newborns (Kharitonov and Barnes 2001).

Other volatiles

Exhaled formaldehyde from women with breast cancer and in the tumor-bearing mice is significantly higher than in healthy subjects, suggesting that these carbonyl compounds may be used as a biomarker (Ebeler et al. 1997). Volatile sulfur compounds like, e.g., carbonyl sulfide (OCS), may be characteristic for hepatic diseases (Risby et al. 2001).

A big number of other molecules have been found in expired air, however there is little information about the excretion of these compounds in the breath; in some cases it is not even clear whether they are endogenous or not.

12.2.2 Other biological sources of gaseous emissions

Biogenic emissions of volatile compounds can be found in a wide variety of biologic systems. For example, it is well-known that the gastro-intestinal tract of animals and humans is a rich source of volatile emissions, mostly originating from metabolization of the food and from the colonic flora. Predominant compounds are methane and sulfur compounds.

Plants emit a large number of volatile compounds; these emissions provide an interesting insight into several physiological processes of plants. For example, ethylene is an important plant hormone which plays an major role during fruit ripening, etc. (Harren et al. 1998). Also, fermentation of fruits, yielding in particular ethanol and acetaldehyde emissions, has been studied. The contribution by Kühnemann (Chap. 16) describes this interesting field in more detail.

Apart from the *in-vivo* measurements, numerous studies investigated volatile emissions from biological samples *in vitro*. For example, in (Morimoto et al. 2001; Kosterev et al. 2002), the biogenic carbon monoxide production rate above cultures of vascular cells has been observed. In that work, an extractive technique was used with gas samples taken from the flask containing the cell culture.

It should be noted that volatile metabolites from animals or humans are also excreted across the skin. For example, Harren and co-workers observed elevated ethylene emissions from the skin after application of UV radiation (Harren 2001).

Table 12.2 summarizes some examples of biogenic emissions.

Table 12.2. Selected examples for biological sources of volatile emissions other than exhaled breath.

Biological source	Volatile emissions
Microorganism (Bacteria, Macrophages)	Methane, Carbon monoxide, Nitric oxide
Plants	Ethylene, Isoprene, Ethanol, Acetaldehyde
Skin of humans and animals	Ethylene
Aortic vascular cells of rats	Carbon monoxide
Gastro-intestinal tract of animals & humans	Methane, Dihydrogensulfide

12.3 Instrumentation for Laser Analytics of Breath and Other Biological Gas Samples

At first glance, the analysis of biological or medical gas samples is not too much different from the analysis of atmospheric air. However, there are some differences that must be paid particular attention. The acquisition of a biogenic gas sample may be much more complicated than the acquistion of an atmospheric air sample. This is obvious in the case of breath samples: the fractions of different volatiles may depend on many parameters, like the exhalation flow rate, etc. and the fractions may change rapidly during the exhalation. Thus, sample preparation and storage must be performed extremely carefully, in particular for analysis of trace constituents in exhaled breath. This topic as well as a number of laser spectroscopic techniques that have proven to be suitable and advantageous for biomedical trace gas analysis are discussed in the following.

12.3.1 Sample collection and preparation

Online vs Offline analysis

The term "online measurement" refers to exhaled breath testing with a real-time display of exhaled gas fractions, whereas „offline measurement" refers to the collection of exhalate into suitable containers for subsequent analysis.

Online methods are characterized by continuous gas sampling; the resulting concentration profile versus time, together with other variables (e. g., airway flow during breath tests) are captured and displayed without significant delay. This allows real-time monitoring of the gas sample; in the case of breath tests, for example, sub-optimal exhalations can be immediately identified and discarded (Baraldi et al. 2000).

Offline methods perform a delayed analysis on gas samples collected in a reservoir. As compared with online techniques, offline collection offers the potential for sample collection at remote sites (e.g. at the bedside in an hospital), and offline collection is independent from the analyzer response times (Paredi et al. 1998; Silkoff et al. 1999). Potential problems with offline methods include error introduced by sample storage, and an inability to allow for instantaneous feedback. Particular attention must be paid to the storage of a breath sample. The reservoir used to collect the exhaled gas sample must be non-reactive and relatively impermeable. Suitable materials include Tedlar and Mylar. It has been demonstrated that new Tedlar or Mylar balloons allow for sample stability for at least 48 hours (American Thoracic Society 1999). Another possibility is the use of evacuated electro-polished canisters (Pleil and Lindstrom 1995). All these sample collection methods show problems: Electro-polished canisters are expensive; bags are potential sources for contamination, depending on the plastic and valves used

Contamination with ambient air during breath testing

A potential problem of both online and offline methods is the contamination of the biological gas sample with ambient air. If the trace gas fractions of interest in the inhaled ambient air are of comparable or even larger magnitude than in the biological sample, contamination with even small amounts of ambient air may render the results unreliable._Currently there exist two approaches to deal with ambient air contaminations. First, a wash-out procedure (approx. 5 min) is applied. The wash-out period can serve to flush ambient-air ethane and pentane from the lungs. Second, the local ambient trace gas levels are recorded and subtracted from the levels in exhaled breath.

Special care must be taken with the measurements of exhaled hydrocarbons. Hydrocarbons present in ambient air are inhaled and may be retained in various body compartments (body fat). Although it takes only a few minutes to wash out the lungs, it may require 90 minutes or longer to wash out the body stores of hydrocarbons (Kharitonov and Barnes 2001).

Flow dependence and dead-space contributions

Exhaled trace gas concentrations may exhibit significant expiratory flow dependence. This problem is particularly important for measurements of exhaled nitric oxide. This flow dependence of eNO fractions has been attributed to faster flows minimizing the transit time of alveolar gas in the airway. Mixing during exhalation between the NO produced by the alveoli and by the conducting airways, explains the flow rate and breath-hold dependencies. The rate of nitric oxide output is greater at higher flow rates, but not in direct proportion. In view of this flow dependence, it is recommended to enforce a constant expiratory flow rates and to monitor the flow rate during the sample exhalation. The flow rate recommended by the American Thoracic Society is 50 ml/s (American Thoracic Society 1999). Additionally, exhaled nitric oxide is usually determined during single breath exhalations against a resistance to avoid nasal contamination of the sample as mouth pressure falls off and the velum opens. This is particularly important for eNO measurements since nasal nitric oxide has high concentrations relative to the lower respiratory tract, which can be excluded by exhalation against a resistance.

For some measurements it may be useful to discarde the dead space during the first part of exhalation. Paredi *et al.* have reported an offline system that incorporates discardment of the initial portion (dead space) of the exhalate (Paredi et al 1998). It should be noted that exhaled CO levels, for example, are less flow rate or breathhold-dependent than exhaled NO, suggesting less airway contribution.

Preconcentration of breath samples

For the analysis of VOCs in breath, the breath samples are usually offline collected, accumulated, and concentrated up to hundred-fold by means of a sorbent trap, since hydrocarbon concentrations in expired air are below the detection limit of most analytical methods. Such trap-and-purge techniques are commonly used before gas-chromatographic analysis of the sample. The trapped components of the collected breath sample are separated in a suitable gas chromatographic column and are then analyzed via mass spectrometry. The sorbent traps have to be carefully conditioned before usage and the chromatographic column must be very carefully selected and prepared (Ebeler et al 1997; Knutson et al. 1999; Hyspler et al. 2000).

This trap-and-purge process plus the GC/MS analysis is very time consuming; usually, it takes one hour to complete the analysis of a single breath sample. Another drawback of this procedure is, that under several circumstances the measurement is prone to errors, as Kneepkens *et al.* pointed out (Kneepkens et al. 1994).

12.3.2 Laser spectroscopic techniques

Most of the laser spectroscopic techniques developed and used for environmental analytical investigations can be adapted to biomedical problems. A number of methods that have been developed for the sensitive in-situ detection of atmospheric trace gases can also be applied to the analysis of exhaled human breath.

Table 12.3. Performance of selected methods applied to gas analysis in life sciences. The compounds given in the second column are selected examples.

Method	Well-suitable for analysis of	Sensitivity	Specificity	Isotope selective	Suitable for online analysis
Laser spectroscopy					
Direct absorption	CH_4, C_2H_6, CO, NO	O	+	+	yes
Photoacoustic	C_2H_6, C_5H_8	+	O	+	Yes
Laser magnetic resonance	NO	+	+	+	Yes
Cavity enhanced	CH_4, C_2H_6, CO, NO	+	+	+	Yes
Other					
GC/MS	VOCs, CO	-	O	-	No
PTRMS, SIFT-MS	C_5H_8	+	+	+	Yes
Electro-chemical	CO	O	-	-	No
NDIR	CO, CO_2, NO	O	+	+	Yes
Chemoluminescence	NO, NO_2	+	+	-	Yes

VOCs: Volatile Organic Compounds, GC/MS: Gas Chromatography/Mass Spectrometry, PTRMS: Proton-Transfer-Reaction Mass Spectrometry, SIFT-MS: Selected Ion Flow Tube Mass Spectrometry, NDIR: Non-Dispersive Infrared.

Spectroscopy in the mid-infrared spectral region is most advantageous since most of the gaseous compounds of biomedical interest are molecular gases that have characteristic, strong ro-vibrational absorption bands in this spectral region. For example, nitric oxide and carbon monoxide have strong fundamental absorption bands in the wavelength region near $\lambda = 5$ µm; for hydrocarbons the most interesting wavelength region is around 3 µm where strong absorption lines according to the CH stretching vibration are located. Infrared spectroscopy of these fingerprint spectra allows sensitive, specific and rapid monitoring of gas mixtures.

Infrared laser sources

Laser spectroscopy generally requires laser sources with outstanding spectral properties, i. e., narrow linewidth, wide wavelength tunability and smooth beam profile. Suitable coherent light sources for mid-infrared spectroscopy are molecular gas lasers, semiconductor lasers, and recently all-solid-state pumped nonlinear conversion devices. In the following, these laser sources are briefly characterized in terms of suitability for laser spectroscopy applied to life sciences.

The CO laser is a line-tunable molecular gas laser which is based on laser transitions between vibrational-rotational transitions in the electronic ground state of the CO molecule. It provides several hundred laser lines in the wavelength region between 4.8 and 8.3 µm (Urban 1995). Moreover, a CO laser can be operated on overtone transitions providing additional laser lines in the wavelength region between 2.6 and 4.0 µm (Bachem et al. 1993). Though molecular gas lasers generally lack continuous tunability, they are powerful instruments for high-resolution infrared spectroscopy due to their relatively narrow linewidth (typically below 1 MHz). Improved tunability of the laser wavelength can be achieved by means of an electro-optic modulator. By mixing the CO laser radiation with microwave radiation in an CdTe crystal tunable microwave sidebands in the range between 8 and 18 GHz above and below each laser line are obtained, providing about 0.1 mWatt of tunable laser radiation (Mürtz et al. 1998). The disadvantage of the CO laser is that the entire laser system is large in size, cannot be handled with ease, and hence is not suitable for usage at clinical locations.

Non-linear frequency conversion is an interesting technique to generate laser radiation in the mid-infrared spectral region. Considerable progress has been made with difference frequency generation (DFG) using the interaction of two near-infrared lasers in a non-linear crystal. A prototype of a portable difference- frequency laser system for trace gas monitoring has been reported by Tittel and co-workers (Lancaster et al. 1998). Moreover, a cw optical parametric oscillator (cw-OPO) has recently been developed that proved to be suitable for trace gas analysis (Kühnemann et al. 1998).

The quantum cascade laser (QCL) is a novel source of coherent light in the mid-infrared spectral region. It combines high output power (up to 10 mW in cw mode below T=100 K) with continuous tunability and narrow linewidth. First results on QCL-based trace gas monitoring have recently been reported (Paldus et al. 2000; Menzel et al. 2001; Kosterev et al. 2001).

Tunable lead-salt diode lasers (TDLs) are widely used for molecular spectroscopy in the mid-infrared. TDLs have many practical advantages including easy wavelength sweeping through the control of the laser diode temperature and drive current. Portable TDL spectrometers for trace gas analysis are commercially available since more than ten years. The major disadvantages of TDLs is the need for cryogenic cooling. More details on TDLs can be found in the contribution by Werle (Chap. 11).

Laser spectroscopic techniques

Of the great variety of laser spectroscopic techniques there are a number of methods which have been demonstrated for the analysis of gas samples in life sciences. Most techniques are based on the principle of absorption spectroscopy: The gas sample of interest is transferred into an absorption cell and the wavelength-dependent attenuation of the laser light which passes the cell is measured.

TDL absorption spectroscopy has been applied to the measurement of exhaled carbon monoxide and nitric oxide (Lee et al. 1991; Stepanov et al. 1993; Lee et al. 1994; Moskalenko et al. 1996; Chuchalin et al. 1999; Giubileo et al. 2001; Roller et al. 2002). In most cases, a multipass absorption cell is utilized which allows multiple passes of the laser beam through the gas sample. In this way, a detection sensitivity for CO on the ppb level is achieved.

Cavity ring-down spectroscopy (CRDS) is a very sensitive technique developed in the past decade. The chapter by Paldus *et al.* (Chap. 15) reviews this technique and describes applications to medical breath testing. Our group has recently demonstrated real-time monitoring of ethane in exhaled human breath by using cavity leak-out spectroscopy (CALOS), a cw variant of cavity ring-down spectroscopy. The spectrometer is based on a tunable CO laser and a high-finesse cavity (length: 50 cm) which provides an optical absorption pathlength of 3.6 km. A minimum detectable absorption of 10^{-9}/cm with a 10 s integration time has been reported (Kleine et al. 2000, Dahnke et al. 2001). This corresponds to sub-ppb fractions of ethane and other relevant trace gases. More details can be found in the contribution by Dahnke *et al.* (Chap. 14).

Faraday laser magnetic resonance spectroscopy (LMRS) is a sensitive, specific and isotope-selective technique which exploits the property of magnetic molecules to rotate light polarization in a magnetic field. The contribution by Urban (Chap. 13) describes this method in more detail. This method allows to determine very low concentrations of gaseous free radicals with high time resolution. This technique has been successfully used to measure exhaled nitric oxide in healthy subjects (Mürtz et al. 1999).

Trace gas detection by means of laser-based photoacoustic spectroscopy (PAS) has proven to provide a sensitive and non-intrusive method of studying physiological processes in biological samples (Bijnen et al. 1996; Martis et al. 1998; Oomens et al. 1998; Dahnke et al. 2000). Using a photoacoustic cell placed inside a CO_2 laser cavity it became possible to detect the plant hormone ethylene (C_2H_4) at ppt levels (Fink et al. 1996; Harren et al. 1998). A continuous flow-through system at atmospheric pressure transfers the trace gases released from the plant chamber to the photoacoustic resonator cells at flow rates where various physiological processes can be studied with high time resolution. The contribution by Kühnemann (Chap. 16) describes the PAS technique in more detail.

Comparison with conventional techniques

A number of conventional analytical techniques exists for the analysis of biogenic gas samples. Most of them are suitable or optimized for one particular compound, like, e.g., the chemoluminescence technique for nitric oxide. In contrast, laser-spectroscopic analysis techniques are of universal character since they are applicable to a large number of different compounds.

Another advantage of laser spectrocopic techniques is their high specificity. Conventional analysis techniques often suffer from cross-interferences between different compounds in the gas mixture, whereas spectroscopic detection is almost free of such complications. For example, in the case of chemoluminescent detection of nitric oxide, there are some substances known besides nitric oxide that are volatile and generate a chemoluminescence signal (e.g., hydrogen sulfide, ethylene, propylene, sulfur dioxide, and dimethyl sulfide) (Dillon et al. 1996).

The standard technique used to quantify hydrocarbons from biogenic sources is gas chromatography combined with a mass spectrometer (GC/MS). Various methods for analysing ethane and pentane in expired air have been reviewed in (Knutson et al. 2000). To detect ppb trace gas levels lare amounts of the gas sample must be collected and preconcentrated (Kneepkens et al. 1994).

Mass spectrometry techniques with a 'soft' ionization method which leads to little or no fragmentation, like the proton-transfer-reaction mass spectrometry (Taucher et al. 1997) or the the „selected ion flow tube" mass spectrometry (Davies et al. 2001) have been demonstrated to allow sensitive detection of isoprene in exhaled breath.

Non-dispersive spectrometers mostly use a broadband infrared source emitting a continuum in the infrared and a detector cell filled with pure gas of interest. The isotopic selectivity is achieved by using two cells filled with isotopically pure gases. These devices are commercially available for $^{13}CO_2$ breath tests.

12.4 Application of Breath Tests

Non-invasive breath monitoring may assist in differential diagnosis of various diseases, assessment of disease severity and response to treatment. Because these techniques are completely non-invasive, they can be used repeatedly (to give information about kinetics), they can be used to monitor patients with severe disease and children, including neonates.

Currently, three groups of breath tests are used for medical diagnosis: First, the analysis of endogenous volatile diseasemarkers which are present in exhaled breath on the ppb level. Second, the analysis of the isotopic ratio $^{13}CO_2/^{12}CO_2$ in exhaled air after ingestion of a ^{13}C-labelled pharmaceutical. Third, the use of breath tests for the study of Lactose intolerance and related tests (H_2 tests) which will not be discussed here (see for example (King and Toskes 1983)).

12.4.1 Monitoring of endogenous volatile diseasemarkers in breath

Table 12.4. Endogenous diseasemarkers found in breath which exhibit potential for clinical use.

Molecule	Origin	Useful for monitoring of
NO	Inflammatory processes	Airway diseases, Asthma
Ethane, pentane	Lipid peroxidation	Oxidative stress status
CO	HO activity	Smoking habits
OCS	?	Hepatic diseases

Recently there has been growing interest in the use of exhaled breath analysis in the diagnosis and monitoring of lung diseases. Many lung diseases, including asthma, chronic obstructive pulmonary disease (COPD), bronchiectasis, cystic fibrosis, and interstitial lung disease, involve chronic inflammation and oxidative stress (Kharitonov and Barnes 2001). Although most studies have focused on exhaled nitric oxide, several other volatile compounds (carbon monoxide, ethane, pentane) are being investigated. The benefit of all these exhaled markers for routine clinical diagnosis is still controversial. To make breath tests reliable and usable in clinical medicine, several problems have to be solved:

1. Standardization: Although there are numerous publications on exhaled nitric oxide or exhaled hydocarbons, the publications about measurements of exhaled diseasemarkers are characterized by a variation in published exhaled levels in health and disease, much of which is attributable to the lack of standardization.
2. Evaluation: Potential applications must be evaluated in extensive clinical studies.

In the following, potential applications of exhaled diseasemarkers are briefly outlined.

Volatile organic compounds (VOCs)

Oxidative stress is a ubiquitous process related to various diseases (Sies 1997; Christen 2000). Numerous methods have been developed to measure oxidative stress and the related oxidative damage in tissues and cells. Lipid peroxidation has gained increasing interest as one of the important features of free-radical-induced damage in biology and medicine. During the process of peroxidation of polyunsaturated fatty acids, volatile hydrocarbons, like ethane and pentane are formed which can be detected in exhaled breath. Since peroxidation of polyunsaturated fatty acids is considered as the major, probably the only, endogenous source of the pentane and ethane in breath these compounds can serve as a specific marker for oxidative damage (Kneepkens et al. 1994). Several studies provide evidence that ethane and pentane in exhaled air are useful markers of in-vivo lipid peroxidation (Bilton et al. 1991; Andreoni et al. 1999; Risby and Sehnert 1999; Risby et al. 1999; Paredi et al. 2000).

Despite the growing number of reports on breath ethane and pentane there exist no widely accepted methods for collecting and analyzing expired air. As a result, most investigators have had to develop their own techniques. The lack of extensive validation of the different analysis techniques is very likely responsible for much of the striking variability in expired-air ethane and pentane values reported by different investigators. The large variability in normal adult values has been reviewed in (Kneepkens et al. 1994; Knutson et al. 1999).

In humans, breath alkane output is currently difficult to measure: sensitivity and accuracy are insufficient, subjects sometimes are breathing room air with variable ambient concentrations of alkanes (Knutson et al. 2000). Several attempts were made to standardize this non-invasive method of measuring breath alkane and validate it as a measure of lipid peroxidation in humans (Pryor et al. 1991; Aghdassi et al. 2000). The results suggest that breath alkanes are good markers for lipid peroxidation (Aghdassi et al. 2000).

Recently, a new breath test for the measurement of free radical activity in the body was suggested, called breath methylated alkane contour. This comprises the alveolar gradients of a wide spectrum of VOCs ranging from C2 to C20 alkanes plotted as function of carbon chain length (Phillips 2001). Also, a breath test was suggested for diagnosing the presence of lung cancer in a mammal, using a group of VOCs as biomarkers (Phillips et al. 1999).

Smoking increases exhaled hydrocarbons, in particular ethane and pentane. This effect may be related to oxidative damage caused by smoking; also, this effect may mirror the high concentrations of hydrocarbons in cigarette smoke (Habib et al. 1995; Miller et al. 1997; Habib et al. 1999). Increased levels of exhaled VOCs could be used as biochemical markers of exposure to cigarette smoke and oxidative damage caused by smoking.

Nitric oxide (NO)

Exhaled nitric oxide (eNO) has been proposed as a non-invasive marker of airway inflammation in asthma and other airway diseases (Alving et al. 1993; Arnal et al. 1997; Arnal et al. 1997; Hart 1999). The standard technique to analyse eNO concentrations is the chemoluminescence analyzer. eNO is high in untreated asthma and falls rapidly after treatment with inhaled corticosteroids (Silkoff et al. 1999). The diagnostic value of exhaled NO measurements to differentiate between healthy subjects and asthma patients has been analyzed with 90% specificity and 95% positive predictive value when an exhaled NO level of more than 15 ppb is used as a cutoff for asthma (Kharitonov and Barnes 2001; Dupont et al. 1999).

Carbon monoxide (CO)

Over the last 20 years, exhaled carbon monoxide (eCO) has been measured to identify current and passive smokers, to monitor patients after CO poisoning, to determine bilirubin production, including hyperbilirubinemia in newborns, and in the assessment of the lung diffusion capacity (Kharitonov and Barnes 2001).

eCO has been measured by a number of different techniques. Most of the measurements in humans have been made using electro-chemical or near-infrared (NDIR) sensors, as used for continuous monitoring of atmospheric carbon monoxide

Carbon monoxide is discussed as indicator for smoking habits A cutoff level of 6 ppm effectively separates non-smokers from smokers (Kharitonov and Barnes 2001). It is currently discussed that exhaled CO may be useful in non-invasive monitoring of pediatric asthma.

Carbonyl sulfide (OCS)

Very recently, it has been reported that the concentration of carbonyl sulfide found via GC/MS in exhaled breath of patients with liver diseases are significantly different from that found in the breath of normal subjects (Risby et al. 2001).

12.4.2 Use of stable isotope markers for medical and pharmaceutical research

Table 12.5. Selection of available ^{13}C breath tests (Wetzel and Fischer 2001).

13C-Test for monitoring of	Labelled pharmaceutical
Helicobacter pylori infection	^{13}C-Urea
Liver function and cirrhosis	^{13}C-Galactose
Liver function	^{13}C-Aminopyrine, ^{13}C-Phenylalanine
Fat malabsorption	^{13}C-Trioctanoin
Gastric emptying	^{13}C-Acetate

For the investigation of a specific metabolic reaction it is advantageous to apply substances that are labelled with a rare isotope. By monitoring the increase of metabolites that contain the rare isotope one can draw conclusions about the metabolic reaction rate etc. Such investigations are called tracer investigations. In early tracer investigations, radio isotopes, like ^{14}C, have been used for this purpose. Nowadays, radio isotopes are only used in special cases; the radio isotope ^{14}C has been almost completely replaced by the stable ^{13}C isotope (Krumbiegel 1991; de Meer et al. 1999).

Tracer techniques are of particular interest in such cases where the studied metabolic reaction yields a volatile compound which can be directly analysed in exhaled breath. For example, if the metabolic process to be studied yields CO_2, i.e. the ^{13}C-labelled substrate is metabolized to CO_2, then the exhaled $^{13}CO_2$ fraction will start to increase and this increase is usually measured through monitoring of the isotopic ratio $^{13}CO_2/^{12}CO_2$ of exhaled breath samples. Thus, the principle for all the ^{13}C breath tests is the same: The substrate contains one or more functional groups labelled with ^{13}C. A certain amount of a compound artificially labelled with ^{13}C is applied to the patient and the kinetics of the $^{13}CO_2$ fraction exhaled is monitored in a certain time period after application of the substrate.

^{13}C breath tests are now widely used for the diagnosis of various diseases. Many diseases may be diagnosed using such breath analysis, and a variety of substrates have been proposed or developed for medical diagnosis. Table 12.5 shows some examples. The most prominent example is the ^{13}C-urea test for *Helicobacter pylori* bacteria in the gastro-intestinal tract, a cause of ulcers and gastritis. The urea breath test has been in clinical practice for a considerable period of time as one of the most important non-invasive methods for verifying a *Helicobacter pylori* infection. The test exploits the hydrolysis of orally administered urea by the enzyme urease, which *H. pylori* produces in large quantities. Urea is hydrolyzed to ammonia and carbon dioxide, which diffuses into the blood and is excreted by the lungs. It is expected that many more such tests with associated substrates may be developed in the near future.

The common device used for the measurement of the isotopic ratio $^{13}CO_2/^{12}CO_2$ of exhaled breath samples is a non-dispersive infrared analyzer (Haisch et al. 1993, Steffen at al. 1993, Braden et al. 1994, Haisch et al. 1994a, Haisch et al. 1994b, Koletzko et al. 1995, Haisch et al. 1996). There have also been proposed analyzers based on laser spectroscopy as new carbon isotope analyzers (Cooper et al. 1993; Sauke et al. 1997; Tanahashi et al. 1998). Such analyzers are based on tunable diode lasers. Carbon dioxide has strong absorption lines in the vicinity of wavelengths of 4.3 μm, a wavelength which can be accessed with lead-salt TDLs. A novel approach using CRDS is described by Paldus *et al.* (see Chap. 15).

12.5 Conclusion and Perspectives

Laser-based analytical methods have proven to be very useful for the investigation of biogenic gas samples. Under certain circumstances laser analytics is superior to conventional analytical techniques, regarding sensitivity, specificity and/or time response. Particularly, this is the case for medical breath testing. Breath analysis is currently a research procedure, but there is increasing evidence that it may have an important place in the diagnosis and management of diseases in the future. The measurement of diseasemarkers in breath, being non-invasive, is well suited for routine use in research and clinical settings. This will drive the development of cheaper and more convenient analyzers, which can be used in a hospital, and then eventually to the development of personal monitoring devices for use by patients.

Laser spectroscopic techniques have an enormous potential to enable more accurate and reproducible online measurements of volatile trace constituents in human breath, opening up new perspectives for exhaled breath analysis. Thus, the investigation of exhaled breath and other biogenic gaseous samples with precise laser-based techniqes is a promising and developing field of research. However, it should be noted, that laser-based analytical devices are still not in a technological stage which allows a simple and robust use in an clinical environment.

There is a pressing need for the evaluation of these techniques in long-term clinical studies. Whether repeated measurements of exhaled markers will help in the clinical management of lung diseases needs to be determined by longitudinal studies relating exhaled markers to other measurements of asthma control. This is most advanced with measurement of exhaled NO, but it is still uncertain whether routine measurement of exhaled NO will improve the clinical control of asthma in a cost-effective way. The value of particular biomarkers will depend on the availability of reliable, fast, and inexpensive detector systems. Advances in technology will result in smaller devices that are cheaper and easier to use. This will increase the availability of the measurements, which will facilitate the evaluation of biomarkers in clinical settings. Moreover, new endogenous substances and markers may be detected in expired breath in the future.

References

Aghdassi E, Allard JP (2000) Breath alkanes as a marker of oxidative stress in different clinical conditions [Review]. Free Radic.Biol.Med. 28:880-886

Alving K, Weitzberg E, Lundberg JM (1993) Increased amount of nitric oxide in exhaled air of asthmatics. Europ.Respir.J. 6:1368-1370

American Thoracic Society (1999) Recommendations for standardized procedures for the online and offline measurement of exhaled lower respiratory nitric oxide and nasal nitric oxide in adults and children. Am.J.Respir.Crit.Care Med. 160:2104-2117

Andreoni KA, Kazui M, Cameron DE, Nyhan D, Sehnert SS, Rohde CA, Bulkley GB, Risby TH (1999) Ethane: a marker of lipid peroxidation during cardiopulmonary bypass in humans. Free Radic.Biol.Med. 26:439-445

Arnal JF, Didier A, M'Rini C, Charlet JP, Serrano E, Besombes JP (1997) Nasal nitric oxide is increased in allergic rhinitis. Clin.Exp.Allergy 27:358-362

Arnal JF, Dinh-Xuan AT, Pueyo M, Darblade B, Rami J (1999) Endothelium-derived nitric oxide and vascular physiology and pathology [Review]. Cell.Mol.Life Sci. 55:1078-1087

Bachem E, Dax A, Fink T, Weidenfeller A, Schneider M, Urban W (1993) Recent progress with the CO-overtone laser. Appl.Phys.B 57:185-191

Baraldi E, Scollo M, Zaramella C, Zanconato S, Zacchello F (2000) A simple flow-driven method for online measurement of exhaled NO starting at the age of 4 to 5 years. Am.J.Respir.Crit.Care Med. 162:1828-1832

Barnes PJ (1993) Nitric oxide and airways. Europ.Respir.J. 6:163-165

Bijnen FGC, Harren FJM, Hackstein JHP, Reuss J (1996) Intracavity CO laser photoacoustic trace gas detection; cyclic CH_4, H_2O and CO_2 emission by cockroaches and scarab beetles. Appl Opt 35:5357-5368

Bilton D, Maddison J, Webb AK, Seabra L, Jones M, Braganza JM (1991) Cystic fibrosis, breath pentane, and lipid peroxidation. Lancet 337:1420

Borland C, Cox Y, Higenbottam T (1993) Measurement of exhaled nitric oxide in man. Thorax 48:1160-1162

Braden B, Haisch M, Duan LP, Lembcke B, Caspary WF, Hering P (1994) Clinically feasible stable isotope technique at a reasonable price: analysis of 13CO2/12CO2-abundance in breath samples with a new isotope selective nondispersive infrared spectrometer. Gastroenerol. 32:612

Christen Y (2000) Oxidative stress and Alzheimer disease [Review]. Am.J.Clin.Nutr. 71:621S-629S

Chuchalin AG, Voznesenskiy N, Dulin K, Sakharova S, Soodaeva E, Stepanov EV (1999) Exhaled nitric oxide and exhaled carbon monoxide in pulmonary diseases. Am J Respir Crit Care Med 159:A410

Coburn RF (1970) Endogenous carbon monoxide production. N Engl J Med 282:207-209

Cooper DE, Martinelli RU, Carlisle CB, Riris H, Bour DB, Menna RJ (1993) Measurement of $^{12}CO_2$:$^{13}CO_2$ ratios for medical diagnostics with 1.6 μm distributed-feedback semiconductor diode lasers. Appl Opt 32:6727-6731

Culotta E, Koshland DE, Jr. (1992) NO news is good news. Science 258:1862

Dahnke H, Kahl J, Schüler G, Boland W, Urban W (2000) On-line monitoring of biogenic isoprene emissions using photacoustic spectroscopy. Appl.Phys.B 70:275-280

Dahnke H, Kleine D, Hering P, Mürtz M (2001) Real-time monitoring of ethane in human breath using mid-infrared cavity leak-out spectroscopy. Appl.Phys.B 72:971-975

Davies S, Spanel P, Smith D (2001) A new 'online' method to measure increased exhaled isoprene in end-stage renal failure. Nephrol Dial Transplant 16:836-839

de Meer K, Roef MJ, Kulik W, Jakobs C (1999) In vivo research with stable isotopes in biochemistry, nutrition and clinical medicine: an overview. Isotop.Env.Health Stud. 35:19-37

Dillon WC, Hampl V, Shultz PJ, Rubins JB, Archer SL (1996) Origins of breath nitric oxide in humans. Chest 110:930-938

Dupont LJ, Demedts MG, Verleden GM (1999) Prospective evaluation of the accuracy of exhaled nitric oxide for the diagnosis of asthma. Am.J.Respir.Crit.Care Med. 159:A861

Ebeler SE, Clifford AJ, Shibamoto T (1997) Quantitative analysis by gas chromatography of volatile carbonyl compounds in expired air from mice and human. J Chromatogr B Biomed Sci Appl 702:211-215

Fink T, Büscher S, Gäbler R, Yu Q, Dax A, Urban W (1996) An improved CO_2 laser intracavity photoacoustic spectrometer for trace analysis. Rev.Sci.Instrum. 67:4000-4004

Giubileo G, Fantoni R, de Dominicis L, Giorgi M, Pulvirenti R, SnelsM (2001) A TDLAS system for the diagnosis of helicobacter pylori infection in humans. Laser Physics Russia 11:154-157

Gustafson LE, Leone AM, Persson MGeal (1991) Endogenous nitric oxide is present in the exhaled air of rabbits, guinea pigs, and humans. Biochem.Biophys.Res.Commun. 181:852-867

Habib MP, Clements NC, Garewal HS (1995) Cigarette smoking and ethane exhalation in humans. Am.J.Respir.Crit.Care Med. 151:1368-1372

Habib MP, Tank LJ, Lane LC, Garewal HS (1999) Effect of vitamin E on exhaled ethane in cigarette smokers. Chest 115:684-690

Haisch M, Hering P, Wendel U, Broesicke H, Schadewaldt P (1993) Determination of 13C-labelled CO2: A new application of non-dispersive infrared spectroscopy. Biological Chemistry Hoppe Seyler 374:688

Haisch M, Hering P, Fabinski W, Fuß W (1994a) A sensitive isotope selective non-dispersive infrared spectrometer for 13CO2 and 12CO2 concentration measurement in breath samples. Isotopenpraxis Isotopes in environmental and health studies 30:247

Haisch M, Hering P, Schadewaldt P, Brösike H, Braden B, Koletzko S, Steffen C (1994b) Biomedical application of an isotope selective nondispersive infrared spectrometer for 13CO2 and 12CO2 concentration measurement in breath samples. Isotopenpraxis Isotopes in environmental and health studies 30:253

Haisch M, Hering P, Fabinski W, Zöchbauer M (1996) Isotopenselektive Konzentrationsmessungen an Atemgasenmit einem NDIR-Spektrometer. Technisches Messen 63:322

Harren FJM, Oomens J, Persijn S, Veltman RH, de-Vries HSM, Parker D (1998) Multicomponent trace gas analysis with a CO laser based photoacoustic detector; emission of ethanol, acetaldehyde, ethane and ethylene from fruit. Proc.SPIE 3405:556-562

Harren FJM (2001) Personal Communication

Hart CM (1999) Nitric oxide in adult lung disease. Chest 115:1407-1417

Hyspler R, Crhova S, Gasparic J, Zadak Z, Cizkova M, Balasova V (2000) Determination of isoprene in human expired breath using solid-phase microextraction and gas chromatography-mass spectrometry. J.Chromatogr.B 739:183-190

Ignarro LJ, Cirino G, Casini A, Napoli C (1999) Nitric oxide as a signaling molecule in the vascular system: an overview [Review]. J.Cardiovasc.Pharmacol. 34:879-886

Kharitonov SA (1999) Exhaled nitric oxide and carbon monoxide in asthma. Europ.Respir.J. 9:212-218

Kharitonov SA, Barnes PJ (2001) Exhaled markers of pulmonary disease. Am.J.Respir.Crit.Care Med. 163:1693-1722

King CE, Toskes PP (1983) The use of breath tests in the study of malabsorption. Clin.Gastroent. 12:591-610

Kleine D, Dahnke H, Urban W, Hering P, Mürtz M (2000) Real-time detection of $^{13}CH_4$ in ambient air by use of mid-infrared cavity leak-out spectroscopy. Opt.Lett. 25:1606-1608

Kneepkens CM, Lepage G, Roy CC (1994) The potential of the hydrocarbon breath test as a measure of lipid peroxidation. Free Radic.Biol.Med. 17:127-160

Knutson MD, Lim AK, Viteri FE (1999) A practical and reliable method for measuring ethane and pentane in expired air from humans. Free Radic.Biol.Med. 27:560-571

Knutson MD, Handelman GJ, Viteri FE (2000) Methods for measuring ethane and pentane in expired air from rats and humans. Free Radic.Biol.Med. 28:514-519

Koletzko S, Haisch M, Seeboth I, Braden B, Hengels K, Koletzko B, Hering P (1995) Isotope-selective non-dispersive infrared spectrometry for detection of Helicobacter pylori infection with 13C-urea breath test. Lancet 345:961-962

Kosterev AA, Malinovsky AL, Tittel FK, Gmachl C, Capasso F, Sivco DL, Baillargeon JN, Hutchinson AL, Cho AY (2001) Cavity ringdown spectroscopic detection of nitric oxide with a continuous-wave quantum-cascade laser. Appl Opt 40:5522-5529

Kosterev AA, Tittel FK, Durante W, Allen M, Gmachl C, Capasso F, Sivco DL, Cho AY (2002) Detection of biogenic CO production above vascular cell cultures using a near-room-temperature QC-DFB laser. Appl Phys B 74:95-99

Krumbiegel P (1991) Stable Isotope Pharmaceuticals for Clinical Research and Diagnosis. Gustav Fischer, Jena

Kühnemann F (1998) Photoacoustic trace gas detection using a cw single-frequency parametric oscillator. Appl Phys B 66:741-745

Lancaster DG, Richter D, Curl RF, Tittel FK (1998) Real-time measurements of trace gases using a compact difference frequency based sensor operating at 3.5 μm. Appl.Phys.B 67:339

Lee PS, Majkowski RF, Perry TA (1991) Tunable diode laser spectroscopy for isotope analysis-detection of isotopic carbon monoxide in exhaled breath. IEEE Trans Biomed Eng 38:966-973

Lee PS, Schreck RM, Hare BA, McGrath JJ (1994) Biomedical applications of tunable diode laser spectrometry: correlation between breath carbon monoxide and low level blood carboxyhemoglobin saturation. Ann Biomed Eng 22:120-125

Martis A, Büscher S, Kühnemann F, Urban W (1998) Simultaneous ethane and ethylene detection using a CO Overtone laser PA spectrometer: A new tool for stress/ damage studies. Instr.Sci.Technol. 26:177-187

Menzel L, Kosterev AA, Curl RF, Tittel FK, Gmachl C, Capasso F, Sivco DL, Baillargeon JN, Huchinson AL, Cho AY, Urban W (2001) Spectroscopic detection of biological NO with a quantum cascade laser. Appl Phys B 72:859-863

Miller ER, Appel LJ, Jiang L, Risby TH (1997) Association between cigarette smoking and lipid peroxidation in a controlled feeding study. Circulation 96:1097-1101

Moncada S, Palmer RM, Higgs EA (1991) Nitric oxide: physiology, pathophysiology, and pharmacology [Review]. Pharmacol.Rev. 43:109-142

Morimoto Y, Durante W, Lancaster DG, Klattenhoff J, Tittel FK (2001) Real-time measurements of endogenous CO production from vascular cells using an ultrasensitive laser sensor. Am J Physiol Heart Circ Physiol 280:H483-H488

Moskalenko K, Nadezhdinskii A, Adamovskaya IA (1996) Human breath trace gas content study by tunable diode laser spectroscopy technique. Infrared Phys Technol 37:181-192

Mürtz M, Frech B, Palm P, Lotze R, Urban W (1998) Tunable carbon monoxide overtone laser sideband system for precision spectroscopy from 2.6 to 4.1 μm. Opt.Lett. 23:58-60

Mürtz P, Menzel L, Bloch W, Hess A, Michel O, Urban W (1999) LMR spectroscopy: a new sensitive method for on-line recording of nitric oxide in breath. J Appl Physiol 86:1075-1080

Oomens J, Zuckermann H, Persijn S, Parker D, Harren FJM (1998) CO-laser-based photo-acoustic trace-gas detection: applications in postharvest physiology. Appl Phys B 67:459-466

Paldus BA, Harb CC, Spence TG, Zare RN, Gmachl C, Capasso F, Sivco DL, Baillargeon JN, Hutchinson AL, Cho AY (2000) Cavity Ring-down Spectroscopy using Mid-Infrared Quantum Cascade Lasers. Opt.Lett. 25:668

Paredi P, Loukides S, Ward S, Fantoni R, et al. (1998) Exhalation flow and pressure-controlled reservoir collection of exhaled nitric oxide for remote and delayed analysis. Thorax 53:775-779

Paredi P, Kharitonov SA, Leak D, Shah PL, Cramer D, Hodson ME, Barnes PJ (2000) Exhaled ethane is elevated in cystic fibrosis and correlates with carbon monoxide levels and airway obstruction. Am.J.Respir.Crit.Care Med. 161:1247-1251

Phillips M (1992) Breath tests in medicine. Sci.Am.(Int.Ed.) July:52

Phillips M (1999) Breath test for detection of lung cancer. US Patent 5996586

Phillips M (2001) Breath methylated alkane contour: a new marker of oxidative stress and disease. US Patent 6254547

Pleil JD, Lindstrom AB (1995) Measurement of volatile organic compounds in exhaled breath as collected in evacuated electropolished canisters. J Chromatogr B Biomed Appl 665:271-279

Pryor WA, Godber SS (1991) Noninvasive measures of oxidative stress status in humans [Review]. Free Radic.Biol.Med. 10:177-184

Riely CA, Cohen G, Lieberman M (1974) Ethane evolution: a new index of lipid peroxidation. Science 183:208-208

Risby TH, Sehnert SS (1999) Clinical application of breath biomarkers of oxidative stress status [Review]. Free Radic.Biol.Med. 27:1182-1192

Risby TH, Jiang L, Stoll S, Ingram D, Spangler E, Heim, Cutler R, Roth GS, Rifkind JM (1999) Breath ethane as a marker of reactive oxygen species during manipulation of diet and oxygen tension in rats. J.Appl.Physiol. 86:617-622

Risby TH, Sehnert SS, Jiang L, Burdick JF (2001) Volatile biomarkers for analysis of hepatic disorders. US Patent 6248078

Roller C, Namjou K, Jeffers J, Potter W, McCann PJ, Grego J (2002) Simultaneous NO and CO_2 measurement in human breath with a single IV–VI mid-infrared laser. Opt.Lett. 27:107-109

Sauke TB, Becker JF, Torre-Bueno J (1997) Laser diode spectrometer for analyzing the ratio of isotopic species in a substance. US Patent 5640014

Sies H (1997) Oxidative stress: oxidants and antioxidants [Review]. Exp.Physiol. 82:291-295

Silkoff PE, Stevens A, Bucher-Bartelson B, Martin RJ (1999) A method for the standardized offline collection of exhaled nitric oxide. Chest 116:754-759

Steffen C, Haisch M, Hering P (1993) The evaluation of drug metabolism capacity by 13C and 14C carbon dioxide exhalation data. Fundamental & Clinical Pharmacology 7:381

Stepanov EV, Moskalenko KL (1993) Gas analysis of human exhalation by tunable diode laser spectroscopy. Opt Eng 32:361-367

Tanahashi T, Kodama T, Yamaoka Y, Sawai N, Tatsumi Y, Kashima K, Higashi Y, Sasaki Y (1998) Analysis of the 13C-urea breath test for detection of *Helicobacter pylori* infection based on the kinetics of delta-$^{13}CO_2$ using laser spectroscopy. J Gastroenterol Hepatol 13:732-737

Taucher J, Hansel A, Jordan A, Fall R, Futrell JH, Lindinger W (1997) Detection of isoprene in expired air from human subjects using proton-transfer-reaction mass spectrometry. Rapid Commun.Mass.Spectrosc. 11:1230-1234

Urban W (1995) Physics and spectroscopic applications of carbon monoxide lasers, a review. Infrared Phys.Technol. 36:465-473

Wetzel K, Fischer H (2001) 13C–breath tests in medical research and clinical diagnosis. Unpublished Report

13 Detection of Nitric Oxide in Human Exhalation Using Laser Magnetic Resonance

Wolfgang Urban

13.1 Free Radical Spectroscopy, a Challenge for Sensitive Detection

We have been heavily involved in the development of spectroscopic methods for the investigation of the IR spectra of gaseous free radicals and open shell molecular ions. Based on our expertise in molecular gas lasers, we have explored Laser Magnetic Resonance (LMR) particularly in combination with carbon monoxide (CO) lasers. LMR was first developed by K.M.Evenson and his co workers in 1968 in combination with FIR lasers for pure rotation spectroscopy of free radicals (Evenson 1968). Since these species are short lived and cannot be generated in high concentrations in general, the necessity of a very sensitive detection scheme is obvious.

In the past, the only powerful IR lasers of high beam quality in the medium IR were the merely stepwise tuneable molecular gas lasers. These lasers use molecular vibration-rotation transitions as gain medium. For spectroscopy the frequency of such a laser cannot be tuned across the molecular transition to be investigated. However, provided the transition frequencies of one particular laser line and of the free radical are close enough, one can tune the free radical transition across the laser line. This is possible due to the magnetic moment inherent to the open shell structure of the radical via its Zeeman effect. Thus *magnetic tuning* is the key to LMR. On top of this, *magnetic modulation* provides a most powerful means for specific detection of the magnetic species by Lock-In techniques. In this way it is possible to discriminate the magnetic molecules from any non magnetic background absorptions. We should point out, however, that we need to reduce the pressure broadening of the transitions and therefore have to operate at reduced total pressure. In our case this would be on the order of 1000 Pa. Then pressure broadening and Doppler width are comparable. The magnetic modulation amplitude, necessary to produce a modulation depth on the order of the linewidth for sensitive detection, is in the order of 100 Gauss.

Since LMR may not be familiar to our readers, we briefly outline its principle of operation. We start with two vibrational levels, v=0 and v=1. In a simplified hypothetical case they should both have J=1/2, so the connecting line would be a

Q(1/2) transition. The energy level diagram is plotted in *Fig. 13.1a.* There are two transitions allowed by the experimental arrangement, ΔM=+1and ΔM=-1. In an external magnetic field the corresponding transition frequencies will tune by the Zeeman effect to higher and lower frequencies respectively. The transition frequency diagram is shown in *Fig. 13.1b.*

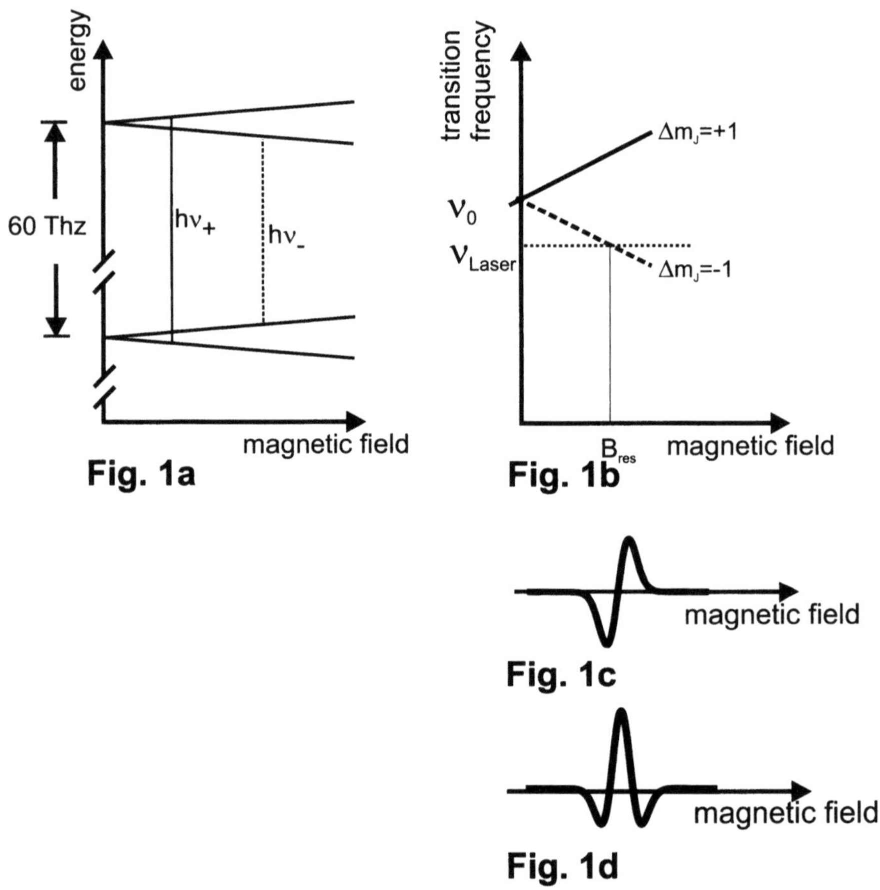

Fig. 13.1. Vibration rotation transition for a hypothetical Q(1/2) transition.

a) Energy level diagram, v=0, J=1/2 and v=1, J=1/2.

b) Transition frequency diagram for the Q(1/2) transition. The laser frequency is indicated at ν_L.

c) First derivative of the absorption line taken by conventional LMR.

d) First derivative of the dispersion line, observable by Faraday-LMR.

If a fixed frequency laser radiation at ν_L is applied that lies within the Zeeman tuning range, we can tune the molecular transition across the laser line and observe a LMR transition. For vib.-rot. transitions, the laser frequency is in the order of 60 THz. The Zeeman tuning covers a range of several GHz, depending on the magnetic moment and the available magnetic field strength B_0.

In *Fig. 13.1c* the observed line shape for a LMR absorption first derivative is plotted. The corresponding experimental setup is shown as a block diagram in *Fig. 13.2.* The main components are a molecular gas laser and an external magnetic field.

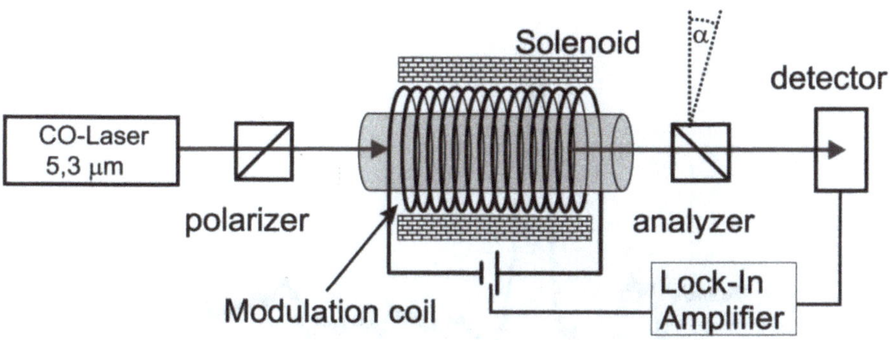

Fig. 13.2. Block Diagram of the LMR setup. (The analyzer in front of the detector is only needed for Faraday detection. The position of the analyzer deviates by a small angle α from the perpendicular orientation)

For free radical spectroscopy in general, we use a variety of molecular gas lasers, e.g. CO_2 (900-1100 cm^{-1}), CO (1200-2100 cm^{-1}) and CO-overtone (2500-3800 cm^{-1}) lasers.

There we have a superconducting solenoid for Zeeman tuning with magnetic field strengths up to 5 Tesla. This solenoid has a room temperature bore (2-5 cm inner diameter) for the sample.The active volume is 30 cm long, the range of high homogeneity that is covered by the modulation coil.

In general the laser radiation is well polarized by itself, however, to make sure a high polarisation quality, a polarizer is set in front of the sample cell. Only in the Faraday set up, to be discussed next, we need the option of an analyser, set at 90° plus a small offset angle δ in front of the detector.

With "conventional" Zeeman absorption LMR (Rohrbeck 1983) we soon reached limitations in our spectroscopic goals. Thus we developed a modification and introduced a polarisation type detection method, the so-called *Faraday-LMR.* Here the Faraday rotation caused by the magnetic species is detected behind an almost crossed analyser put behind the sample. The corresponding Faraday LMR signal shape produced by Zeeman modulation, is the first derivative of the dispersion line rather than the absorption line shape as in conventional LMR. This lineshape is shown in *Fig. 13.1d.*

We have verified that Faraday-LMR in the mid-IR is more sensitive by two orders of magnitude compared with conventional LMR (Hinz 1985). The block diagram is that of *Fig. 13.2* including the analyser. Setting the direction of the offset angle δ we determine the sign of the signal. In contrast to conventional LMR signals, in Faraday LMR the sign of the signal depends on whether the transition is $\Delta M=+1$ or $\Delta M=-1$.

When testing the sensitivity of our equipment we often used the absorption features of nitric oxide (NO). This "radical" falls into the easy range of the CO laser, and it was obvious, when testing the detection sensitivity of our equipment, to see whether NO was present in the atmosphere. *Fig. 13.3* shows the Faraday LMR signal obtained, when we were pumping air from the Bonn atmosphere through the apparatus (Bohle 1988).

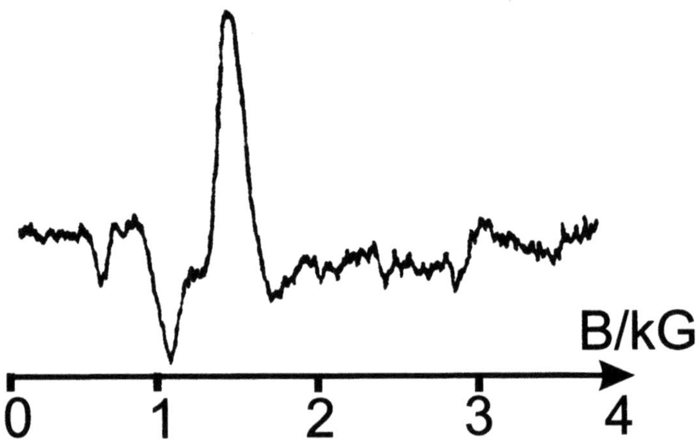

Fig. 13.3. Faraday LMR signal taken by pumping air from the Bonn atmosphere through the sample (After (Bohle 1988))

To complete this introductory chapter we show the tuning features of two relevant NO transitions. This is the P(3/2) for $^{14}N^{16}O$, observable by the CO laser line $P(13)_{v=9\text{-}8}$ at 1884.3493 cm^{-1} (*Fig. 13.4*), and the Q(3/2) for $^{15}N^{16}O$ accessible with the CO laser line $P(17)_{v=10\text{-}9}$ (*Fig. 13.5*). This shows one additional fine feature of the LMR method. It is not only sensitive and specific but inherently selective for different isotopomers.

For NO detection we only need magnetic field strengths of up to 0.2 Tesla, which can be generated in a conventional copper solenoid. The apparatus has a solenoid 50 cm long and uses a liquid nitrogen cooled CO laser, delivering ca. 1 Watt of single line power. The achieved detection sensitivity is below 1 ppb of NO in N_2 at a total pressure of 1000 Pa. The time constant was set to 0.3 s. For further details we refer to two previous articles from our group [5, 6].

There are other methods to detect NO in the gas phase. In particular the one based on chemiluminescence has reached a high degree of perfection which may be more sensitive than LMR (Sievers 1998). Therefore we would not recommend LMR for atmospheric monitoring of NO. However, when time resolution is required, LMR may be an alternative, and as soon as isotopic selectivity is required, LMR may be the method of choice.

Other IR laser detection methods are available that may be used for quantitative NO analysis, e.g. Photoacoustics and Cavity Ring Down or Cavity Leak Out Spectroscopy [8, 9].

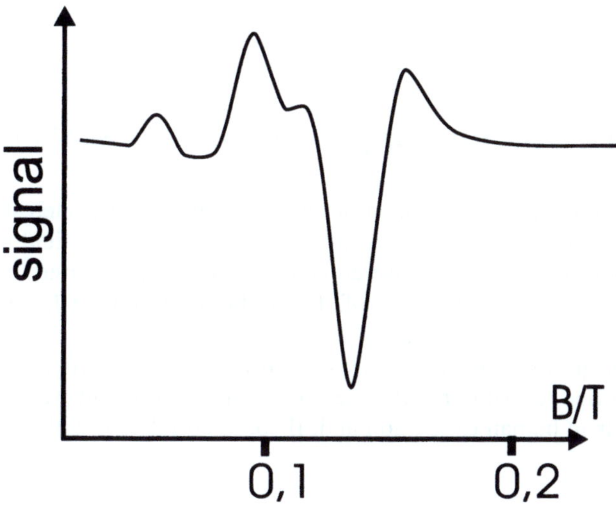

Fig. 13.4. Plot of the Faraday LMR Signal of $^{14}N^{16}O$ R(3/2) transition taken with the CO Laser Line P(13) $_{v=9\text{-}8}$

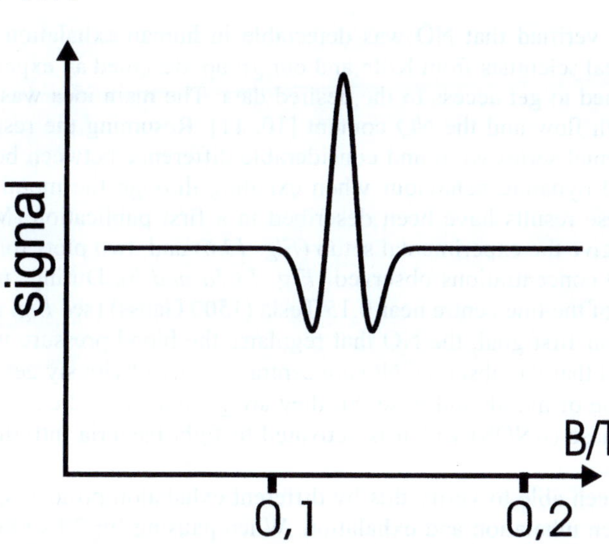

Fig. 13.5. Plot of the Faraday LMR Signal of $^{15}N^{16}O$ Q(3/2) Transition with CO P(17) $_{v=10\text{-}9}$

As far as Photoacoustics (PA) is concerned, there are good coincidences with CO laser transitions available. To produce a good PA signal, however, the diatomic molecule NO does not provide a fast enough VT relaxation channel to be a good candidate for sensitive detection.

Cavity Leak Out Spectroscopy (CALOS), on the other hand, may suffer from strong background absorption by nonmagnetic species. At the very least, it will need gas processing of the sample before it can be used.

13.2 Applied Spectroscopy using the LMR Method

At a rather early stage it was pointed out to us that it could be interesting to use a sensitive detection method for NO to investigate its role in the field of medicine. However a physicist needs a precise enough question to get started in the field of interdisciplinary research. It was Prof. K. Addicks (Anatomie I, Universität zu Köln) who helped us in that respect.

The general question of interest was the role of NO as a regulator for blood pressure. As a first test of our LMR method we had to find out whether we could see NO signals in human exhalation and, if so, to correlate it with the respiratory phases.

13.2.1 Dynamic Behaviour of NO in Exhalation

Once we had verified that NO was detectable in human exhalation by LMR, a team of medical scientists from Köln and our group designed an experiment, from which we hoped to get access to the desired data. The main idea was to correlate both the breath flow and the NO content [10, 11]. Resuming the results of these first experimental series we found considerable difference between both NO concentration and dynamic behaviour when exhaling through the mouth or through the nose. These results have been described in a first publication (Mürtz 1999). Here we just give the experimental setup (*Fig. 13.6*) and two plots for the dynamics of the NO concentrations observed (*Fig. 13.7a and b*). During measurement we sit on top of the line centre near 0.15 Tesla (1500 Gauss) (see *Fig. 13.4*).

As far as our first goal, the NO that regulates the blood pressure is concerned, we must admit that the observed NO concentrations are obviously generated in the mucosal tissue of mouth and nose. So they are generated by the so-called *inducible* NO synthase (i-NOS) which is activated to fight bacteria intruding from the inhaled breath.

We have been able to verify this by different exhalation protocols, varying the pauses between inhalation and exhalation. When pausing for 20 s after inhalation the exhaled peak concentration went up to 500 ppb (*Fig. 13.8*). The mucosal tissue, which contains the NO sources have different NO-release in mouth and nose. The NO-relase in nose and mouth are dependent from mucosal structure and surface as well as NOS expression..

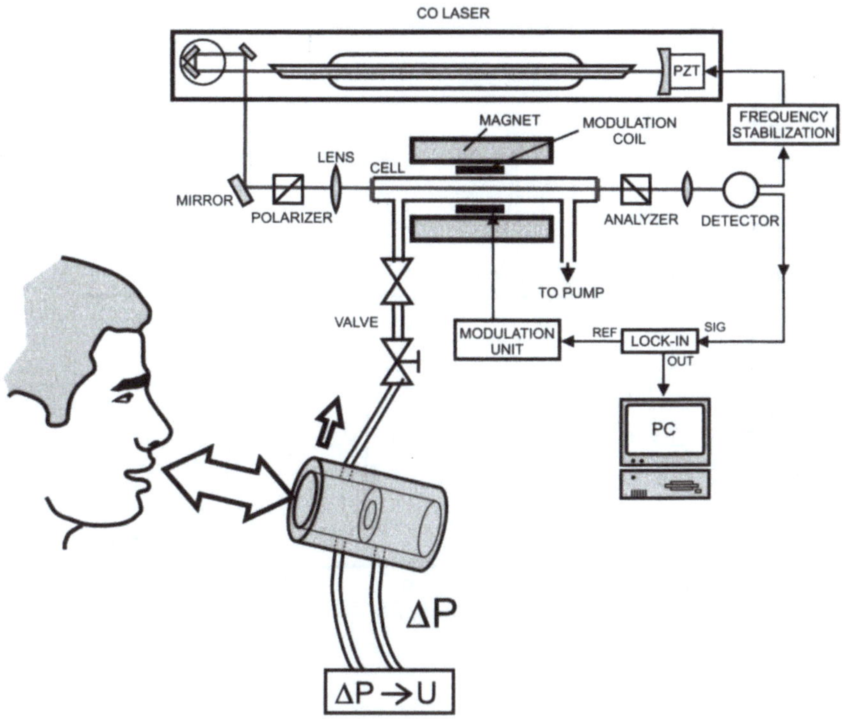

Fig. 13.6. Block diagram of the LMR equipment for exhalation analysis. (After Menzel (Menzel 2000))

There is one interesting observation we made more or less accidentally [12, 13]. The NO concentration, when exhaling through the nose will increase dramatically as soon as the person starts humming. The effect is plotted in *Fig. 13.9.* Discussions with colleagues from the Klinik und Polyklinik für Hals- Nasen- Ohrenheilkunde der Universität zu Köln has led us to the conclusion that the observed NO comes from the paranasal sinuses. It is flushed out by the sound vibrations.

There are characteristic time constants, the decay time for the flushing out, which indicates the coupling of the paranasal cavity to the exterior and the building up time constant that is a measure for the activity of the mucosal tissue cladding the sinuses. The latter is accessible by repetitive humming after different time intervals. Some data are shown in *Fig. 13.10.* Systematic investigations are underway. We may have found a non-invasive method to characterise the NOS-activity of the paranasal sinuses. A publication on this subject is in preparation (Menzel 2002).

So far we can say that we have made interesting observations for the NO dynamics in human exhalation, but these observations do not allow to distinguish NO-origin from lower and upper airways. A detection of NO-released in the lower airways, which is mainly derived from endothelial NO synthase (eNOS) in blood vessels would allow to correlate NO-exhalation with blood pressure regulation.

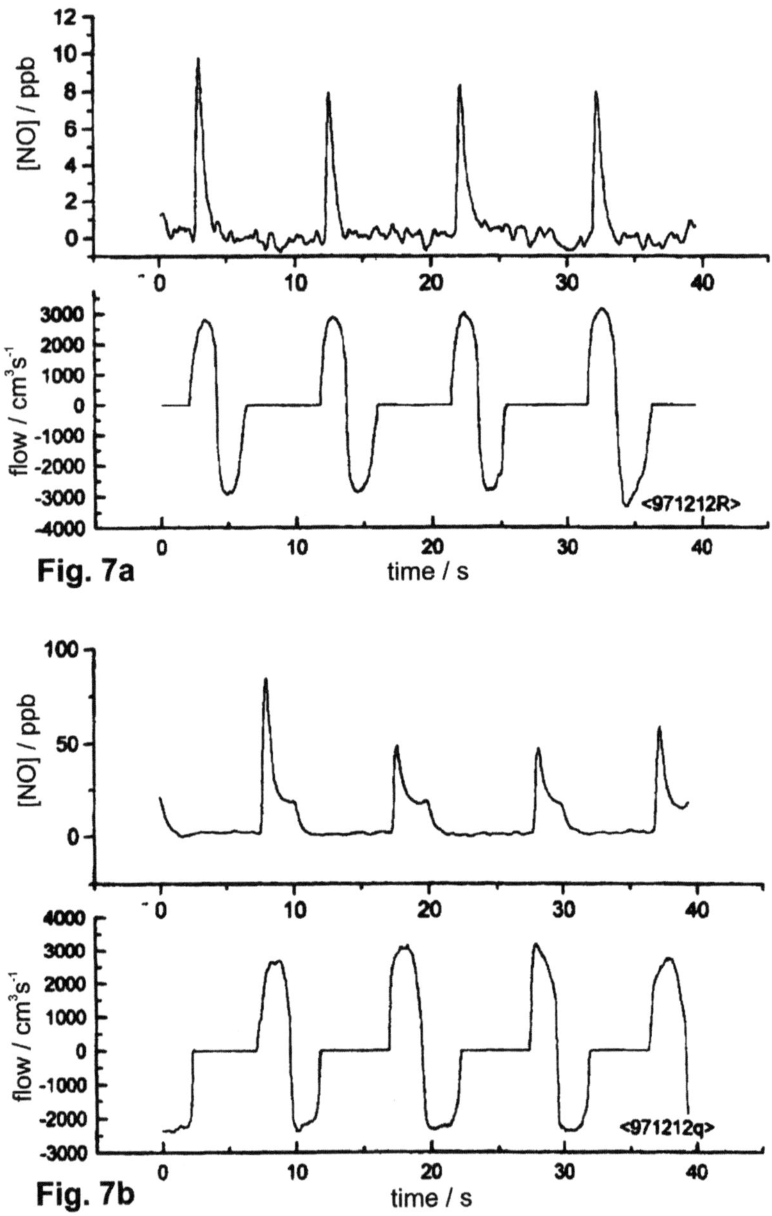

Fig. 13.7. Breath flow and NO dynamics.
a) Exhalation through the mouth.
b) Exhalation through the nose.

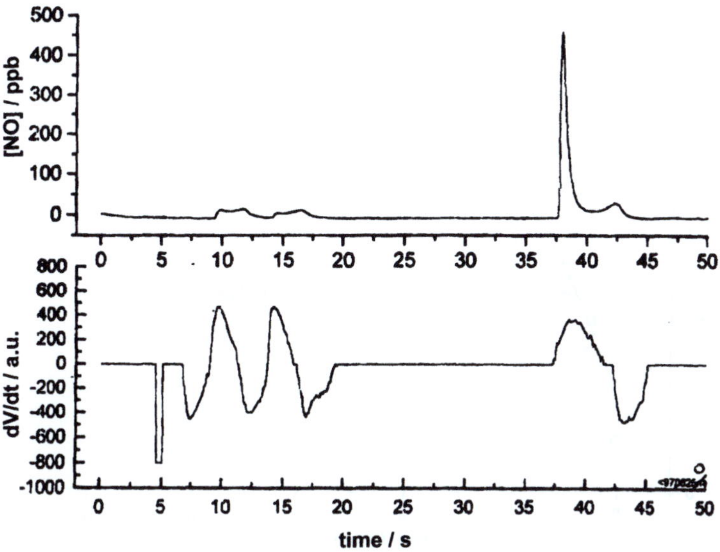

Fig. 13.8. Exhalation through the nose, buildup of NO peak concentration after 20 s pause

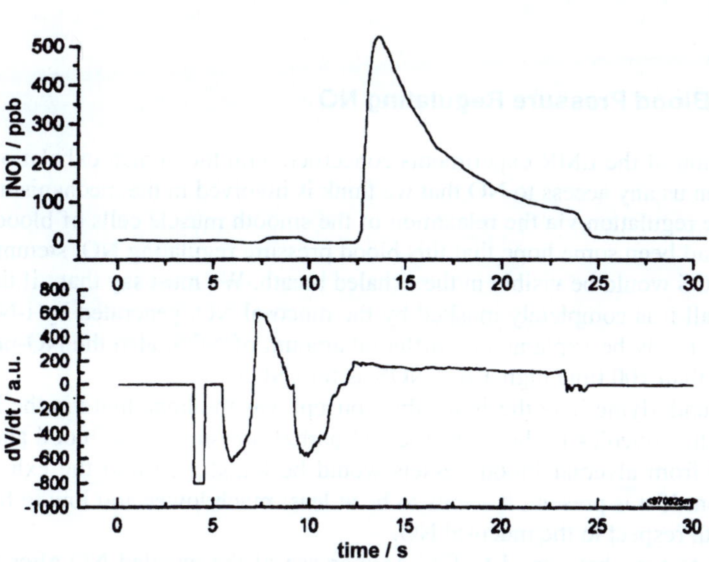

Fig. 13.9. Washing out of NO by humming when exhaling trough the nose. Peak Value and Decay time constant.

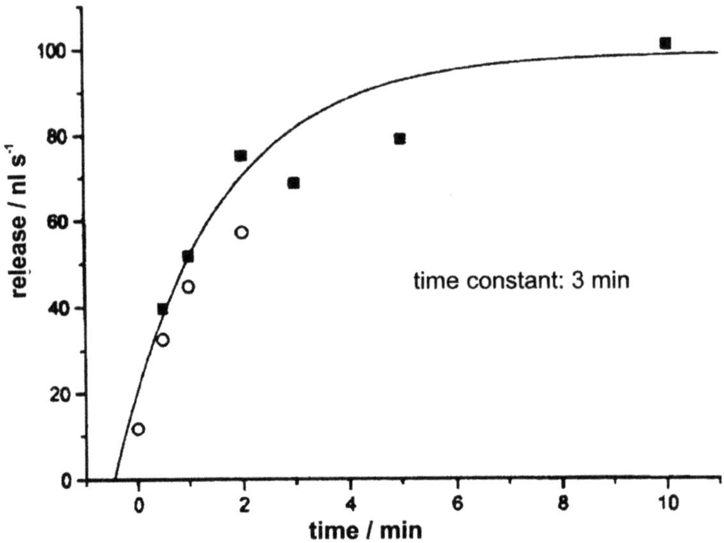

Fig. 13.10. Restoration of the NO concentration after washing out by humming. The restoration time constant may be an indicator for the status of the mucosal tissue cladding inside the paranasal sinuses.

13.2.2 Blood Pressure Regulating NO

Evaluation of the LMR experiments concerned with the human exhalation has not yet given us any access to NO that we think is involved in the mechanism of blood pressure regulation via the relaxation of the smooth muscle cells of blood vessels. There had been some hope that this blood pressure regulating NO, stemming from the e-NOS would be visible in the exhaled breath. We must say that , if this is present at all it is completely masked by the mucosal NO, generated by i-NOS. This could not only be explained by different amount of NOS, also the NO-production is more than 100 time higher for iNOS as for eNOS.

The underlying hypothesis for this concept was the hope that via the huge surface of the alveoles in the lungs the NO partial pressure in the blood and the NO derived from alveolar blood vessels would be transferred into the exhalation by diffusion. If it is present, it seems to be at least much lower, too low to be observable with respect to the mucosal NO.

As a test we have tried to find an increase of the exhaled NO after treatment with a blood pressure regulating drug. This experiment was done in collaboration with *Hoechst Marion Roussell,* who provided their drug *Corvaton*^{R}, containing

Molsidomin, an NO-donor, as active compound. Our observations, although negative, have been reported (Urban 2000).

How could we increase the sensitivity particularly for that NO that was stemming from the drug? Since with our LMR method we can switch to $^{15}N^{16}O$ we can use the rare isotopomer as tracer for the metabolised drug. So we needed Molsidomine specially prepared with the rare Isotope ^{15}N.

In the next step of collaboration colleagues from Institut für Pharmazeutische Chemie der Universität Bonn prepared the isotopomeric version of Molsidomine and we switched to the $^{15}N^{16}O$ LMR transition. The background signal of this isotopomer should be lower by a factor of 300, the natural abundance of ^{15}N. Therefore we should become more sensitive to the metabolisation product by this same factor.

Nevertheless there was no clear positive result in a first test of the exhalation. Either our method was not sensitive enough or the metabolisation pathway was not as we had anticipated. We should mention, however, that we are able to find traces of $^{15}N^{16}O$ in a sample containing 1 ppm of natural NO (*Fig. 13.11*). It seems, however, that gas phase investigations were not suitable to get the sought for information.

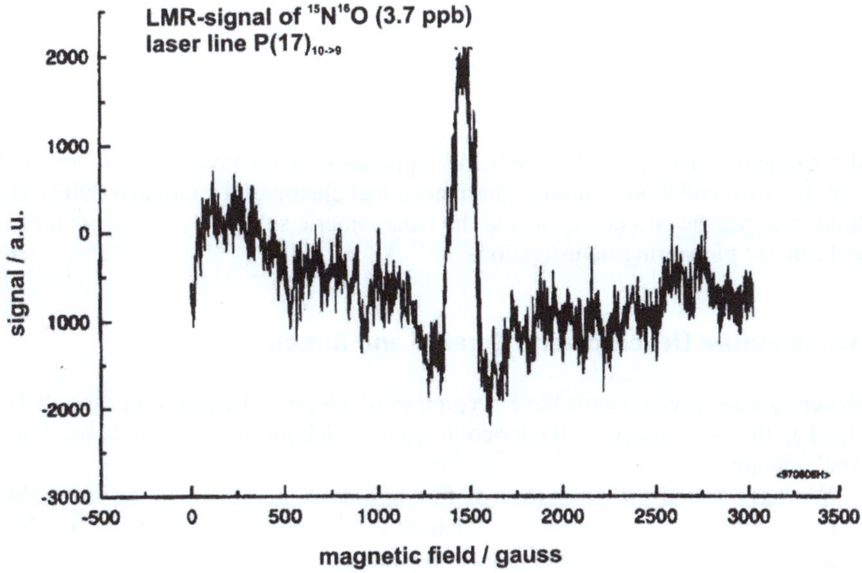

Fig. 13.11. Faraday LMR signal of $^{15}N^{16}O$. The natural content in 1ppm NO. The corresponding concentration of the rare Isotopomere is 3.7 ppb

13.2.3 In-Vitro Investigations using LMR

There are situations where NO that is initially produced as a diatomic is immediately converted to a reaction product, such as nitrite or nitrate ions. These products are then present in a liquid and therefore not accessible to detection by LMR. The question is, how to get these reaction products back to set free NO in the gas phase in order to become accessible by LMR?

In a pilot experiment L.Menzel (Menzel 2000) has measured the NO concentration above an aqueous solution containing nitrites by a reduction reaction. These experiments were quite successful and have opened up a series of new investigations, including isotopomeric selectivity.

Coming back to our first goal, the pathway of NO in the human body, one possible way for transport of endothelial NO is its chemical reaction with the blood. It is well known that NO has a high affinity to hemoglobin. It will replace the O_2 quickly and thus not be able to get out via the lungs by diffusion. This would explain our negative results quite easily. Very promising experiments in this direction are under way (Jentsch 2000). Furthermore, an enrichment of $^{15}N^{16}O$ in urine from the probands that had taken the ^{15}N doped Molsidomine has successfully been observed (Lehmann 2000).

13.2.4 Applications in Pharmacology

Expression rates from chemicals that might be useful as new NO donors can be determined rather easily by our LMR apparatus. Such investigations are under way in close collaboration with pharmaceutical chemists (Horstmann 2002). Here both, the specific NO detection and the isotopomeric selectivity can play an essential role for pioneering investigations.

13.2.5 Future Development, Smaller and Simpler

Recently new laser systems have become available, e.g. the quantum cascade laser (QCL), first developed in Federico Capasso´s laboratory at Bell Labs, Lucent Technologies.

We have tested the properties of this new laser and have found its qualities most promising for our purposes (Ganser 2001). Since it is tunable, we do no longer need the bulky solenoid to tune the molecular transition into resonance with the laser, we can just sit on the molecular transition frequency. Our test of the beam quality was positive, with the necessary precautions one can run this laser to give sufficient power for high sensitivity detection of NO (Ganser 2002). This new development will provide us with a much smaller experimental setup, much more appropriate for the field of applied spectroscopy for medical sciences, maybe even for everyday use.

Nevertheless, the bulky LMR system has given us the key to a new field of applied spectroscopy in medicine. The evolution of this branch is showing its own dynamics and develops in unexpected directions. But this is how progress in science takes place.

Acknowledgements

First of all I need to thank my previous co-workers at the Institut für Angewandte Physik der Universität Bonn. Dr. rer. nat. Petra Mürtz was actively supervising the construction of the first LMR set-up for medical application and got the medical program running. Dr. rer. nat. Lars Menzel did both his Diploma Thesis and Doctoral Thesis in the field of "Applied Laser Spectroscopy" for medical and pharmacological research, and so has contributed very much to the results of this article. Dr. rer. nat. Christian Schmidt and Dipl. Phys Andreas Jentsch started with great enthusiasm the isotopomeric NO measurements and finally Dr. rer. nat. Ralph Gäbler is trying to find out whether the LMR is a useful work horse in the real world of applications.

Then I am indebted to all medical colleagues who have encouraged us to get this new branch of activities started. Prof. Dr. med. Klaus Addicks, who initialised the first NO project and PD Dr. med. Wilhelm Bloch, who has been the most actively involved medical counterpart for us. Their Anatomisches Institut I der Universität zu Köln has always been our main source for relevant information. Prof Dr. med. Olaf Michel as well as Dr. med. Alexander Hess, Klinik und Poliklinik für Hals- Nasen- Ohrenheilkunde der Universität zu Köln have given us support in setting up and understanding the exhalation measurements.

We had many illuminating discussions concerning the role of NO as regulator for blood pressure and active support for our project with Dr. rer. nat. Rolf Grewe, Hoechst Marion Roussell, in Frankfurt/Main.

Prof. Dr. rer. nat Jochen Lehmann and Dr. rer. nat. Axel Horstmann, Institut für Pharmazeutische Chemie der Universität Bonn, have not only prepared the ^{15}N doped Molsidomine, they also brought in new aspects and new ideas for further applications of the LMR method for NO research.

Most of the equipment used for the experiment was set up for fundamental research in the SFB 334, and thus was provided through funding by the Deutsche Forschungsgemeinschaft (DFG).

Finally I want to thank Heiko Ganser for preparing the figures for this text.

References

Bohle W, private commun. Bonn (1988)

Dahnke H, Kleine D, Urban W, Hering P, Mürtz M (2001) Isotopic ratio measurement of methane in ambient air using mid-infrared cavity leak-out spectroscopy. Appl. Phys. B **72** 121-125

Evenson KM, Broida HP, Wells JS, Mahler RJ, Mizushima M (1968) Electron Paramagnetic Resonance Absorption with the HCN Laser. Phys.Rev.Letters **21**: 1038ff

Ganser H, .Frech B, Jentsch A, Mürtz M, Gmachl C, Capasso C, Sivco DL, Ballargeon JN, Hutchison AL, Cho AY, Urban W (2001) Investigation of the spectral width of quantum cascade laser emission near 5.2 μm by a heterodyne experiment. : Opt. Commun. **197**, 127-130

Ganser H, Urban W, Brown J M (2002) Sensitive Detection of NO by Faraday Modulation Spectroscopy with a Quantum Cascade Laser. Molec. Phys. In print

Hinz A, Zeitz D, Bohle W, Urban W (1985) A Faraday Laser Magnetic Resonance Spectrometer for Spectroscopy of molecular Radical Ions. Appl. Phys.**B 36**: 1-4

Hinz A, Pfeiffer J, Bohle W, Urban W (1982) Mid-infrared laser magnetic resonance using the Faraday and Voigt effects for sensitive detection. Molec. Phys. **45**: 1131-1139

Horstmann A, Menzel L, Gäbler R, Jentsch A, Urban W, Lehmann J (2002) Release of Nitric Oxide from novel Diazeniumdiolates Monitored by Laser Magnetic Resonance Spectroscopy. Nitric Oxide Biol. Chem.**6,** *In print*

Jentsch A (2000) Isotopomerenselektiver Stickstoffmonoxid-Nachweis aus biologischen Quellen, Diploma Thesis, University of Bonn

Koch M, Luo X, Mürtz P, Urban W, Mörike K (1997) Detection of small traces of $^{15}N_2$ and $^{14}N_2$ by Faraday LMR spectroscopy of the corresponding isotopomers of nitric oxide. Appl. Phys. **B 64**: 683-688

Kühnemann F (2002) Photoacoustic trace gas detection in plant biology. This Book (to be updated by the editors)

Lehmann J (2000) private commun. Bonn

Menzel L (1997) Zeitaufgelöste NO-Messung im menschlichen Atem mit einem LMR-Spektrometer . Diploma Thesis, University of Bonn

Menzel L (2000) Infrarotspektroskopischer Nachweis von Stickstoffmonoxid aus biologischen Quellen. Doctoral Thesis, University of Bonn

Menzel L, Hess A, Bloch W, Urban W (2002) in preparation

Mürtz P, Menzel L, Bloch W,Hess A, Michel O, Urban W (1999) LMR spectroscopy: a new sensitive method for on-line recording of nitric oxide in breath. J.Appl.Physiol. **86**, 1075-1080

Rohrbeck W, Hinz A, Nelle P, Gondal MA, Urban W (1983) A Broadband Mid-Infrared Laser Magnetic Resonance Spectrometer for the Spectral Range of 1200-2000 Wavenumbers. Appl. Phys **B 31**: 139-144

Sievers Instruments (1998) NOA 280 NO Analyzer, Boulder CO:

Urban W, Menzel L, Bloch W, Hess A, Grewe R (2000) Die Laser-Magnet-Resonanz-Spektroskopie (LMRS). Ein neues, Stickstoffisotopen-selektives Verfahren zur „NO"-Forschung in Physiologie und Pharmakologie, in: Aktuelle Trends in der invasiven Kardiologie, R.Bach und S.Spitzer (Hrsg.) AKA: 262-267

14 Medical Trace Gas Detection by Means of Mid-Infrared Cavity Leak-Out Spectroscopy

Hannes Dahnke, Sandra Stry and Golo von Basum

14.1 Introduction

Ambient air and the human breath both contain many different volatile organic and inorganic compounds. Most of them are present in very low concentrations. The analysis of these trace gases leads in case of ambient air to a better understanding of the atmospheric chemistry and in the case of breath tests to a deeper knowledge of the physiological processes in the human body. Amongst other sources these compounds are produced in the human organism and find their way via the blood through the lungs into the human breath. Most of the volume fractions of these trace gases are on the order of some ppb (parts per billion: $1{:}10^9$) down to several ppt (parts per trillion: $1{:}10^{12}$). This shows the need for ultra sensitive analytical methods for the in-vivo monitoring of human breath, which helps to understand various physiological and pathophysiological processes in the human organism.

Many of these volatile compounds are hydrocarbons, such as ethane or methane. The standard technique to analyse these hydrocarbons in human breath is gas chromatography (GC). Due to the low concentration and the limited sensitivity of the GC method the breath sample must go through a trap-and-purge process to be pre-concentrated. This is very time consuming and prone to errors. Due to these disadvantages other methods for the analysis of breath samples are currently developed.

One way to overcome these problems is to use the spectral fingerprint of the molecules, which can be observed, in the mid-infrared spectral region between 1.5 and 10 µm. This can be done by means of absorption spectroscopy with infrared lasers. This technique has proven to be a sensitive, selective and fast way for the detection of many hydrocarbons and other volatile compounds. More details on the principle of absorption spectroscopy can be found in the contribution of Werle in this book. It has been demonstrated that photoacoustic spectroscopy, which is one way to detect the fingerprint spectra of molecules in the IR, is applicable to sensitive analysis of hydrocarbons [see contribution of Kühnemann in this book].

In the last decade another absorption spectroscopic technique, the cavity ring-down spectroscopy (CRDS), has been developed and successfully applied to trace gas detection. Using the visible and near-infrared spectral fingerprint region it has been shown that extremely high sensitivities up to 10^{-9}/cm can be reached (Engeln et al. 1996; Paldus et al. 1998; Romanini et al. 1997). The principle of CRDS is described in more detail by Paldus and Lauterbach et. al in their contributions in this book (chapter 15 and chapter 7).

Conventional CRDS uses pulsed lasers to excite an external optical cavity where the subsequent decay of the pulsed ring-down signal is observed. Cavity-leak-out spectroscopy (CALOS) is a continuous-wave variant of CRDS. Compared to pulsed CRDS, the excitation of the cavity with a continuous-wave laser has the advantage of a higher spectral resolution. Moreover it allows the use of low-power laser sources, since the transmission through the ring-down-cell is up to six magnitudes higher than with pulsed CRDS.

With IR-laser spectroscopy it is furthermore possible to distinguish different isotopomeres, since they show slightly different absorption spectra due to the difference in mass contribution. This feature gives rise to the monitoring of different pathways in the human metabolism and in atmospheric chemistry (Dahnke et al. 2001b; Kleine et al. 2000).

Also, multigas-analysis can be performed, for example the simultaneous detection of methane and ethane in a breath sample (Dahnke et al. 2001a). As there is no pre-processing of the gas sample needed, this detection method offers a new way to real-time monitoring of various volatile compounds.

The medical applications presented here were performed using a CO-overtone-laser-sideband-spectrometer. As this configuration is fixed to the laboratory, an all solid-state laser source, based on difference-frequency generation (DFG), has been developed, which is small in size and leads to a transportable CALO spectrometer. In this article the application of CALO spectroscopy for medical breath tests will be described. These experiments have been carried out with the stationary CO-overtone-laser spectrometer. Furthermore the development and first results of the transportable DFG spectrometer will be described (Stry et al. 2002).

14.2 The CO-Overtone Spectrometer

The light source of the stationary CALO spectrometer is a continuous wave CO-laser (Urban 1995) in an overtone configuration that emits radiation on about 300 rovibrational $\Delta v=2$ transitions in the wavelength region between 2.6 and 4.0 μm (Bachem et al. 1993). The laser has a single line output power in the range between 50 and 500 mW.

Figure 14.1 shows a schematic of the stationary CALOS setup. The main parts are the CO laser, the ring-down cell (RDC), a detector and a computer. The CO-overtone laser is formed by a grating in Littrow configuration, a liquid nitrogen cooled discharge tube and an outcoupler (R=98 %) mounted on a piezotubus (PZT). In heterodyne experiments the spectral linewidth of the laser has been determined to be 100 kHz (integration time: 1 s) (Mürtz et al. 1998; Mürtz et al. 1999a). By mixing the laser light with microwave radiation in an electro-optic

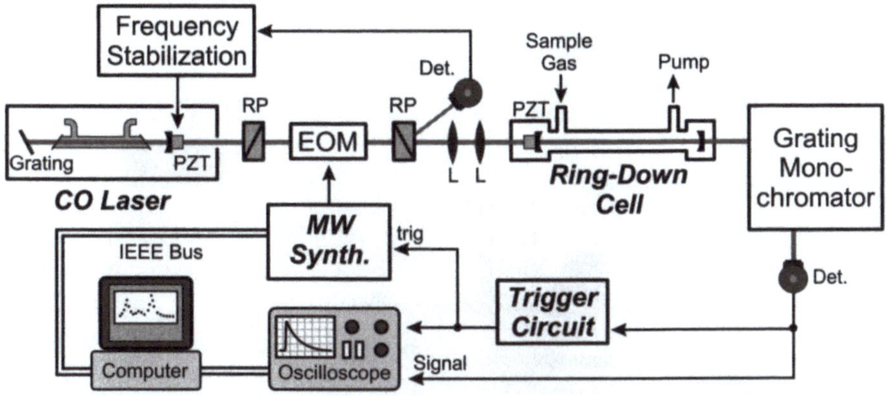

Fig. 14.1. Setup scheme of the stationary CALO spectrometer. *PZT*: piezoceramic transducer; *RP*: rochon polarizer; *L*: Lens; *EOM*: electro-optic modulator; *MW Synth*: microwave synthesizer

modulator, tunable laser sidebands are generated (power: ca. 100 – 200 μW), in the spectral range of 8 to 18 GHz above and below each laser line. To distinguish the two sidebands a grating monochromator is used, which only transmits the sideband wavelengths of interest onto the InSb detector.

The beam profile is mode matched by means of two lenses to excite the fundamental transverse mode (TEM_{00}) of the external high finesse resonator. This ring-down cell (RDC) consists of two high reflective mirrors (R=99.987 % at λ=3.3 μm) that are mounted in a stainless steel tube with a length of 50 cm (figure 14.2). The high reflectivity of the mirrors leads to an effective absorption path-length of 3.6 km. After exciting the resonator the light is turned off quickly and the decay time is determined via an exponential fit. Form the difference of the decay time of the empty cavity and the cavity filled with the gas sample the absorption coefficient can be directly determined in absolute values.

The ring down cell provides two gas supplies that enable a continuous flow through the cell. The complete gas setup is shown in figure 14.3. All parts in contact to the gas are made of stainless steel, glass or Teflon in order to prevent any

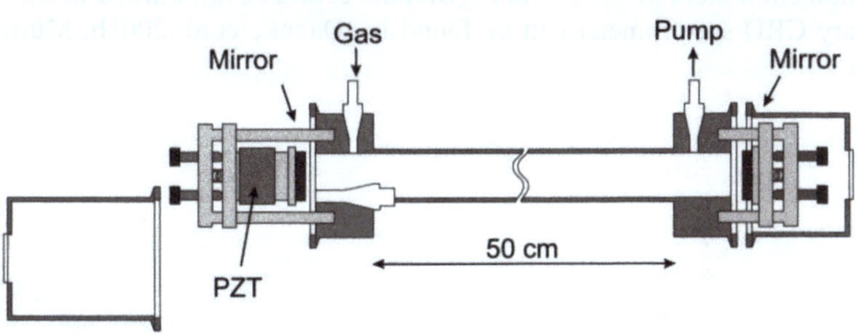

Fig. 14.2. Schematic of the CALOS absorption cell. The cell is made entirely of stainless steel. The piezoceramic tubus (PZT) allows a fine-tuning of the cavity length. One of the end pieces is demounted.

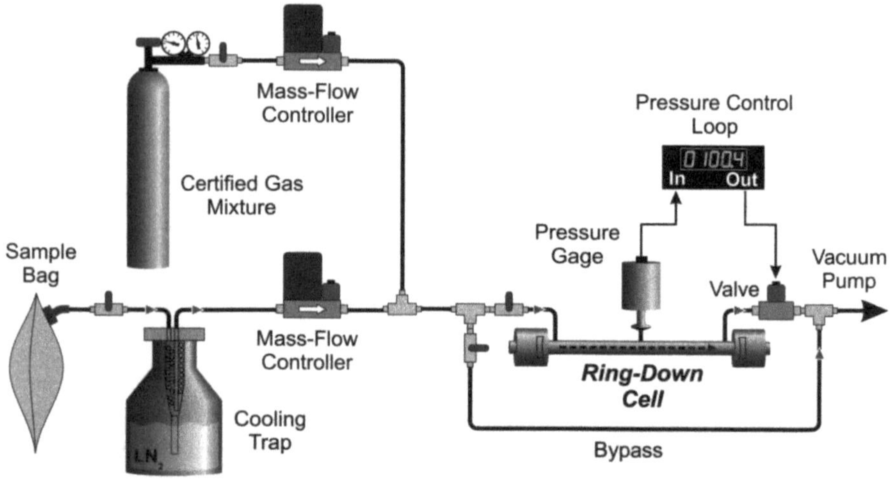

Fig. 14.3. Schematic of the gas-flow setup. The pressure control loop keeps the cell pressure at a fixed value (e.g. 100 mbar), while the flow (e.g. 100 sccm) can be controlled independently. The breath samples go through a LN_2 cooling trap to eliminate disturbing gas fractions (e.g. water). The certified gas mixture is used to calibrate the system.

interaction between the spectrometer and the sample gas. A control loop is used to stabilize the pressure inside the cell.

As the human breath contains many volatile compounds, which might disturb the analysis of ethane, it is necessary to filter the gas sample by a liquid nitrogen-cooling trap (temperature: 120 K).

In order to determine the instrumental detection limit, a certified gas mixture was used. Figure 14.4a) shows a spectrum of ethane near 3000.3 cm^{-1} at a concentration of 62.5 ppb and a pressure of 100 mbar. The solid line shows the corresponding spectrum of ethane obtained from a FTIR spectrometer (Popp et al. 2000) of 5 % ethane and scaled to the appropriate absorption for 62.5 ppb. The detection limit can be determined by the noise equivalent signal and is about 100 ppt with an integration time of 5 seconds. This is equivalent to a minimum detectable absorption coefficient of 1×10^{-9} cm^{-1} (Dahnke et al. 2001a). Further details on the stationary CRD spectrometer can be found in (Dahnke et al. 2001b; Mürtz et al. 1999b).

14.3 Demonstration of Medical Applications with the CO-Overtone Spectrometer

14.3.1 Oxidative Stress

The investigation of cell damage evoked by free radicals is a current major research field in biological and physiological medicine. The human organism is continuously affected by free radicals whereas the balance of sinks and sources of free radicals is called the oxidative stress status (Sies 1985). Several factors are believed to affect the oxidative stress status like hyperbaric oxygen, cigarette smoke or ultraviolet light (Knutson et al. 2000). Furthermore the pathogenesis of a number of chronic and acute diseases seems to be influenced by the cell damage induced by free radicals (Aghdassi and Allard 2000).

Free radicals, which are present in the human organism, can oxidise polyunsaturated fatty acids. This lipid peroxidation (Gardner 1989) can cause cell damage or even cell death and became one of the main targets in the investigation of oxidative damage of tissue (Pryor and Godber 1991). For this reason it is necessary to in-vivo monitor the lipid peroxidation. Several methods have been developed. The most common method used in clinical routine is the analysis of blood plasma for secondary reaction products such as malondialdehyde (MDA) (Kneepkens et al. 1994). Many of these tests show a low specificity and they are invasive since they need blood samples. The only available non-invasive way to monitor lipid peroxidation is the analysis of ethane (C_2H_6) or pentane (C_5H_{12}) in exhaled breath (Esterbauer 1996). These hydrocarbons are exclusively produced via a minor degeneration pathway of the lipid peroxidation, which ensures a high specificity. As they are poorly soluble in water or tissue, these by-products can be found in human breath as an indicator for the oxidative stress status in the human organism. The concentration found in breath ranges from hundred ppt up to several ppb. Ethane is considered to be the most specific volatile marker for the investigation of lipid peroxidation (Knutson et al. 2000).

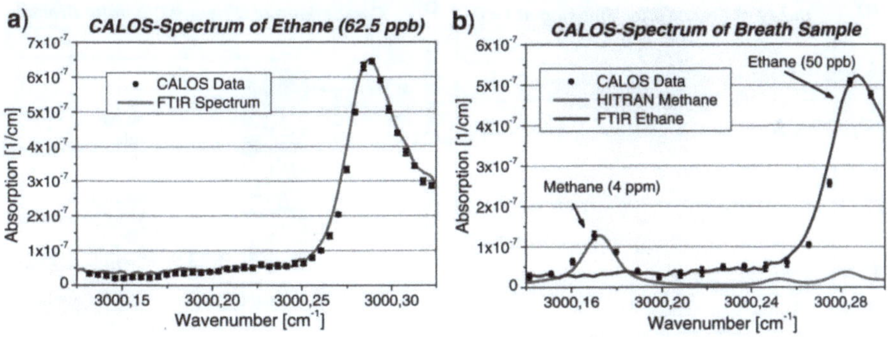

Fig. 14.4. a) Observed CALOS spectrum of ethane near 3000.3 cm^{-1}. The concentration of ethane was 62.5 ppb at a pressure of 100 mbar.
b) Observed CALOS spectrum of a breath sample (pressure: 100 mbar). The left arrow indicates the methane line that was used for the correction of the ethane data.

14.3.2 Measurements on Smokers

To demonstrate the suitability of the CALO spectrometer for precise time-resolved breath testing, investigations on exhaled breath after smoking a cigarette were performed. The breath of smokers was analyzed during a period of several hours after smoking a cigarette.

A typical spectrum of a breath sample is shown in figure 14.4b). The peak at 3000.29 cm^{-1} belongs to the absorption of 50 ppb ethane. Furthermore there is a second absorption line at 3000.172 cm^{-1}, which arises from methane absorption, present in ambient air. Using the HITRAN database (HITRAN 1996) the concentration of methane can be calculated to 4 ppm. This example shows that even a simultaneous detection of ethane and methane is possible.

The diagram in figure 14.5a) shows the result of a measurement that displays a strong increase and a subsequent decay of the ethane concentration in the exhaled breath after smoking a cigarette. For the analysis of the breath ethane fraction, the concentration was determined from the peak absorption of the ethane line. For the investigation breath samples were taken every 30 minutes in a period of 4 hours after smoking a cigarette. The test person (29-year-old male) normally smokes a cigarette every 45 min during the day and volunteered not to smoke during the 4-hour period. The first breath sample was taken 30 minutes after smoking the cigarette to ensure that no cigarette smoke was present in the lungs.

It is known that the ethane fraction is enhanced in the breath of smokers due to the free-radical induced damage of the respiratory system (Habib et al. 1995). Height and decay time strongly depend on the condition of the individual, on the smoking frequency and even on the cigarette brand. However, one should consider that part of the ethane observed is not endogenously generated but originates from the cigarette smoke, which has been absorbed in the respiratory system and is then slowly released.

Even the influence of different cigarette brands was examined. For this purpose an irregular smoker was asked to smoke a light and a strong cigarette. The result is

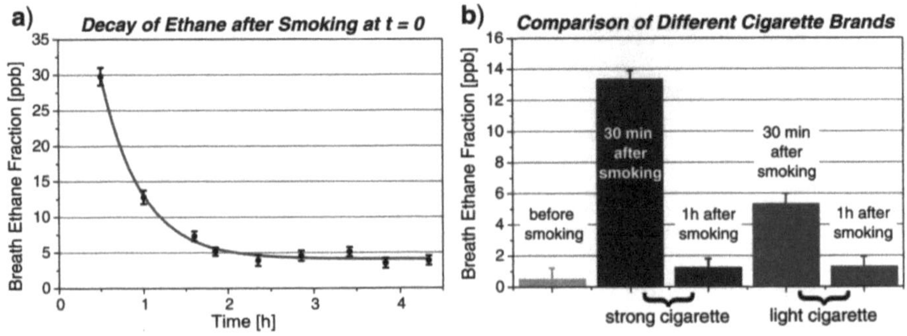

Fig. 14.5. a) Typical decay of the breath ethane fraction after smoking a cigarette. The cigarette was smoked at $t = 0$. Breath test were made every 30 minutes.
b) Observed influence of the cigarette brand on the ethane expiration.

shown in diagram 14.5b).

It reveals the possibility to distinguish a strong and a weak cigarette 30 min after smoking. A significantly increased level of exhaled breath ethane was found with the strong cigarette (0.9 mg nicotine, 12 mg tar condensate), which was nearly twice as high as the level of the light cigarette (0.6 mg nicotine, 9 mg tar condensate). One hour after smoking the ethane fraction had dropped to about 1 ppb, this is so close to the detection limit that the difference of the cigarette brands becomes unobservable.

These measurements are not considered to be a systematic study. They were performed to demonstrate the capability of the CALO spectroscopy for a highly sensitive fast and selective detection of ethane in human breath.

14.4 Further Applications

The CALO spectroscopy is also suitable for the detection of trace gases in environmental or atmospheric applications or even industrial matters. One example of an atmospheric measurement is the analysis of the greenhouse gas methane (CH_4), which is the most abundant hydrocarbon in the atmosphere and also plays an important role in complex feedback mechanisms in the troposphere and stratospheric chemistry. Within the last 300 years its average fraction in the atmosphere has risen from 0.8 to 1.7 ppm (IPCC 1995; Kennet et al. 2000). The measurement of the abundance of the stable isotopes is perceived as a key for understanding the sinks and sources of methane in the atmosphere. By knowing the exact isotopic ratio of $^{13}CH_4/^{12}CH_4$ it is possible to obtain information about the relative source strength of the diverse biochemical and biogenic as well as the anthropogenic sources of methane. For example, emissions caused by biomass burning are enriched in $^{13}CH_4$. Also other sources like rice fields or rumen of cattle can be characterized by their $^{13}CH_4/^{12}CH_4$ ratio, which varies from 10‰ to 80%The CO-overtone spectrometer was used for the detection of the $^{13}CH_4/^{12}CH_4$ ratio with precision of 11‰ at a detection limit of 100 ppt of methane (Dahnke et al. 2001b; Kleine et al. 2000; Kleine et al. 2001). This detection method has the great advantage of a high time resolution (>10 s). The standard method for isotope detection is gas chromatography, which reaches a higher precision but at the cost of a large sample volume and a process time of several hours.

Also other hydrocarbons show fingerprint spectra in the mid-infrared region and hence can be analyzed with the CALO spectroscopy. For atmospheric chemistry investigations and industrial matters the rapid online detection of formaldehyde (H_2CO) is an important application. Formaldehyde is a well-known volatile hydrocarbon, which is toxic for humans in even small concentrations and plays an important role in various atmospheric processes, e.g. the ozone formation in the troposphere. It could be shown that formaldehyde can be detected at the low ppb level by means of the CO-overtone spectrometer with a time resolution of a few seconds (Dahnke et al. 2002).

14.5 Transportable Setup (DFG Laser)

The presented applications have been tested in the lab because of the immobility of the CO-overtone laser. This shows the need for a transportable CALO spectrometer that can be taken to the patient's bedside in a hospital or to the field for *in-situ* atmospheric measurements. The only cw laser sources for the 3 μm wavelength region to date which are applicable for CALO spectroscopy are devices based on nonlinear frequency conversion, like optical parametric oscillation (OPO) or difference-frequency generation (DFG). An application of OPO's is described in the contribution of Kühnemann in this book (Popp et al. 2002). The transportable spectrometer described here is based on difference frequency generation.

By mixing two near-infrared laser sources in an efficient non-linear crystal made of periodically poled lithium niobate (PPLN), a tunable spectroscopic light source has been realised. The pump source is an external-cavity grating-tuned diode laser (EOSI) in Littmann configuration with a single-mode emission between 794 nm and 820 nm and an output power of 20 mW. The other light source is a single frequency, diode-pumped monolithic Nd:YAG ring laser with an output power of 670 mW at 1064 nm. Difference-frequency mixing is obtained in a bulk PPLN crystal (length: 5 cm, width 500 μm). The nonlinear crystal has 21 different gratings with poling periods varying from 20.6 μm to 22.6 μm. Translating the crystal to the adjacent grating results in a frequency shift of 20 cm^{-1} to 75 cm^{-1} depending on the utilized poling period. The fine-tuning is performed by altering the crystal temperature. The infrared output powers amount to about 27 μW in the 3 μm region which corresponds to a conversion efficiency of 0.042 %/(W·cm); the modehop-free tuning range is about 50 GHz. The accessible wavelength range is between 2997 cm^{-1} and 3340 cm^{-1}. The optical set-up of the complete spectrometer has a size of 60 x 80 cm. For both lasers a nearly diffraction limited spatial beam

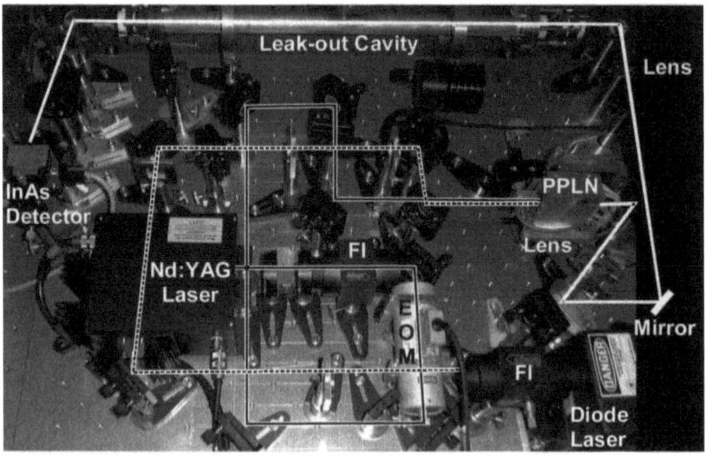

Fig. 14.6. Experimental setup (60 x 80 cm) of a DFG laser combined with a high-finesse cavity. *FI*: Faraday Isolator, *EOM*: Electrooptical Modulator, *PPLN*: Periodically Poled Lithium Niobate. The lines show the different laserbeams.

shape ($M^2 = 1.15$) was observed which is needed for a high efficiency of the conversion process. Furthermore, in order to achieve a good coupling to the high-finesse cavity, a pure TEM_{00} mode profile is necessary. Our measurements of the beam parameters showed that the DFG-laser is also nearly diffraction limited with a beam quality factor of $M^2 = 1.1 \pm 0.1$. In heterodyne experiments with the CO-overtone laser the linewidth of the DFG spectrometer has been determined to be 1 MHz in 20 ms.

DFG light sources have found application for trace gas detection by means of conventional absorption spectroscopy (Richter et al. 2000). The system presented here shows the potential for very sensitive trace gas detection since this DFG setup is used for CALO spectroscopy. This gives rise to the high absorption path length described above which is a factor of 100 longer than the path length of conventional absorption spectroscopy. Besides the advantages of the compact setup this spectrometer is not restricted to the sideband tunability of the CO-overtone spectrometer. A fast alteration of the DFG wavelength with a rate of 0.013 cm^{-1}/V can be achieved by tuning the diode laser wavelength. This enables the use of the strongest absorption features within the 3 μm range which leads to an optimal detection limit and very low cross interference of other gases. The complete setup is shown in Fig. 14.6. The ring-down cell and the gas system are equivalent to the CO-overtone setup described above.

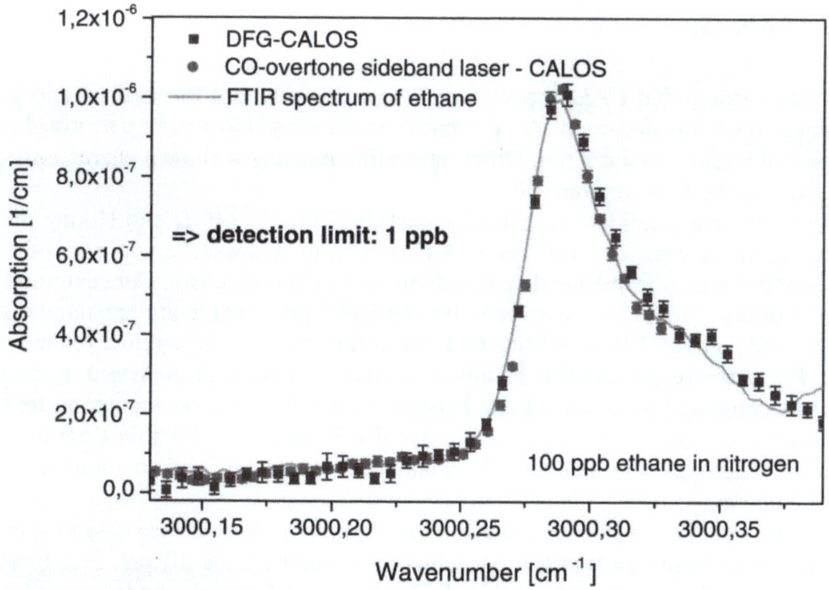

Fig. 14.7. High-resolution cavity leak-out spectrum of 100 ppb ethane in nitrogen at a pressure of 100 mbar. The squares show the measurement using the DFG-laser system as light source for the spectrometer, the dots the measurement done with the CO-overtone sideband laser. Solid line: data from an FTIR ethane spectrum (spectral resolution: 0.005 cm^{-1}, scaled to the corresponding concentration). The error bars indicate the 1σ uncertainty.

The capability of the spectrometer was evaluated for the detection of ethane. Fig. 14.7 shows the CALO spectrum of a gas mixture containing 100 ppb ethane in grade 5 nitrogen. The same absorption feature as shown in Fig. 14.4 was chosen to compare the DFG spectrometer with the CO-overtone spectrometer. For comparison both DFG and CO-overtone spectra are plotted in figure 14.7.

Since the power of the DFG laser is about one order of magnitude below the power of the CO-overtone sideband power the error bars of the DFG based measurement are bigger. The detection limit of the DFG spectrometer is 1 ppb for ethane. The comparison with CO-overtone spectrum shows that a higher power results in a directly higher sensitivity. A promising way to improve the detection sensitivity is to increase the power of the pump laser by use of a tapered amplifier. For this a MOPA system (Master Oscillator Power Amplifier) is currently integrated into the spectrometer that enlarges the power of the pump laser by a factor of 20. This power amplification will directly enlarge the power of the DFG laser by the same factor, since the DFG signal scales with the product of the power of the two mixed lasers.

This shows that the DFG spectrometer, which has proven to be capable for CALO spectroscopic application, has the potential to become a very sensitive tool for trace gas detection. Since it is transportable it enlarges the number of applications and allows real time trace gas detection outside the lab.

14.6 Outlook

It has been shown that CALO spectroscopy is a versatile tool for medical and atmospheric trace gas detection. All presented applications have been performed in the spectral region around 3 μm. Other interesting molecules show a characteristic fingerprint in the 5 μm region.

One prominent candidate is carbon monoxide (CO). As Mürtz and Hering discuss in their review article in this book, CO is not only a noxious component of air pollution, but is also formed endogenously in the human organism. Concentrations found in human breath vary between 500 ppb and 5 ppm. There are presumptions that the endogenous CO is related to a protection mechanism against oxidative stress. Furthermore the exhaled breath of smokers contains an increased fraction of CO as compared to non-smokers, because they inhale carbon monoxide from the cigarette smoke, which is deposed in the blood and released within a period of 24 hours. Therefore the analysis of exhaled CO could be used to monitor the smoking habits.

Another volatile organic compound, which could be detected via CALO spectroscopy, is carbonyl sulfide (OCS), which also shows a significant fingerprint spectrum around 5 μm. As reported by Risby et al., OCS can be used as a marker for hepatic disorders (Risby et al. 2001).

For the CALO spectroscopy in the 5 μm region, a CO laser working on the $\Delta v=1$ transition is an available light source which will be applied for CALOS in the near future. Another promising laser available for CALOS spectroscopy in this spectral region is the quantum cascade laser (QCL) (Ganser et al. 2001). It has been applied for cavity ring-down spectroscopy by (Paldus et al. 2000). Since the QCL is a small solid-state laser it is currently investigated to integrate it into the transportable setup.

Acknowledgement

The Authors are grateful to group leaders Prof. Dr. P. Hering, Dr. M. Mürtz for their helpful advice and organisation. We would also like to thank Dr. D. Kleine for his contribution to this work and Dr. J.-P. Meyn, University of Kaiserslautern, for providing us several PPLN crystals.

The presented works were financially supported by the German Research Foundation (Deutsch Forschungsgemeinschaft – DFG) and the German Federal Foundation for the Environment (Deutsche Bundesstiftung Umwelt – DBU).

References

Aghdassi E and Allard JP (2000) Breath alkanes as a marker of oxidative stress in different clinical conditions, Free Radic. Biol. Med. **28**:880-886

Bachem E, Dax A, Fink T, Weidenfeller A, Schneider M, and Urban W (1993) Recent progress with the CO-overtone $\Delta v=2$ laser, Appl. Phys. B. **57**:185-191

Dahnke H, Basum Gv, Kleinermanns K, Hering P, and Mürtz M (2002) Rapid formaldehyde monitoring in ambient air by means of cavity leak-out spectroscopy, Appl. Phys. B **75**:311-316

Dahnke H, Kleine D, Hering P, and Mürtz M (2001a) Real-time monitoring of ethane in human breath using mid-infrared cavity leak-out spectroscopy, Appl. Phys. B **72**:971-975

Dahnke H, Kleine D, Urban C, Hering P, and Mürtz M (2001b) Isotopic ratio measurement of methane in ambient air using mid-infrared cavity leak-out spectroscopy, Appl. Phys. B **72**:121-125

Engeln R, vonHelden G, Berden G, and Meijer G (1996) Phase shift cavity ring down absorption spectroscopy, Chem. Phys. Lett. **262**:105-109

Esterbauer H (1996) Estimation of peroxidative damage. A critical review, Pathol. Biol. (Paris) **44**:25-28

Ganser H, Frech B, Jentsch A, Mürtz M, Gmachl C, Capasso F, Sivco DL, Baillargeon JN, Hutchinson AL, Cho AY, and Urban W (2001) Investigation of the spectral width of quantum cascade laser emission near 5.2 μm by a heterodyne experiment, Opt. Commun. **197**:127-130

Gardner HW (1989) Oxygen radical chemistry of polyunsaturated fatty acids, Free Radic. Biol. Med. **7**:65-86

Habib MP, Clements NC, and Garewal HS (1995) Cigarette smoking and ethane exhalation in humans, Am. J. Respir. Crit Care Med. **151**:1368-1372

HITRAN (1996) Database available at http://www.hitran.com

IPCC (1995) Second assesment report: climate change 1995; available at http://www.ipcc.ch

Kennet JP, Cannariato KG, Hendy IL, and Behl RJ (2000) Carbon isotopic evidence for methane hydrate instability during quaternary interstadials, Science **288**:128-133

Kleine D, Dahnke H, Urban W, Hering P, and Mürtz M (2000) Real-time detection of (CH_4)-C-13 in ambient air by use of mid-infrared cavity leak-out spectroscopy, Opt. Lett. **25**:1606-1608

Kleine D, Mürtz M, Lauterbach J, Dahnke H, Urban WG, Hering P, and Kleinermanns K (2001) Atmospheric trace gas analysis with cavity ring-down spectroscopy, Isr. J. Chem. **41**:111-116

Kneepkens CM, Lepage G, and Roy CC (1994) The potential of the hydrocarbon breath test as a measure of lipid peroxidation, Free Radic. Biol. Med. **17**:127-160

Knutson MD, Handelman GJ, and Viteri FE (2000) Methods for measuring ethane and pentane in expired air from rats and humans, Free Radic. Biol. Med. **28**:514-519

Mürtz M, Frech B, Palm P, Lotze R, and Urban W (1998) Tunable carbon monoxide overtone laser sideband system for precision spectroscopy from 2.6 to 4.1 μm, Opt. Lett. **23**:58-60

Mürtz M, Frech B, and Urban W (1999a) High-resolution cavity leak-out absorption spectroscopy in the 10 μm region, Appl. Phys. B **68**:243-249

Mürtz M, Kayser T, Kleine D, Stry S, Hering P, and Urban W (1999b) Recent developments on cavity ring-down spectroscopy with tunable cw lasers in the mid-infrared, Proc SPIE **3758**:53-62

Paldus BA, Harb CC, Spence TG, Wilke B, Xie J, Harris JS, and Zare RN (1998) Cavity-locked ring-down spectroscopy, J. Appl. Phys. **83**:3991-3997

Paldus BA, Harb CC, Spence TG, Zare RN, Gmachl C, Capasso F, Sivco DL, Baillargeon JN, Hutchinson AL, and Cho AY (2000) Cavity ringdown spectroscopy using mid-infrared quantum-cascade lasers, Opt. Lett. **25**:666-668

Popp A, Dahnke H, Orphal J, Basum Gv, Burrows JP, and Kühnemann F (2000) HASIBO- a database for high-resolution absorption cross-sections of VOCs in the mid-IR

Popp A, Müller F, Kühnemann F, Schiller S, Basum Gv, Dahnke H, Hering P, and Mürtz M (2002) Ultra-sensitive mid-infrared cavity leak-out spectroscopy using a cw optical parametric oscillator, Appl. Phys. B (submitted)

Pryor WA and Godber SS (1991) Noninvasive measures of oxidative stress status in humans, Free Radic. Biol. Med. **10**:177-184

Richter D, Lancaster DG, and Tittel FK (2000) Development of an automated diode-laser-based multicomponent gas sensor, Appl. Opt. **39**:4444-4450

Risby TH, Sehnert SS, Jiang L, and Burdick JF (2001) Volatile biomarkers for analysis of hepatic disorders **Patent**:6248078-

Romanini D, Kachanov AA, Sadeghi N, and Stoeckel F (1997) CW cavity ring down spectroscopy, Chem. Phys. Lett. **264**:316-322

Sies H (1985) Oxidative stress: oxidants and antioxidants

Stry S, Hering P, and Mürtz M (2002) Portable difference-frequency laser based cavity leak-out spectrometer for trace gas analysis, Appl. Phys. B **75**:297-303

Urban W (1995) Physics and spectroscopic applications of carbon-monoxide lasers, a review, Infrared Phys. Techn. **36**:465-473

15 Practical Applications of CRDS in Medical Diagnostics

B. A. Paldus, E. R. Crosson and H. Dahnke

15.1 Introduction

Cavity ring-down spectroscopy (CRDS) is revolutionizing the sensitivity, speed, ease of use, robustness, and portability of trace chemical species detection. Today, CRDS is being commercialized across a broad range of application areas, including medical diagnostics, industrial process control, environmental monitoring, and civilian and military security. In this monograph we will address the application of CRDS to medical breath testing, a new field in medical diagnostics that promises to usher in a new era of testing methodology with improved specificity, sensitivity, and painless administration compared to the current generation of medical tests.

A CRDS instrument measures the concentration of a trace gas by observing its effect on a beam of light circulating inside a small sample chamber (Busch 1997). The rate of decay of the light intensity indicates the concentration of the targeted chemical species with sensitivities typically in the parts per billion range and in some cases to better than one part per trillion. Though well established within the scientific community for over ten years, CRDS has historically been too complex for use outside the laboratory. We will describe several of the pioneering innovations that enable CRDS to be deployed in commercial environments, such as doctor's offices or clinics where the staff have minimal technical training and expert technicians are not available on a regular basis.

15.2 Cavity Ring-down Spectroscopy

Anderson and coworkers discovered CRDS in 1985 while developing a technique to measure the reflectivity of very high quality mirrors (Anderson 1984). "[We] found our results depended on the ambient Pasadena air quality." They observed CRDS as an annoying experimental artifact. The first reported use of CRDS as an analytical technique appeared in the literature in 1988 in a paper by O'Keefe and Deacon, who used a pulsed dye laser to investigate the overtone transitions of molecular oxygen (O'Keefe 1988). O'Keefe began a collaboration with U.C. Berkeley which endures to this day. In 1990, the Berkeley group reported CRDS experiments with a pulsed dye laser (P-CRDS) to investigate the spectroscopic constants of jet-cooled metal clusters in a molecular beam (O'Keefe 1990).

From 1993 through 1996, K.K. Lehmann and D. Romanini at Princeton applied CRDS to spectroscopic studies of line intensities, transition wavelengths, and coupling effects in the vibrational and rotational spectra of gas phase species (Romanini 1993). In 1995, a group at the University of Nijmegen demonstrated the utility of P-CRDS studies for trace gas detection in the ultraviolet. Using a pulsed Nd:YAG pumped dye laser followed by a frequency doubler to obtain radiation at 300 nm, the Nijmegen group demonstrated P-CRDS sensitivities of 10 parts per billion to ammonia and 1 part per trillion to elemental mercury in air (Jongma 1995). In 1996, Hodges, Looney, and van Zee at NIST conducted detailed studies of laser source effects on the noise and sensitivity of P-CRDS (Hodges 1996).

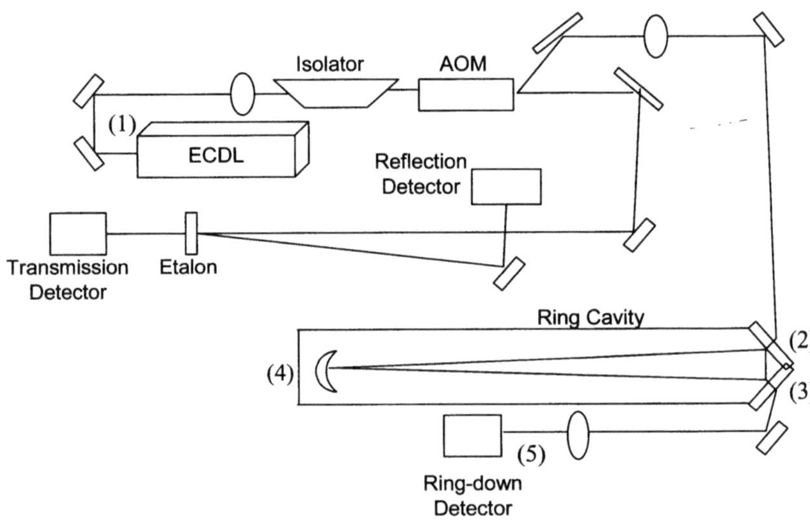

Figure 15.1. Schematic of CW-CRDS implementation using an external cavity diode laser.

In 1996, Lehmann proposed coupling a narrowband continuous wave diode laser to a ring-down cavity (CW-CRDS) in order to build up enough energy in the cavity to perform shot noise limited measurements. In 1997, Romanini and colleagues demonstrated that in fact a continuous-wave laser (CW-CRDS) provided more light on the sensor, thus improving the signal-to-noise ratio (Romanini 1997). In 1999, Paldus and Zare achieved a record equivalent sensitivity of 10^{-12} $cm^{-1}/Hz^{1/2}$ (Spence 2000). Applications of CRDS include simple molecular spectroscopy, detection of concentration profiles of ions in flames and plasmas, to reaction dynamics measurements, and detection of isotopic ratios. In 2001 Dahnke et al published the first use of CRDS for medical breath analysis (Dahnke 2001). In 2002, Crosson and Paldus achieved a reproducibility of 3 parts in 10,000 in measurements of isotopic CO_2 in human breath with a sensitivity of 10^{-11} $cm^{-1}/Hz^{1/2}$ (Crosson 2002).

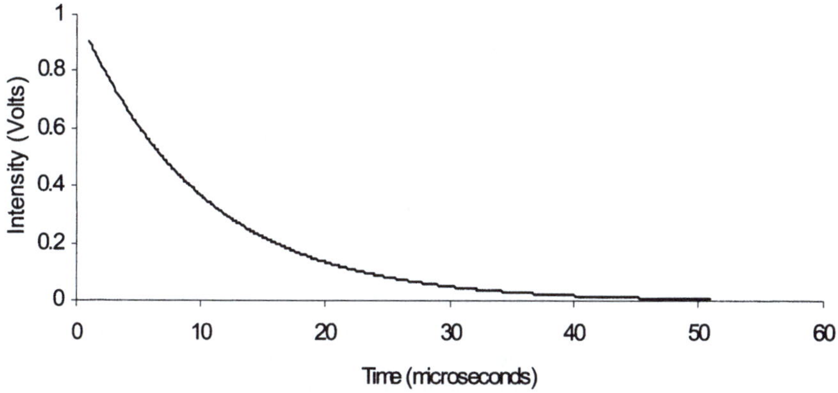

Figure 15.2. Typical Ring-down trace showing exponential decay.

Early attempts to conduct CRDS experiments with continuous wave diode laser sources revealed a problem: whenever a beam generated by a diode laser reflects directly back into the laser, as it does in a linear cavity CRD configuration, the optical feedback will result in phase fluctuations and mode hopping in the laser. Paldus and Zare solved this problem in 1997 by placing an acousto-optic modulator (AOM) between the laser and the cavity (Paldus 1997). Results indicated that the AOM, isolated the laser diode from the cavity thus leaving the frequency and linewidth of the laser diode unperturbed. They reported a sensitivity of 20 ppm in measurements of water vapor. Later in 1998, the group demonstrated they could greatly reduce feedback into the diode laser by using a ring resonator cavity (such as a triangle or bow tie) rather than a linear cavity (Paldus 1998). This improvement eliminated the need for optical or frequency isolation for many applications, including the use of mid-infrared quantum cascade lasers (Paldus 2000).

Although continuous wave laser-diode based CRDS can assume many different forms, they all share the same basic principle of operation. For definiteness, consider our implementation, illustrated in Figure 15.1. A continuous wave tunable diode laser *(1)* shines into an optical cavity, which consists of three high reflectivity mirrors (typically 99.99% reflectance). The light circulates around the cavity *(2-3-4-2-3-4-2-3....)*, executing thousands of loops. At every reflection, a small fraction of the light, typically 0.01%, escapes through one of the mirrors. A photodetector *(5)* captures and records the light that escapes from one end of the cavity. At the beginning of a measurement, the laser injects light into the cavity for a few microseconds, then shuts off. Without replenishment, the intensity of the circulating light in the cavity decays exponentially. Figure 15.2 shows a typical decay curve as recorded by the photodetector. The characteristic decay time—the "ringdown time"—depends on the size and configuration of the cavity and the quality of the mirrors.

Now assume that the cavity contains trace amounts of the target species to be detected. Suppose also that the laser wavelength corresponds to a peak in this target molecule's absorbance. The target species in the cavity will then absorb some of the circulating light and accelerate the ringdown time. The decrease in the ringdown time correlates directly with the concentration of the target gas in the cavity. The actual measurement procedure is slightly more complicated. Instead of measuring at just one wavelength, the laser is tuned across the peak, as shown in Figure 15.3. The instrument makes a series of consecutive ringdown measurements at closely spaced wavelengths. The measured ringdown time off the absorbance peak gives the baseline value that tells how quickly the light decays without influence from the gas. Thus, the instrument intrinsically auto-calibrates itself at the beginning of every measurement sequence without the need for reference or calibration samples. Subsequent measurements subtract this baseline value to profile the absorption peak of the target species. The instrument measures about 1000 points in less than one second and calculates the total area under the peak.

CRDS offers advantages over traditional absorption spectroscopies in all areas that to date have posed limitations on the use spectroscopic techniques for diagnostics applications. Primarily, it addresses the issues of stability, allowing instruments to be used by unskilled operators; sensitivity, allowing the detection of trace levels of gases as typically occur in human breath; and selectivity, allowing the direct identification of markers and correlation of such markers with disease.

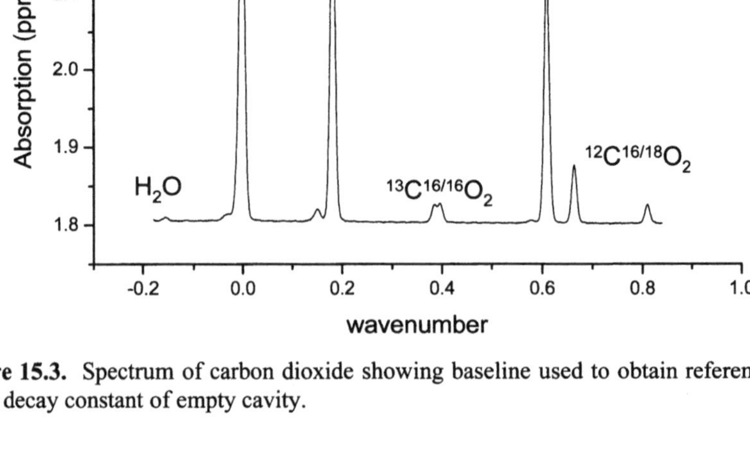

Figure 15.3. Spectrum of carbon dioxide showing baseline used to obtain reference ringdown decay constant of empty cavity.

The strengths of CRD are summarized below.

Stability. CRDS measures the characteristic time of the decay of light exiting the cavity after the source laser has been shut off. As such CRDS is insensitive to fluctuations in the intensity of the source laser. The extremely high tolerance of CRDS to laser fluctuations reduces laser requirements, thus allowing the use of less expensive laser sources, while still vastly outperforming traditional spectroscopic techniques.

Sensitivity. The extremely long effective path length, up to 10 km, translates directly into extremely high sensitivity because the light beam has more opportunity to interact with the sample gas. Also, the measurement of a characteristic time is far less prone to systematic errors than the determination of an absolute signal intensity or two relative signal intensities, further enhancing performance.

In automated, inexpensive instruments, we have demonstrated sensitivities that are comparable to mass spectrometry.

Specificity. CRDS's extreme sensitivity also enhances its potential selectivity. The limited sensitivity of competing techniques generally requires them to target the strongest spectral features of the target molecule; the constrained choice makes finding a suitable peak that is free from interferences difficult or even impossible. CRDS enables us to be more selective and, if necessary, to choose weaker peaks that avoid interferences. In effect, a small trade-off in detection sensitivity, which CRDS has in abundance, can vastly improve specificity.

Absolute Measurement. CRDS is an absolute measurement. Unlike most other spectroscopic techniques, it requires no calibration or gas standard.

Self-calibration. As described above, every measurement sequence incorporates an intrinsic recalibration by tuning the laser first to the baseline and then through the absorption peak. No special calibration procedure or reference gases are required.

Situation. Non-optical techniques such as gas chromatography, mass spectrometry, and ion mobility spectroscopy all force the physical separation of the target analyte from the matrix and then count up the results. But accounting for the motion of individual molecules is a difficult proposition: even in an ultra-high vacuum chamber, quintillions of stray molecules bounce around and can potentially knock a molecule into the wrong bin. CRDS measures the sample as it is, *in situ*, and accepts a full sample directly into the measurement cavity.

Simplicity. CRDS requires only a few basic optical components and no moving parts. We can fuse all of the components together into one monolithic, vibration-resistant body. We are reducing all of the electronics, intelligence and diagnostics needed to run the measurement onto a single logic card, so that the instrument simply reports the concentration of the target gas. Anyone can operate the instrument with a few minutes of training. Sample preparation and introduction are trivial: for example, in a breath diagnostics application, the patient will simply breathe through a straw into a test tube and the test tube will then be inserted into the instrument. The concentrations of the analytes of interest will be measured within at most a few minutes.

Though CRDS has gained rapid acceptance as a high-end laboratory technique, it is not widely known outside of an academic setting. We have pioneered several key innovations that simplify the technique from a laboratory optical table filled with parts to a simple, manufacturable prototype with very high accuracy and precision (Figure 15.4). These innovations cover the four principal subsystems of a CRDS instrument: the laser/laser controller, the cavity, the detection/signal processing system, and the stabilization of the temperature and pressure of the sample:

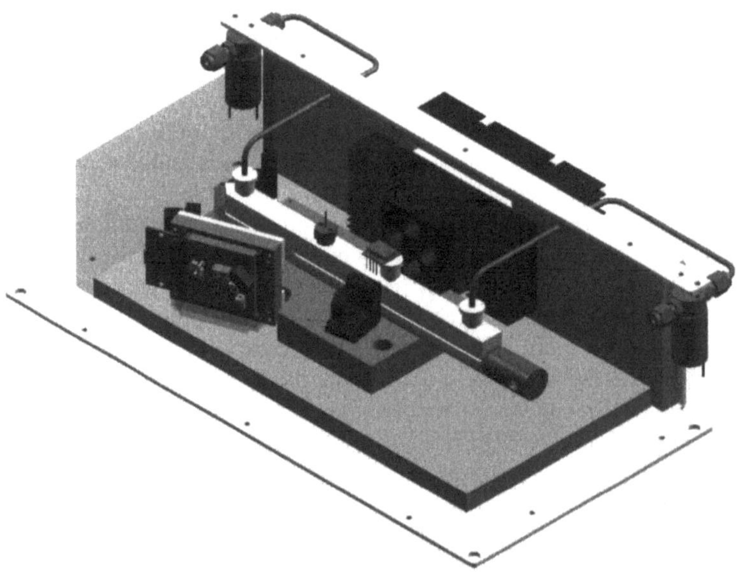

Figure 15.4. Cavity ring-down universal prototype.

Continuous wave tunable diode lasers (TDL). Instead of bulky, pulsed laser systems, our implementations use ordinary TDLs mass-produced for the telecommunications industry, such as distributed feedback lasers (DFB) for single species detection systems, or broadly tunable external cavity lasers (ECL) for multi-species detection. Continuous wave operation increases the measurement bandwidth and ultimate instrument performance. Proprietary electronics allow us to switch the laser on and off to perform the measurement without destabilizing the laser. These electronics also allow extremely accurate control of laser wavelength to generate high precision in the frequency axis data collected for each spectrum.

Novel cavity configurations. We have designed ring cavities that require far fewer supporting optical components to interface to the laser and the detector. Typically for a linear ring-down cavity the incident light from the laser is retroreflected into it, so that isolation and feedback stabilization schemes need to be implemented in order to accurately control its wavelength. For a ring cavity, however, the incident light is reflected off the incoming mirror into a detector or a beam dump and never returns to the source, thereby allowing much simpler coupling optics. By designing the laser and the cavity to be optically matched it is also possible to simplify further any mode matching optics and maximize the light throughput from the ring-down cavity.

Detection electronics and algorithms. We have optimized the processing of the detector signal using proprietary analog circuitry to extract a 100X increase in signal-to-noise performance with inexpensive A/D converters. Simple on-board logic handles all of the subsequent data processing and the extraction of a analyte concentration from the ring-down curves. This circuitry also allow the precise and accurate measurement of the temperature and pressure of the sample which is critical when making absolute measurements of concentration.

Temperature and Pressure Stabilization. In applications requiring absolute concentration measurements or isotopic ratio measurements, the pressure and temperature of the sample become important parameters in determining measurement accuracy and precision. We have developed simple, rapid, and inexpensive means for stabilizing the temperature of the sample to within 4 mK and the pressure to within 0.2 Torr. This reduces the error from environmental fluctuations during measurement to less than 1 part in 10,000.

15.3 Applications

15.3.1 Helicobacter Pylori Detection

In 1983, Warren and Marshall first isolated Helicobacter pylori, a spiral shaped bacterium that lives in the lining of the stomach and duodenum. Approximately 2/3 of the world's population is infected with H. pylori. The incidence of H. pylori in the United States is about 35%, and it increases by about 1% with each year of age. The incidence approaches 35% in Europe and 90% in Africa and Asia.

H. pylori is an important pathogen in the upper GI tract of humans. Research has revealed strong associations between H. pylori infection and peptic ulcer disease and between H pylori infection and gastric cancer. In this section we discuss the relationships between H pylori and those diseases and the conditions in which the medical community is testing for H pylori.

Since the discovery of H. pylori as a gastrointestinal pathogen, the medical community has come to recognize the majority of peptic ulcer disease as an infectious disease readily cured with oral antibiotic therapy. Research has associated H. pylori with more than 90% of cases of duodenal ulcer and 60% to 80% of cases of gastric ulcer. In fact, tests for H. pylori infection are the only laboratory tests useful in the diagnosic workup of peptic ulcer disease. For peptic ulcer patients who undergo antibiotic therapy to eradicate H. pylori and then experience recurrence of their symptoms, the consensus is that the physician should repeat the H. pylori test to confirm a cure. One study showed that many patients who do not experience recurrence of their symptoms nevertheless request repeat testing for confirmation. Four alternatives to 13C isotopic breath testing exist today: endoscopy, serology, stool testing, and a radioactive carbon-14 breath test.

Endoscopy/biopsy is the gold standard for patients with alarm symptoms for gastric cancer including weight loss, age, and family history. The endoscopy takes place in the office of the gastroenterologist, not the office of the primary care physician. This procedure entails some discomfort and some risk to the patient. The gastroenterologist inserts a small tube into the mouth, through the esophagus, and into the stomach of the patient to procure a sample of tissue. A laboratory conducts the analysis of the biopsy by histological examination of stained tissue, microbiological culture of the organism, or direct detection or urease activity in the tissue. The physician typically learns the results 1 or 2 days after the endoscopy.

Most often, gastroenterologists, internists, and primary care doctors typically employ a serology blood test which measures the H. Pylori antibody. Tests with office-based serologic kits can be conducted in less than 10 minutes. They offer sensitivity greater than 80% and specificity near 90%, but they cannot distinguish between a present infection and a past infection during the preceding 3 years. Therefore they cannot be used in follow-up tests to check for recurrence or successful eradication of the infection. The serology test does not require that the patient refrain from the medications listed above which would adversely affect the endoscopy/biopsy results.

The FDA recently approved a stool test for H. pylori in the United States. Although experience with this test is limited, it seems to offer accuracy comparable to that of the other available tests. There are inherent disadvantages in stool diagnostics as collection of the sample is much less appealing to the patient, less timely, and the laboratory handling of the specimen is more complicated.

In the radioactive H. pylori test, the patient ingests a capsule of urea labeled with one microCurie of carbon-14, a radioactive isotope. If the patient is infected with H. pylori, then urease, an enzyme secreted by the bacteria, will break down the urea to form labeled carbon dioxide ($^{14}CO_2$) and ammonia (NH_3) at the interface between the gastric epithelium and the lumen. $^{14}CO_2$ dissolves into the bloodstream. In the lungs, the $^{14}CO_2$ diffuses across the alveolar-capillary membrane to be exhaled in the breath. The physician can collect the breath in a plastic bag, then send it to a nuclear medicine laboratory for analysis with a scintillation counter. The physician typically receives the results the next day. The radioactive material renders the test unsuitable for children and pregnant women. Furthermore, some patients resist the idea of ingesting a radioactive material.

The carbon-13 isotopic breath is similar to the carbon-14 test, but does not involve administration of a radioactive material. First the patient exhales into a plastic bag to provide a baseline measurement. The patient then ingests urea labeled with carbon-13, a stable (non-radioactive) isotope. If H. pylori is present in the stomach, urease will break down the urea and produce $^{13}CO_2$. This extra $^{13}CO_2$ causes the ratio of $^{13}CO_2/^{12}CO_2$ to rise by about 5 parts per thousand compared to its baseline value. The patient exhales into collection bags 30 minutes after ingesting the urea. This places the burden of detection on the instrument measuring the isotopic ratio change. Mass spectrometry has been the incumbent technology for at least a decade owing to its high sensitivity and accuracy. However, it is expensive and requires frequency maintenance by skilled technicians.

To date, isotope ratio mass spectrometry (IRMS) has been the method of choice for measuring changes in isotopic ratio. However, over the past decade, optical alternatives, such as non-dispersive infrared spectroscopy (IRMS) have made inroads into medical isotopic measurements. However, most non-coherent methods lack the ultimately sensitivity and precision required to compete head-on with IRMS. CRDS, however, has shown the potential not only to approach single collector IRMS in performance, but to offer a more compact, less expensive and more portable alternative. The precision of CRDS is about five times less than that of the best commercially available IRMS systems, but we anticipate that as CRDS technology evolves, it will rival, if not surpass IRMS in the coming decade. Moreover, CRDS can measure a single sample repeatedly and non-destructively.

The light isotope of carbon, ^{12}C, has a natural abundance of 98.89 % whereas ^{13}C has a natural abundance of 1.11 %. Carbon isotope ratios are usually expressed in parts per thousand (or per mil, $^{o}/oo$) relative to a standard, using the common notation

$$\delta^{13}C = \frac{R_{\text{sample}} - R_{\text{standard}}}{R_{\text{standard}}} \times 1000,$$

where

$$R = {}^{13}C \big/ {}^{12}C$$

is the ratio of the heavier to the lighter stable isotope of carbon. For example, the naturally occurring $\delta^{13}C$ values for biologically interesting carbon compounds range from roughly 0 $^{o}/oo$ to ~ -110 $^{o}/oo$ relative to the Pee Dee Belemnite (PDB) standard. The common reference for $\delta^{13}C$ was obtained from a cretaceous marine fossil, Belemnitella americana, from the Pee Dee formation in South Carolina. This material has a higher ^{13}C / ^{12}C ratio than nearly all other natural carbon-based substances. For convenience, it is assigned a $\delta^{13}C$ value of zero, giving almost all other naturally occurring samples negative $\delta^{13}C$ values. All original supplies of PDB have been essentially exhausted and replaced by secondary carbonate standards calibrated against those prepared by the U.S. National Institute of Standards and Technology (NIST).[3]

A double collector IRMS dedicated to isotopic ratio determinations of carbon dioxide gas has the capability to achieve an instrumental precision of 0.01 $^{o}/oo$ in the determination ($n > 10$) of $\delta^{13}C$ for carbon dioxide.[13] These instruments, however, are confined to specialized analytical laboratories and are not practical for routine isotope ratio measurements, especially in a traditional medical applications environment such as a doctor's office. Traditional single-collector IRMS instruments have a precision in the determination of $\delta^{13}C$ of about 0.04 $^{o}/oo$ under the most favorable conditions, with a typical precision of 0.2 $^{o}/oo$. This can be compared to the CRDS presented here which has a precision of 0.22 $^{o}/oo$.

We have demonstrated the power of the CRDS method for determining isotope ratios by examining 1.0 mL of 5% CO_2 in N_2 in which the ratio of $^{13}CO_2$ is compared to $^{12}CO_2$ with a delta value of –27.76 ‰ ± 0.22 ‰. The design of the optical components of the ring-down spectrometer used to determine ^{13}C / ^{12}C isotope ratios, was similar to previous instruments and used a tunable external-cavity diode laser light source. The total internal volume of the ring-down cavity was 11 mL, allowing the use of standard Excetainer tubes to collect breath samples with a straw, and the round trip path length of light within the optical cavity is 42 cm. The ring-down decay constant of the empty cavity was 18.5 μs with a measured deviation of 0.0005 μs giving a relative error (σ_τ/τ) of 2.7×10^{-4}. With a tracking circuit engaged, the ring-down acquisition rate was 300 Hz, allowing rapid generation of spectra. Figure 15.5 shows three absorption spectra of CO_2 in N_2 obtained with CO_2 concentrations of 79.7 ppmv, 41.2 ppmv, and 18.3 ppmv with a constant total pressure of 100 Torr. By comparing the noise on the baseline to the integrated intensity of the observed spectral features, the minimum detectable CO_2 concentration was determined to be 3 ppmv. This limit indicates a minimum detectable absorption loss of 3.2×10^{-11} cm^{-1} Hz$^{-1/2}$.

Table 15.1. $\delta^{13}C$ values of CO_2 samples determined using IRMS and CRDS.

$\delta^{13}C$ [‰ ±0.2 ‰] by IRMS	$\delta^{13}C$ [‰ ±0.2 ‰] by CRDS
-20.49	-20.43
-14.91	-14.68
-9.38	-9.35
-4.19	-3.54
0.99	0.64
5.98	5.81
11.28	11.37
15.05	15.10
19.89	19.31

The absorption features of $^{12}C^{16}O^{16}O$ and $^{13}C^{16}O^{16}O$ were chosen so that the lines have similar peak absorptions despite the 100:1 abundancy discrepancy. Because these features arise from different rotational/vibrational bands, their ratio is dependent on temperature. To ensure that temperature did not affect the measurement precision, the instrument was designed to stabilize sample temperature to 2 mK standard deviation over several weeks. In order to minimize sample cross-contamination, the cavity was evacuated to 2 Torr from a sample pressure of 70 Torr and purge times were adjusted until there was no observed cross contamization between consecutive samples having –25 ‰ and –50 ‰. About 1 mL of standard temperature and pressure (STP) sample was therefore required the cavity.

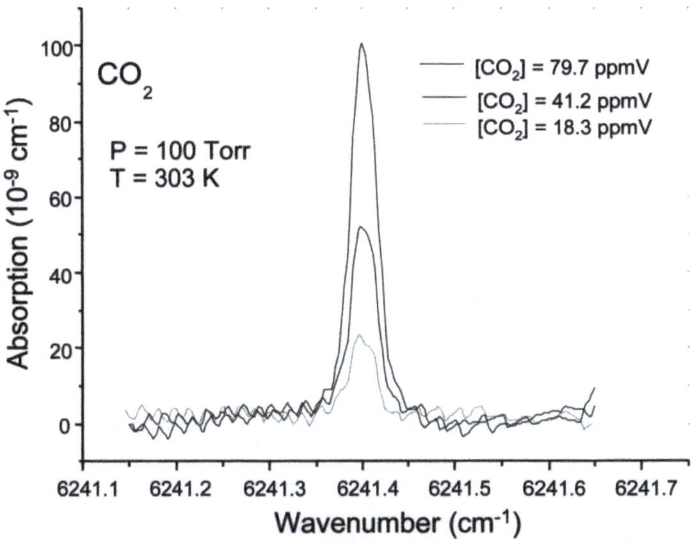

Figure 15.5. Absorption spectra of CO_2 in N_2 obtained with CO_2 concentrations of 79.7 ppmv, 41.2 ppmv, and 18.3 ppmv with a constant total pressure of 100 Torr

Reference samples having a range of isotope rations and characterized using IRMS (precision of 0.2 ‰) were used to test the instrument performance. It is interesting to note that the absorption features were fit to Galatry rather than Voigt profiles in order to improve the measurement precision. Galatry profiles include Dicke narrowing effects in addition to the standard collisional broadening treated by Voigt profiles as a convolution of Gaussian and Lorentzian lineshapes. In fact, the Galatry fit reduced the uncertainty in the peak area fit by a factor of three. The spectral baseline was used as a measure of zero absorption losses within the cavity. The areas under the absorption peaks were used to measure the absolute concentration of the two isotopic species. Unfortunately, the ^{13}C / ^{12}C ratio obtained from absorption spectra disagreed with the known relative concentrations of $^{12}C^{16}O^{16}O$ and $^{13}C^{16}O^{16}O$ in standard samples indicating a relative error in the absorption line strengths of the two isotopes. However, as in IRMS, this systematic error can be eliminated by measuring the ratio of ^{13}C / ^{12}C in a sample relative to the ^{13}C / ^{12}C ratio of a standard reference material as mentioned above. One CO_2 sample generated from a standard carbonate sample and characterized by IRMS was used as the reference standard. Determining $\delta^{13}C$ for eight replicate samples of CO_2 with a large negative delta produced a measure of the precision of the instrument. The mean $\delta^{13}C$ was –27.76 ‰ with a standard deviation of 0.22 ‰. Table 15.1 shows $\delta^{13}C$ values for nine duplicate samples having $\delta^{13}C$ values ranging from approximately –20 ‰ to +20 ‰ characterized by both IRMS and CRDS. Excellent agreement exists between the two measurements. Linear regression analysis of $\delta^{13}C$ (CRDS) vs. $\delta^{13}C$ (IRMS) data gives a slope of 0.989 ± 0.008 and a y-intercept of 7.47×10^{-3} ‰ ± 0.11 ‰.

For H. pylori detection applications, the typical sensitivity required of an instrument is a precision better than 1 °/oo. This requirement illustrates that CRDS could be routinely applied to H. pylori breath testing. A small pilot study illustrating the use of CRDS in H. pylori detection was carried out. Four members of the research team, two of who were known to test positive for the bacterium and two of whom tested negative, were assayed for the presence of H. pylori by measuring $\delta^{13}C$ in samples of their breath using CRDS before and thirty minutes after consuming ^{13}C-enriched urea. Table 15.2 lists the results of this study. The larger error in the breath tests (0.30 °/oo) is caused by the slightly lower concentration of CO_2 in the real breath samples as opposed to that in simulated samples (2% – 3% vs. 5%, respectively). The precision of the CRDS instrument is clearly sufficient to identify the increase in $\delta^{13}C$ observed for subjects known to be infected with Helicobacter pylori.

Table 15.2. CRDS Helicobacter pylori breath test results for known positive (+) and known negative (-) subjects.

Subject	$\delta^{13}C$ prior to ingestion [°/oo ± 0.3 °/oo]	$\delta^{13}C$ following ingestion [°/oo ± 0.3 °/oo]	DOB [°/oo ± 0.4 °/oo]
A (-)	-21.5	-21.8	-0.3
B (-)	-23.7	-23.0	0.7
C (+)	-25.8	-10.4	15.4
D (+)	-23.5	29.6	53.1

15.3.2 Analysis of the Exhaled Breath in Smokers

Carbon monoxide can affect the human body at concentrations above 30 ppm by inhibiting the ability of blood to carry oxygen to body tissues including vital organs such as the heart and brain. Following inhalation, CO displaces oxygen in the erythrocyte to form carboxyhemoglobin (COHb). In this form, CO has a half-life of about 5 to 6 h and may remain in the blood for up to 24 h. A healthy non-smoker shows a CO breath fraction of 1 to 2 ppm. Smokers show a raised CO breath level due to the inhalation of CO via cigarette smoke. In this way CO can identify smoking habits during a 24-hour period. The measurement of the breath CO level may provide an immediate, noninvasive method of assessing the smoking status of a person. In this section, we compare the CO breath faction of smokers and non-smokers. The sensitivity and selectivity of CRDS was sufficient to perform precise measurements of the CO fraction in human breath.

Detection of CO in human breath first required finding an interference free absorption line independent of carbon dioxide or water vapor which are abundant in human breath and dominate the CO absorption features in the near-infrared. Such a set of CO lines is located in the 6392 to 6400 cm^{-1} range. Again, the instrument concentration readings were calibrated by reference gas samples as well as to the HITRAN database. The CO concentration for breath samples was then computed from the area of the absorption feature with the spectral baseline serving as the zero (empty cavity) reference.

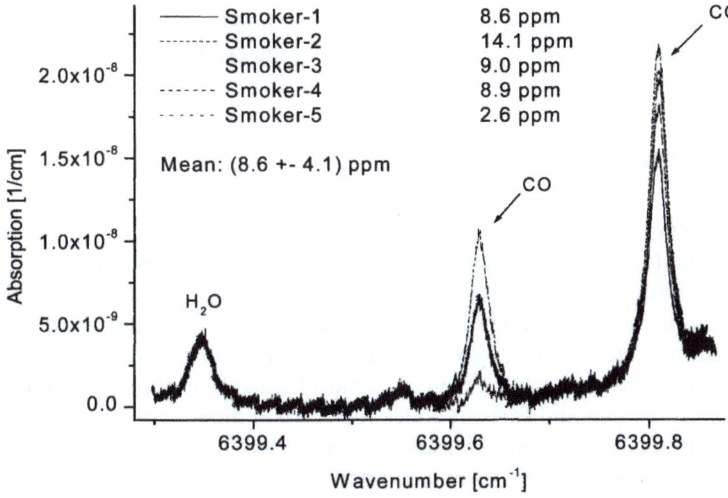

Figure 15.6. Spectra of CO in human breath illustration variation in CO levels between smokers.

In order to test CO in breath samples, breath was first collected from smokers who typically exhibit the highest concentrations. In this instance, the strongest CO absorption feature overlapped, even at reduced sample pressure with neighboring interferences, so that a weaker line (R(16)), at 6399.61 cm^{-1}, was chosen. This is the second vibrational overtone and its weaker peak absorption limited the detection sensitivity of the CO measurements to 500 ppb. This corresponds to a noise equivalent absorption of 4×10^{-10} cm^{-1}.Breath samples were collected from 5 smokers and 5 non-smokers. The subjects were to hold their breath for 15 seconds and then exhaled through a Teflon tube into an exetainer vial that was closed after complete exhalation. This collection method ensured that the gas sample contained in the tube was not mixed with room air and was truly alveolar, i.e., did not originate from the upper parts of the respiratory system.

The breath sample spectra were measured over the 6399.3 to 6399.9 cm-1 range so that two additional absorption features could be measured. A water line is detected at 6399.35 cm^{-1} while a carbon dioxide feature is observed at 6399.8 cm^{-1}. As is illustrated in Figure 15.6, the water line is identical for all subjects, in agreement with the fact that human breath has a saturated humidity. The carbon dioxide concentration varies from 4% to 6%, again in agreement with the known human range. However, the CO concentration shows a strong bimodal distribution between smokers and non-smokers: its mean value is 1 .8 ± 0 .5 ppm for non-smokers while it is 8 .6 ± 4 ppm for smokers. These results are in good agreement with literature, which found a mean value of 6 ppm CO in smokers, or a cut-off point of 8 to 9 ppm between smokers and non-smokers. The variation in CO concentration in the smoker group can be attributed to the small sample set used, especially since one of the smokers was exercising regularly and physical exertions tends to lower the CO level in smokers significantly because CO is released more efficiently from hemoglobin during higher metabolic rates. His CO concentration was 2.5 ppm, much closer to the mean established for non-smokers.

15.4 Summary

Medical diagnosis can only be as good as the tools that the physician has at his or her disposal. In the time of Hippocrates, the highly insensitive human nose was the only breath testing instrument available, yet through the use of smell physicians were able to qualitatively identify many diseases. In the time of Lavoisier, quantification in chemical analysis was in its infancy and allowed for the first time science to demonstrate metabolism and the chemical reactions involved in the conversion of oxygen and carbohydrates into carbon dioxide. Near the end of 20th century, Pauling using sensitive mass spectrometry, identified for the first time with the presence of volatile organic compounds in human breath, which led to detailed studies identifying hundreds of species. For the 21st century, the difficult task of linking specific chemical species to specific disease states remains. Yet, without the means to measure these species in a controlled, quantitative, rapid, and inexpensive manner, progress will not be made, nor will commercial diagnostic instruments come into existence. We hope it is this last challenge that cavity ring-down spectroscopy will address. With its sensitivity, specificity, low cost and small size, CRDS will enable the next generation of breath diagnostics in the next century.

References

Anderson DZ, Frisch JC, Masser CS. (1984) Mirror Reflectometer Based on Optical Cavity Decay Time, Appl. Opt. **23**:1238-1245

Busch KW, Busch MA. (1997) Cavity Ringdown Spectroscopy: An Ultratrace Absorption Measurement Technique, ACS Symposium Series 720, Oxford

Crosson ER, Ricci KN, Richman BA, Chilese FC, Owano TG, Provencal RA, Todd MW, Glasser J, Kachanov AA, Paldus BA, Spence TG, Zare RN. (2002) Stable Isotope Ratios Using Cavity Ringdown Spectroscopy: Determination of $^{13}C/^{12}C$ for Carbon Dioxide in Human Breath, Anal. Chem. **74**(9):2003-7

Dahnke H, Hering P, Kleine D, Murtz M. (2001) Real-time monitoring of ethane in human breath using mid-infrared cavity leak-out spectroscopy Appl. Phys. B **72**:971-975

Hodges JT, Looney JP, Zee RDv (1996) Response of a Ringdown Cavity to an Arbitrary Excitation, J. Chem. Phys. **105**(23):10278-10288

Jongma RT, Boogart MGH, Holleman I, Meijer G. (1995) Trace Gas Detection with Cavity Ringdown Spectroscopy, Rev Sci Instrum **66**(4):2821-2828

O'Keefe A, Deacon DAG (1988) Cavity ring-down optical spectrometer for absorption measurements using pulsed laser sources, Rev. Sci. Inst. **59**:2544-2551

O'Keefe A, Scherer JJ, Cooksy AL, Sheeks R, Heath J, Saykally RJ. (1990) Cavity ring-down dye laser spectroscopy of jet-cooled metal clusters: Cu2 and Cu3, Chem. Phys. Lett. **172**:214

Paldus BA, Harris JS, Martin J, Xie J, Zare RN. (1997) Laser diode cavity ringdown spectroscopy using acousto-optic modulator stabilization, J Appl. Phys. **82**(7):3199-3204

Paldus BA, Harb CC, Spence TG, Wilke B, Xie J, Harris JS, Zare RN (1998) Cavity-locked ring-down spectroscopy, J Appl. Phys. **83**(8) 3991-3997

Paldus BA, Harb CC, Spence TG, Zare RN, Gmachl C, Capasso F, Sivco DL, Bailargeon JN, Hutchinson AL, Cho AY (2000) Cavity ringdown spectroscopy using mid-infrared quantum-cascade lasers, Opt. Lett. **25**(9):666-668

Romanini D, Lehmann KK (1993) Ringdown Cavity Absorption Spectroscopy of the Very Weak HCN Overtone Bands with Six, Seven, and Eight Stretching Quanta, J. Chem. Phys. **99**(9):6287-6301

Romanini D, Kachanov AA, Sadeghi N, Stoeckel F (1997) CW Cavity Ringdown Spectroscopy, Chem. Phys. Lett. **264**:316-322

Spence TG, Harb CC, Paldus BA, Zare RN, Willke B, Byer RL (2000) A laser-locked cavity ringdown spectrometer employing an analog detection scheme. Rev. Sci. Instrum. **71**:347-353

16 Photoacoustic Trace Gas Detection in Plant Biology

Frank Kühnemann

16.1 Introduction

When talking about trace gas detection the usual association is that of monitoring environmental pollution, either as emission (from stack, car exhaust etc.) or immission control (ambient concentrations). A different field, but of growing importance, is the analysis of trace gas emissions from living organisms, either plants, animals or humans. These organisms synthesise different volatile compounds as part of their metabolism, in connection with developmental processes or as a reaction of the organism to external influence. Monitoring the emitted compounds in the air may thus yield important information about the organism and its interaction with the environment in a non-invasive way.

For many years such analysis had been the domain of classical analytical chemistry, using different chromatographic and mass-spectroscopic methods. Infrared spectroscopic techniques offer a different approach to sensitive trace gas detection. Since the mid 80's photoacoustic (PA) spectroscopy with the CO_2 laser has been applied for the detection of ethylene in plant physiological experiments. This particular application had been stimulated by detection limits in the sub-ppb range (Fink et al. 1996, Harren et al. 1990) and the pivotal role of this molecule in plant physiology as a plant hormone and stress-related compound (Abeles et al. 1992). PA detection allows fast, sensitive and continuous analysis. This yields valuable information about the plant which was not accessible with earlier techniques due to limited sensitivity or time resolution .

Besides ethylene, there are numerous other volatile organic compounds (VOCs) synthesised and emitted by plants either as a part of their metabolism or in response to external influence like abiotic stress or biogenic attacks from pathogens or herbivors. Especially the smaller ones among them (up to 5 carbon atoms) show characteristic infrared fingerprints, which can be used for a sensitive detection. This holds especially for the 3 – 4 μm region, where strong fundamental bands due to C-H, N-H and O-H vibrations in the molecules are found.

Photoacoustic trace gas detection requires powerful laser sources with emission frequencies in coincidence with the absorption features of the molecules. Until very recently, the CO overtone laser was the only such source in the 3-4 μm range.

Based on the pioneering work by Urban and co-workers (Bachem et al. 1993) a photoacoustic spectrometer was set up for trace gas detection in plant physiological experiments.

Alternative light sources are emerging with the development of optical parametric oscillators (OPOs) based on frequency conversion in periodically poled nonlinear optical materials, pumped by powerful solid-state lasers. We have conducted first experiments to explore the potential of these sources.

The present contribution summarises the work of our group on both the application of the CO overtone laser PA spectrometer in plant physiological experiments and the use of an OPO for PA trace gas detection.

16.2 The CO Overtone Laser Photoacoustic Spectrometer

In absorption spectroscopy the information is derived from changes in the transmission of the radiation through the sample. Photoacoustic spectroscopy provides an indirect measurement of the amount of radiation which is absorbed by the sample using the photoacoustic effect: For vibrationally excited molecules at atmospheric pressure, inelastic collisions with other molecules provide the dominant channel for relaxation, resulting in a temperature increase in the sample. Periodic amplitude modulation of the laser beam results in temperature oscillations, which, in turn, give rise to periodic pressure fluctuations. A microphone is used to measure the pressure amplitude which is proportional to the number density of absorbing molecules and the intensity of the laser beam.

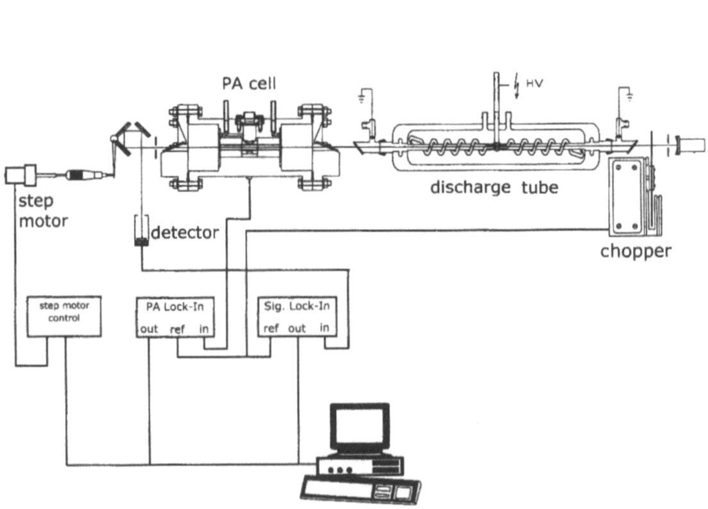

Fig. 16.1. Setup of the CO overtone laser photoacoustic spectrometer

The CO overtone laser (Bachem et al. 1993) operates on $\Delta v=2$ P-transitions in the CO molecule between 2.8 and 4 µm. The laser consists of a liquid nitrogen cooled discharge tube in a resonator, which is formed by a high-reflection end mirror and a grating in Littrow configuration (see fig. 16.1). The photoacoustic cell is placed inside the laser cavity in order to use the high circulating laser power (up to 5W). The laser beam enters and exits the cell through CaF_2 Brewster windows. Amplitude modulation is achieved with an intra-cavity chopper operating on the first longitudinal resonance of the cell. The acoustic signal is detected with an electret microphone. Only a small percentage of the laser power is coupled out through the zero-th order of the grating. It is used for power monitoring with a pyroelectric detector. Photoacoustic (PA) and power signals are processed with dual-phase lock-in amplifiers. Their ratio gives the normalised PA signal (NPAS), which is proportional to the absorber's number density. Operation of the spectrometer is controlled by a computer.

16.2.1 Characterization of the spectrometer

To identify the best laser lines for the detection of particular gases their PA fingerprint spectra were recorded at the CO overtone laser lines using certified gas mixtures. In addition, FTIR spectra of the trace gases served as a reference (Popp et al. 2000). Until now the list of gases for the CO overtone laser includes methane, ethane, ethylene, butane, pentane, acetaldehyde, methanol, ethanol, acetone and isoprene.

Figure 16.2. shows examples for such PA fingerprint spectra. For ethylene and ethane only accidental coincidences are found between the laser lines and the narrow absorption structures. Isoprene and pentane, in contrast, show broad, continuous absorption spectra in this range. Here the wide emission range of the CO overtone laser helps to record the shape of the bands and identify the molecules.

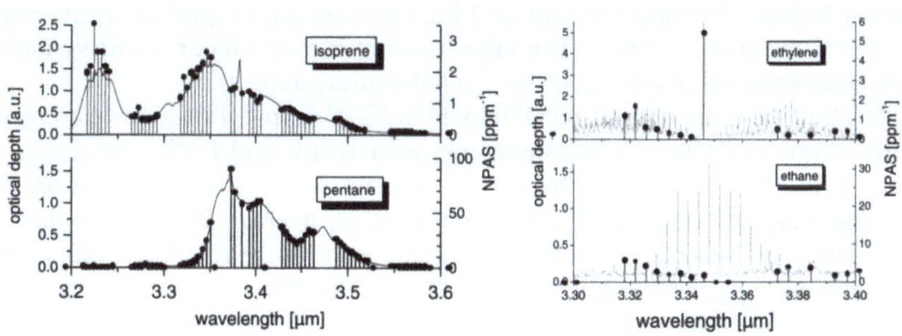

Fig. 16.2. Comparison of FTIR (solid lines) and photoacoustic fingerprints (symbols) in the range of the CO overtone laser for isoprene (C_5H_8), pentane (C_5H_{10}), ethylene (C_2H_4) and ethane (C_2H_6)

Detection limits for individual gases are determined by comparison of the NPA signals with the background signal for clean air. The latter results from wall and window signals. A conservative estimate for the detection limit is the concentration which gives a NPA signal equivalent to the background (background equivalent concentration - BEC). Typical values are around 1ppb, depending on the absorption strength of the individual gas.

In plant biological emission studies the air contains a mixture of volatiles, which may interfere with the compounds to be analysed. Larger, less volatile molecules are removed from the gas flow with a cooling trap. Its operation temperature is selected to ensure full transmission of the desired compounds while the others are removed to a maximum extent. For isoprene concentrations below 1 ppm the trap temperature is 185 K. The trapping efficiency was verified with a standard gas chromatographic analysis of the air stream behind the trap. In isoprene emission studies on *Eucalyptus globulus* leaves the trap effectively removed the heavier hydrocarbons which otherwise would interfere with the PA analysis (Kröner 2000).

If several compounds have to be analysed simultaneously or interfering species may not be eliminated completely, a multi-gas analysis is required. The NPA signal S_i at laser line *i* is given as the superposition $S_i = \Sigma a_i c_i$ of the contributions from *m* individual gases with the concentrations c_i and NPA coefficient a_i. The background signal is introduced as an additional component. The gas concentrations are determined through a non-linear least-squares fit. This requires at least m+2 lines which have to be selected for maximum sensitivity and minimum cross-interference.

16.2.2 Comparison of acetaldehyde detection with PAS and HPLC

As mentioned above different analytical techniques are applied to monitor trace gas emissions by plants, and new methods have to be validated against established ones. In the case of acetaldehyde detection we conducted an intercomparison experiment between PA spectroscopy and high performance liquid chromatography (HPLC) (Kuwata et al. 1983). The latter method is very labour intensive and requires large amounts of aldehydes for a reliable measurement.

Using certified mixtures of acetaldehyde in N_2 (7.7 ppm, Praxair) we compared the data from simultaneous measurements with HPLC and PAS. The results are shown in fig. 16.3. Both data show a very good proportional reading over the concentration range of interest. The PAS data are on average 20% larger than the HPLC data. This was attributed to a systematic error in the absolute calibration of both methods using independent mixtures (Kröner 2000).

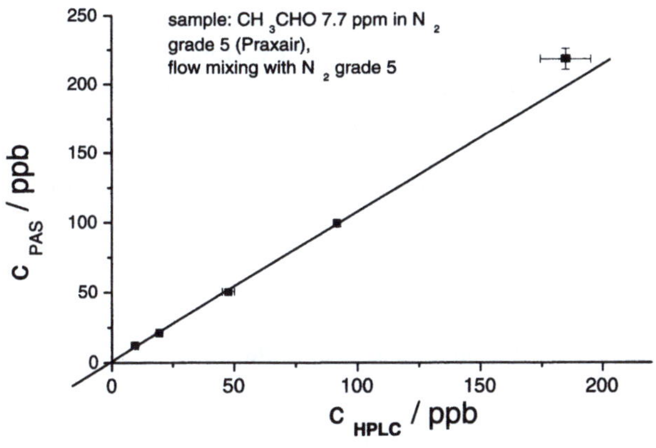

Fig. 16.3. Comparison between acetaldehyde concentration measurements with HPLC (x axis) and PA spectroscopy (y axis) using flow mixtures of a certified sample

16.3 New Radiation Sources for PA Detection

With the infrared gas lasers, detection of trace gases depends on accidental coincidences between laser emission frequency and an absorption feature of the trace gas. While this works well for ethylene and the CO overtone laser, it fails completely in the case of ethane (see fig 16.2.) In such cases light sources are desired which combine the power of the gas lasers with a free selection of the emission wavelength.

One of the new laser sources are quantum cascade lasers, which are capable of delivering several tens of mW in the mid-ir range of interest. Their tunability, however, is limited to a few wavenumbers only. That's why they are suitable for the detection of single gases rather than for the analysis of mixtures. Their potential for PA detection has been demonstrated by Harren, Capasso and co-workers (Paldus et al. 1999).

16.3.1 Photoacoustic detection with an optical parametric oscillator

Optical parametric oscillators offer the potential to deliver both, power and tunability. In 1998 we demonstrated what was, to our knowledge, the first application of an infrared cw OPO for photoacoustic trace gas detection. The OPO, developed at the University of Constance, was a pump-enhanced singly resonant oscillator based on periodically poled $LiNbO_3$, pumped by a narrow-linewidth Nd:YAG laser (800 mW) (Schneider et al. 1998). The wavelength agility of the

OPO allowed to determine detection limits for ethane, ethylene and methane using the strongest absorption features in the 3 µm band. Table 16.1. gives a comparison of the detection limits achieved with the CO overtone laser and with the OPO. The CO laser column shows the laser lines with the best coincidence for the respective gas, while in the OPO column the position of the strongest absorption peak in the 3 – 4 µm region is given. For ethane and methane the frequency agility of the OPO yields a considerable increase in absorption strength, while the coincidence between the 24p13 line of the laser and the ethylene spectrum is already close to the maximum.

Table 16.1. Detection limits for ethane. methane and ethylene using the CO overtone laser and a single-frequency OPO (Kühnemann et al. 1998)

	CO ($\Delta v=2$) laser ($\tau = 300$ ms)				OPO spectrometer ($\tau = 10$ sec)		
molecule	line	ν [cm^{-1}]	α [cm^{-1}atm^{-1}]	c_{min} [ppb]	ν [cm^{-1}]	α [cm^{-1}atm^{-1}]	c_{min} [ppb]
ethane	24p8	3006.95	4.2	**1**	2986.7	36	**0.46**
ethylene	24p13	2988.38	5.27	**2**	2988.4	5.3	**4**
methane	24p5	3017.22	28	**0.1**	3067.3	69	**0.26**

Up to now the OPO delivers considerably less power than the CO overtone laser. The PA signal is thus still noise-limited. An increased lock-in integration time (10 seconds vs. 300 milliseconds) and the gain in absorption coefficient, however, result in lower detection limits for ethane and methane. Based on this first application work is now under way to develop an OPO with increased power and tunability.

16.4 Application to Plant Physiology

In order to carry out plant physiological experiments the PA detection system had to be combined with an experimental setup which ensured adequate conditions for the plant and fulfilled the requirements for trace gas detection at very low concentrations.

The basic setup is shown in fig. 16.4. It consists of the PA spectrometer with electronics and computer, the plant chamber and the gas flow system. Air flow is directed through the system by a small membrane pump and controlled by a mass flow controller. A platinum catalyst removes any hydrocarbons from the air before it enters the plant chamber. The air takes up the volatiles emitted by the plant and passes the temperature-regulated cooling trap prior to entering the PA cell. Behind the PA cell the air flow is monitored as a control for possible leaks or blocked tubings.

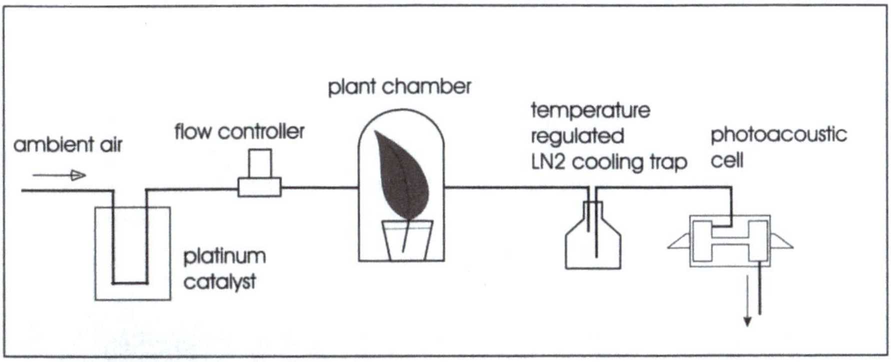

Fig. 16.4. Basic setup for plant physiological trace gas emission studies. The photoacoustic cell is placed inside the cavity of the CO overtone laser

Since 1996 the CO overtone laser PA spectrometer has been used in numerous studies including the detection of ethylene and ethane (Martis et al. 1998), ethanol and acetaldehyde (Kreuzwieser et al. 2000), isoprene (Dahnke et al. 2000), and methanol. Here only a few examples are given in order to illustrate the capabilities of the technique and to demonstrate the links between the detection of the volatiles and other physiological data.

16.4.1 Ethylene and ethane from freezing damage

While ethylene is an indicator of plant stress, severe negative conditions may lead to a damage of the plant. One of these damages is the oxidative degradation of cell membrane lipids, resulting in the formation of ethane. With the CO overtone laser both compounds may be monitored simultaneously, allowing to give a stress / damage rating for the plant (Martis et al. 1998).

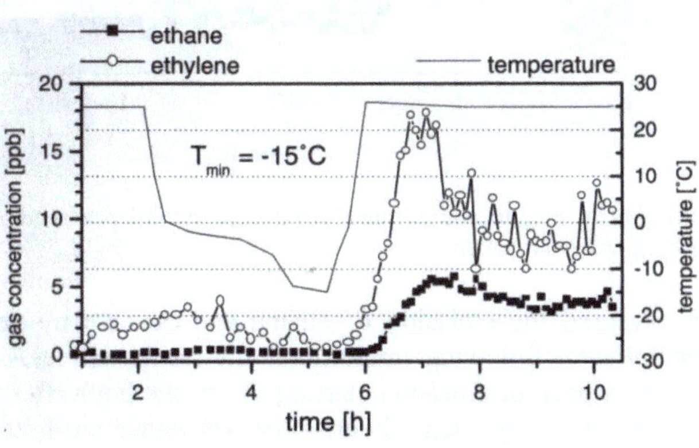

Fig. 16.5. Emission from cut rhododendron leaves during and after freezing to -15°C

To demonstrate the technique, rhododendron leaves were cooled down to different freezing temperatures, and the emission during and after frost stress was monitored. At room temperature, only ethylene is emitted as reaction to the cutting of the leaves. At temperatures down to –15 °C, both ethylene and ethane are emitted after re-warming (shown in fig. 16.5.). Freezing down to –20°C results in a complete breakdown of the cell structure and hence the ethylene synthesis and a strong, short-term ethane emission due to the severe damage.

16.4.2 Ethane and pentane from germinating peas

During storage the seeds are in a glassy state slowing down transport and metabolism in the cells. Germination of seeds starts with imbibition, the rapid and massive water uptake, during which the seed transforms into a liquid state. This transition increases drastically the mobility of compounds in the seed, causes strong morphological changes and starts a complex system of metabolic processes, finally turning the quiescent seed into a seedling. All these steps occur in just a few hours, placing high demands on any technique used to monitor the related emission of volatiles by the seedlings.

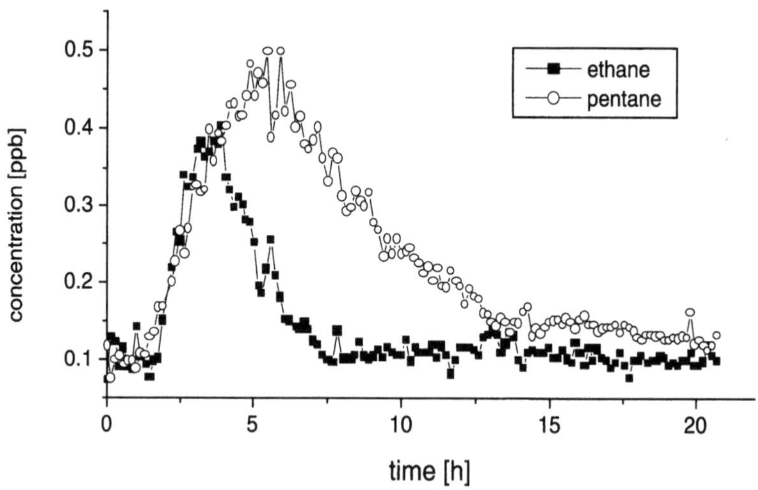

Fig. 16.6. Emission of ethane and pentane from germinating pea seeds. Imbibition was started by applying water at t = 0h

Here we measured the emission of ethane and pentane by germinating pea seeds (*Pisum sativum*) following imbibition. The results are shown in fig. 16.6. Both gases show a peak of emission starting about one hour after imbibition and peaking 2 (ethane) and 5 (pentane) hours later. The peaks are followed by an exponential decay with τ = 3 h (ethane) and 8 h (pentane). The ratio r of the time constants (r =2.7) is close to the ratio of the between the effusion coefficients for pentane and ethane in water (*r* =3) (Reuss and Oomens 1997), suggesting a pure transport effect. That's why it was checked, whether the emission was simply due to the release of previously stored gases upon the phase transition or due to a *de-novo* synthesis after the onset of the metabolism. For this, the glass-liquid phase transition was initiated by heating the seeds without addition of water. No significant emission was observed, pointing towards a *de-novo* synthesis.

16.4.3 Acetaldehyde emission from flooded poplar trees

Acetaldehyde is found in plants as an intermediate of alcoholic fermentation during anaerobic respiration. Anaerobic conditions may occur during fruit storage in an oxygen-free atmosphere or in the roots of flooded trees, when oxygen supply to the roots is impaired. Here the emission of ethanol and acetaldehyde from flooded poplar (*Populus tremula x alba*) trees was studied. The shoots of 3 month old, 1m high poplar trees were placed in a 0.5 m^3 teflon chamber to measure the emission of the gases by the leaves. For the measurement of acetaldehyde, ethanol and ethylene, the cooling trap had to be operated at – 50°C. In order to take into account cross-interferences from unwanted molecules 12 laser lines were used for the reliable determination of the three gases, yielding one data set every 8 minutes.

Switching the conditions between day and night allowed to study the dependence of the emission from photosynthesis and transpiration stream. During the night, the emission was reduced, but still enhanced in comparison to non-flooded trees. Upon sunrise, bursts of both acetaldehyde (see fig.16.7) and ethanol emission could be observed, while ethylene showed only a moderate increase (Kreuzwieser et al. 2000).

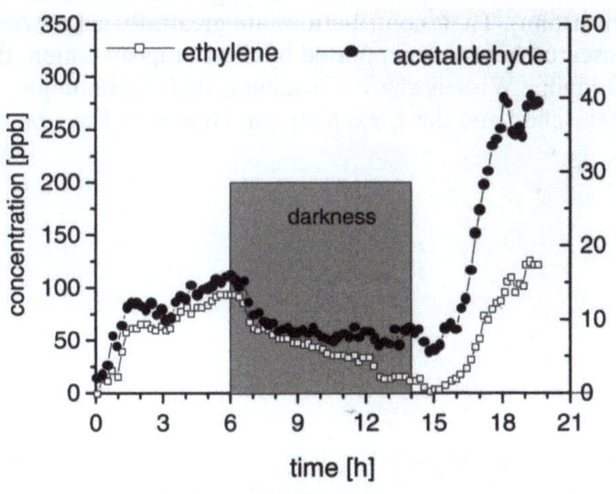

Fig. 16.7. Emission of acetaldehyde and ethylene from the leaves of a flooded, three month old poplar tree. Flooding started at t = 0h, the night period is indicated by the grey area.

16.5 Summary

Photoacoustic spectroscopy allows a sensitive, fast and continuous monitoring of trace gas emissions by organisms. It was successfully applied to study the reaction of plants to changing environmental conditions through the measurement of different volatile organic compounds as ethylene, ethane, pentane, acetaldehyde and ethanol. It has been demonstrated, too, that understanding these results requires a profound analysis of the underlying physical, chemical, metabolic and developmental processes and hence a close interdisciplinary cooperation. Our research group has been successful over the last years in establishing such cooperations with numerous research groups from botany and agriculture.

Progress in infrared trace gas analysis and photoacoustic analysis in particular is brought by the development of optical parametric oscillators as both powerful and tunable sources in the mid-IR.

Acknowledgement

The author would like to thank Wolfgang Urban for many years of encouraging support and Jörg Reuss for stimulating discussions. Numerous postdocs, Ph.D. and diploma students contributed to the work presented here: Stefan Büscher, Albert Martis, Ralph Gäbler, Andreas Hecker, Hannes Dahnke, Arno Kröner, and Michael Wolfertz. The OPO study was a collaboration with Stephan Schiller and Klaus Schneider from the University of Constance. The biological studies presented here were carried out together with Jürgen Kreuzwieser (Freiburg) and Norbert Keutgen (Bonn). Their contributions are greatfully acknowledged.

The work presented here was supported by the European Union, the Bundesministerium für Bildung, Wissenschaft, Forschung und Technologie, the Deutsche Forschungsgemeinschaft and the Alexander von Humboldt Foundation.

References

Abeles F, Morgan P, Saltveit Jr M (1992) Ethylene in Plant Biology, 2nd edition Academic Press, San Diego, New York, Boston

Bachem E, Dax A, Fink T, Weidenfeller A, Schneider M, Urban W (1993) Recent Progress with the CO Overtone Laser. Appl. Phys B57:185-191

Dahnke H, Kahl J, Schüler G, Boland W, Urban W, Kühnemann F (2000) On-line monitoring of biogenic isoprene emissions using photoacoustic spectroscopy. Appl. Phys. B70:275-280

Elstner EF, Konze JR (1976) Effects of point freezing on ethylene and ethane production by sugar beet leaf disks. Nature 263:351-352

Fink T, Büscher S, Gäbler R, Yu Q, Dax A, Urban W (1996) An improved CO_2 laser intracavity photoacoustic spectrometer for trace analysis. Rev. Sci. Instruments 67:4000-4004

Harren F, Bijnen F, Reuss J, Voesenek L, Blom C (1990) Sensitive intracavity photoacoustic measurements with a CO_2 waveguide laser. Appl. Phys B50:137-144

Kreuzwieser J, Kühnemann F, Martis A, Urban W, Rennenberg H (2000) Diurnal pattern of acetaldehyde emission by flooded poplar trees. Physiol Plant. 108:79-86

Kröner A (2000) Photoakustischer Spurengasnachweis mit dem CO-Oberton-Spektrometer. Diploma thesis, Institut für Angewandte Physik der Universität Bonn.

Kühnemann F, Schneider K, Hecker A, Martis A, Urban W, Schiller S, Mlynek J (1998) Photoacoustic trace gas detection using a cw single-frequency parametric oscillator. Appl. Phys. B 66:741-745

Kuwata K, Uebori M, Yamasaki H, Kuge H (1983) Determination of aliphatic aldehydes in air by liquid chromatography. Anal. Chem. 55:2013-2016.

Martis A, Büscher S, Kühnemann F, Urban W (1998) Simultaneous ethane and ethylene detection using a CO Overtone laser PA spectrometer: A new tool for stress/ damage studies. Instrumentation Science and Technology 26:177-187

Paldus BA, Spence TG, Zare RN, Oomens J, Harren FJM, Parker DH, Gmachl C, Cappasso F, Sivco DL, Baillargeon JN, Hutchinson AL, Cho AY (1999) Photoacoustic spectroscopy using quantum-cascade lasers. Optics Letters 24:178-80

Popp A, Dahnke H, Orphal J, Basum Gv, Burrows JP, Kühnemann F (2000) HASIBO- a database for high-resolution absorption cross-sections of VOCs in the mid-IR. Work report. Institut für Angewandte Physik Bonn and Institut für Umweltphysik Bremen.

Reuss J, Oomens J (1997) personal communication.

Schneider K, Kramper P, Schiller S, Mlynek J (1998) Towards an optical synthesizer: Single-frequency parametric oscillator using periodically poled $LiNbO_3$. Opt. Lett. 22:1293-1295

17 DNA Adducts as Biomarkers for Carcinogenesis Analysed by Capillary Electrophoresis and Laser-Induced-Fluorescence Detection

Oliver J. Schmitz

17.1 Significance of DNA Adducts

DNA adducts are the direct products of damage by endogenous or exogenous reactive agents to a critical macromolecular target such as DNA (Figure 17.1). Approximately 90 % of the chemicals considered carcinogenic for humans form DNA adducts (Stiborovà et al. 1998). Although the majority of such DNA damage is eliminated by DNA repair processes, some persistent adducts often cause permanent mutations in important growth-controlling genes or loci, resulting in aberrant cellular growth and cancer (Hemminki et al. 2000). The amount of these reflects an integration of both the toxicokinetic properties of a genotoxic compound and the cellular DNA repair. Measurements of endogenous and exogenous DNA adducts are thus of interest in that they provide molecular, mechanism-based bridges between carcinogen exposure and disease end-points and can serve as biomarkers for carcinogenesis. In practice, investigations of such DNA damage can be viewed as a supplementary test that may, in some instances, help to facilitate the assessment of a compound with an unusual or puzzling profile of activity in statutory genotoxicity assays, animal bioassays, or both (Phillips et al. 2000).

Embryogenesis and differentiation, which are characterised by specific patterns of gene expression in particular tissues and organs, proceed on the other hand without any alterations in DNA sequences. It is known that the irreversibility of phenotypes in differentiated organs is due not to any genetic alteration, but rather to "epigenetic change" (Jones and Buckley 1990). Recent work has shown that hypo- and hypermethylation of cytosine plays a key role in such changes (Baylin and Herman 2000; Costello and Plass 2001). Moreover, cancers often exhibit an aberrant methylation of gene promoter regions that is associated with a loss of gene function. Therefore, methylation of DNA, specifically methylation of cytosine in CpG islands of promoter or enhancer regions of genes, has been discussed as one plausible epigenetic mechanism in carcinogenesis (Sugimura and Ushijima 2000).

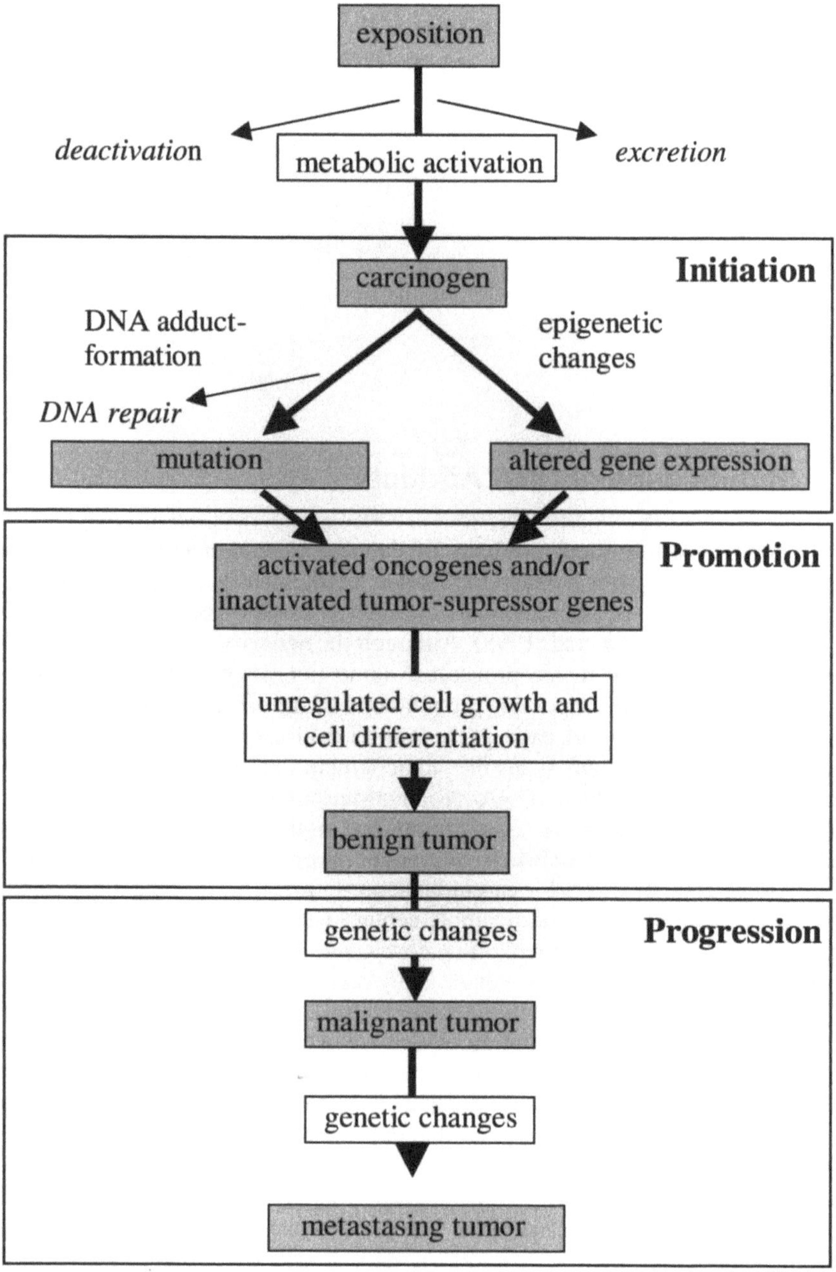

Fig. 17.1. Multi-hit model of chemical carcinogenesis

17.2 Methods for Analyzing DNA Adducts

Various methods are used for the detection of DNA adducts, including the ^{32}P-postlabeling method, LC-MS and immunoassays (Reddy 2000). The immunoassay (Poirier 1993; Poirier et al. 2000) requires the preparation of highly adducted DNA in the range of 1 % for the production of antibodies, which is generally accomplished with either direct-binding chemicals or chemicals whose reactive metabolites have been identified and can be synthesised in large amounts. This is only possible for a few carcinogens. For the determination of DNA adducts by LC-MS (Apruzzese and Vouros 1998; Esmans et al. 1998; Sweetman et al. 1998; Phillips et al. 2000; Poirier et al. 2000), approximately 300 µg DNA or more are necessary (Soglia et al. 2001). This method is very useful for characterizing unknown DNA adducts, but, because of the expensive hydrolysis of such high amounts of DNA, the great problem is how to obtain so much DNA for studies and that the DNA adducts separated in a single analytical run must be similar, LC-MS not being practical for routine analysis. Only the ^{32}P-postlabeling method is thus applicable to screening chemicals for their ability to produce DNA adducts which have not been characterized chemically or to determining low concentrations of adducts. The ^{32}P-postlabeling technique for DNA adduct analysis was introduced in 1981 (Randerath et al. 1981; Gupta et al. 1982; Reddy et al. 1984). Since then ^{32}P-postlabeling has emerged as a major tool for the detection and quantitation of DNA adducts. The method is based on the enzymatic hydrolysis of non-radioactive carcinogen-modified DNA to 2'-deoxyribonucleoside-3'-monophosphates, subsequent [^{32}P]-phosphorylation at the free 5'-OH group by [γ-^{32}P]-ATP and polynucleotide kinase, and chromatographic separation of carcinogen-nucleotide adducts from non-modified (normal) nucleotides (Phillips and Castegnaro 1999). The strengths of the ^{32}P-postlabeling procedure are its ability to determine unknown adducts and the high sensitivity which is made possible by the enrichment of modified nucleotides by butanol extraction or digestion with nuclease P1 before the [^{32}P]-phosphorylation. These enrichments lead to a detectable relative adduct level of 1 adduct in 10^9 nucleotides in only 10 µg DNA. Adduct levels are calculated as relative adduct labeling (RAL) values, which represent the ratio of count rates of adducted nucleotides to count rates of total (adducted and normal) nucleotides.

The use of the ultra-sensitive ^{32}P-postlabeling method is, however, hampered by a number of drawbacks such as the lack of automation, the use of a strong β-emitter (^{32}P) and the fact that the simultaneous detection of DNA adducts derived from several classes of carcinogens is not possible.

Another analytical method with a high sample throughput is therefore necessary to determine various classes of DNA adducts in only 10 µg DNA simultaneously.

Since 1983, fluorescence derivatisation of nucleotides for the analysis of DNA adducts has been pioneered by several research groups to replace the ^{32}P-postlabeling method (Chu et al. 1983; Kelman et al. 1988; Al-Deen et al. 1990;

Sharma et al. 1991; Lee et al. 1991; Li et al. 1992; Wang and Giese 1993, 1998; Lan et al. 1999a; Lan et al. 1999b).

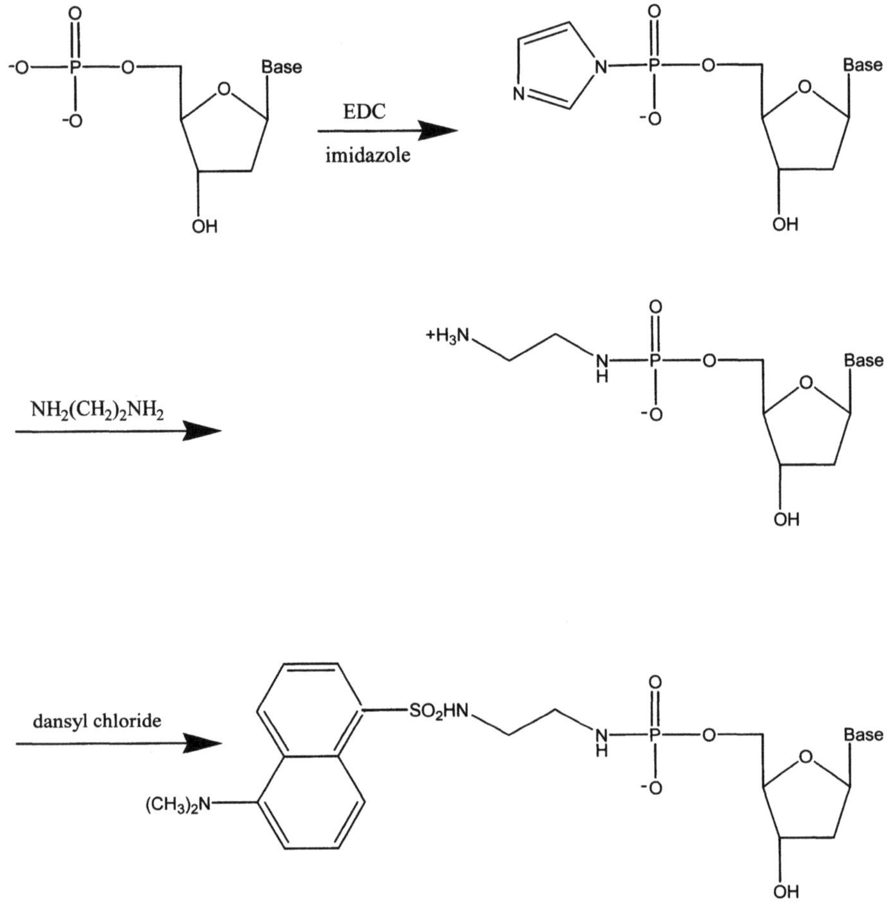

Fig. 17.2. Synthetic scheme for conjugation of mononucleotides with dansyl chloride in the presence of 1-ethyl-3-(3′-*N,N*′-dimethylaminopropyl)-carbodiimide (EDC) [Chu et al. 1983; Kelman et al. 1988; Al-Deen et al. 1990; Sharma et al. 1991; Lee et al. 1991; Li et al. 1992]

Ethylene diamine is used in large excess (Figure 17.2), and the unreacted portion has to be removed. The purification step is not a problem when a relatively small number of analyses are to be carried out, but it is a disadvantage to high-throughput analyses. The one-step derivatization presented in Figure 17.3 is quite elegant, but it may lead to two isomeric derivatives, which sometimes complicates the interpretation.

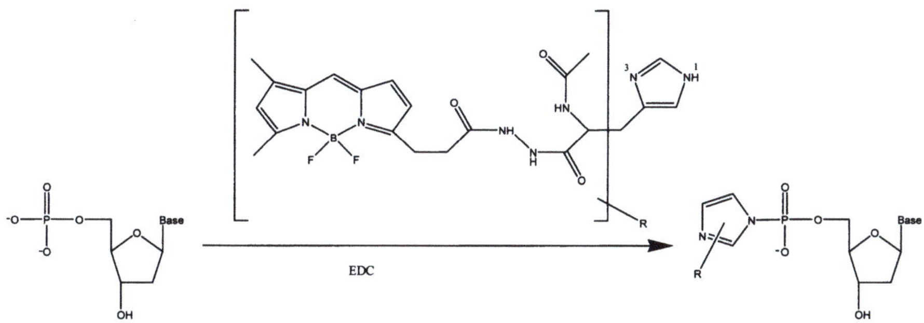

Fig. 17.3. Synthetic scheme for conjugation of mononucleotides with BO-IMI (R) [Wang and Giese 1993, 1998; Lan et al. 1999a; Lan et al. 1999b]

C. Wörth, D. Stach, M. Wießler and I have developed a new method to determine DNA adducts. It involves the hydrolysis of DNA, fluorescence labeling of modified and unmodified nucleotides, micellar electrokinetic chromatography (MEKC) and laser-induced-fluorescence detection (CE-LIF) with 4,4-difluoro-5,7-dimethyl-4-bora-3a,4a-diaza-*s*-indacene-3-propionyl ethylenediamine (BODIPY® FL EDA, Figure 17.4) as fluorescence marker (Schmitz et al. 2002).

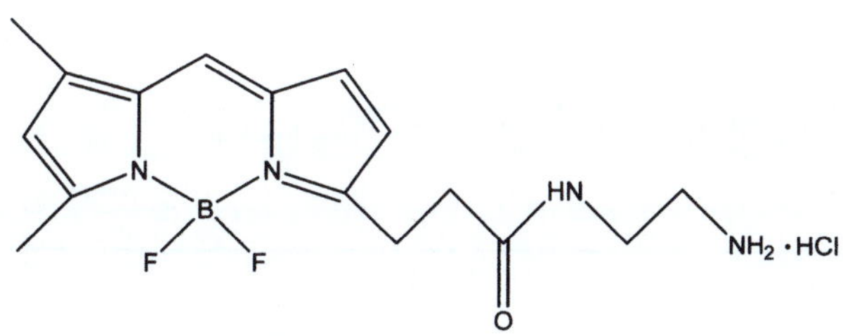

Fig. 17.4. Structure of BODIPY® FL EDA

As in the work mentioned above, especially shown by the research groups of Chu, Sharma and Giese, selective conjugation of 2'-deoxynucleoside-3'-monophosphates (dNMP) via the phosphate moiety to the amino linker of BODIPY® FL EDA was conducted in the presence of a water-soluble carbodiimide, EDC, to activate the phosphate moiety of the dNMPs. This approach to derivatization excludes the use of fluorescence markers or buffer systems with carboxylic acid groups, primary amino groups or phosphate groups. The only suitable buffer system for the derivatization was *N*-(2-hydroxyethyl)-piperazine-*N'*-2-ethane sulfonic acid (HEPES) at pH 6.5. To avoid a purification step, the enzymatic hydrolysis of DNA was also performed in HEPES buffer at pH 6.0. Comparison with ^{32}P-postlabeling analysis showed that digestion of calf-thymus DNA (CT-DNA) in HEPES buffer resulted in the same amounts of dNMPs as the routinely used procedure with sodium succinate. Therefore, the entire sample preparation, including hydrolysis of DNA and derivatization of nucleotides, was carried out sequentially in the same vial. Since samples from the same vial were directly subjected to capillary electrophoresis with laser-induced-fluorescence detection, automation of the analytical sequence with high sample throughput has been achieved (Figure 17.5).

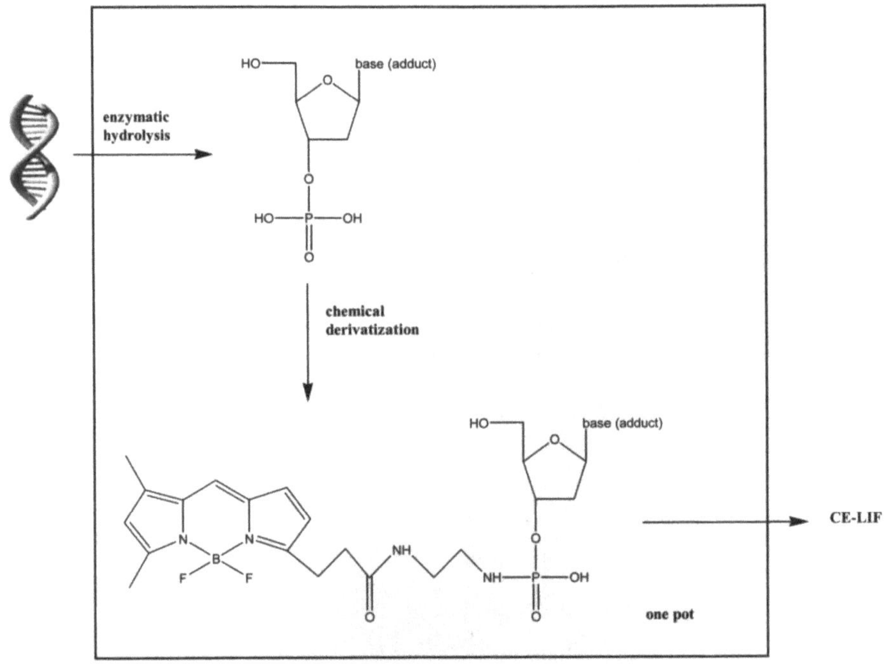

Fig. 17.5. Scheme of the DNA-adduct analysis

With a background electrolyte containing 17 mM sodium phosphate (pH 9.0), 75 mM sodium dodecyl sulfate (SDS) and 15 % methanol, the four unmodified nucleotides and several modified nucleotides (DNA adducts), listed in table 17.1, could be separated. In capillary electrophoresis, however, the reproducibility of the migration time (MT) is often poor because of the influence of the electroosmotic flow in uncoated fused-silica capillaries. To characterize DNA adducts in unknown samples, the MT of all unmodified and modified nucleotides were standardized to the MT of dAMP; this gave excellent reproducibility (standard deviation less than 3% by the use of different charges of capillaries and buffers). Table 17.1 lists the normalized MT of several analyzed nucleotides. A difference in normalized MT of 0.02 between two nucleotides is necessary to ensure a baseline separation. The values are given as the mean and one standard deviation.

Table 17.1. The corrected migration time of modified and unmodified nucleotides (n = number of analysis)

Nucleotide	normalized Migration Time (dAMP/Nucleotide)
1,N^2-propano-2'-deoxyguanosine-3'-monophosphate (Hex-dGMP)	0.78; 0.75 (n = 2)
5-methyl-2'-deoxycytidine-3'-monophosphate (5-Me-dCMP)	0.79 ± 0.01 (n = 32)
2'-deoxycytidine-3'-monophosphate (dCMP)	0.81 ± 0.01 (n = 34)
8-hydroxy-2'-deoxyguanosine-3'-monophosphate (8-HO-dGMP)	0.83 ± 0.01 (n = 6)
2'-deoxythymidine-3'-monophosphate (dTMP)	0.938 ± 0.006 (n = 42)
2'-deoxyguanosine-3'-monophosphate (dGMP)	0.962 ± 0.004 (n = 42)
2'-deoxyadenosine-3'-monophosphate (dAMP)	1.000
apurinic site (AP site)	1.011 ± 0.002 (n = 6)
adducts of benzo[a]pyrene diol epoxide (B[a]P-dGMP, 2nd isomer)	1.17 ± 0.01 (n = 5)
adducts of aristolochic acid I (dA-AAI, dA-AAII, dG-AAI)	1.23; 1.19 (n = 2) 1.20; 1.17 (n = 2) 1.13; 1.11 (n = 2)
adducts of benzo[a]pyrene diol epoxide (B[a]P-dGMP, 1st isomer)	1.29 ± 0.01 (n = 5)
1,N^6-etheno-2'-deoxyadenosine-3'-monophosphate (etheno-dAMP)	1.38; 1.36 (n = 2)

Figure 17.6 shows the electropherogram of an oligonucleotide after hydrolysis and derivatization of the 2'-deoxribonucleoside-3'-monophosphates with BODIPY-EDA. Because of the strong interaction between the fluorescence marker and SDS the great surplus of BODIPY-EDA migrates to the detection window a long time after the analytes of interest.

As an example this method can be used for the analysis of DNA adducts of aristolochic acid, which plays a key role in Chinese-herb nephropathy (Vanherwegham et al. 1993; Vanhaelen et al. 1994). This is a progressive form of renal fibrosis that develops in some patients who take weight-reducing pills containing Chinese herbs. Because of a manufacturing error, one of the herbs in these pills (*Stephania tetrandra*) was inadvertently replaced by *Aristolochia fangchi*, which is nephrotoxic and carcinogenic (Schmeiser et al. 1994; Nortier et al. 2000). Figure 17.7 shows the analysis by CE-LIF and ^{32}P-postlabeling of CT-DNA incubated with aristolochic acid I (AAI) under zinc activation. 10 µg of normal CT-DNA (upper electropherogram) and 10 µg of a CT-DNA incubated with AAI (lower electropherogram) were analyzed, and the electropherograms were compared with the results of the ^{32}P-postlabeling method (inset of the lower electropherogram). The samples were diluted 100-fold with water. In contrast, the CE-LIF method allowed the simultaneous determination of unmodified nucleotides as well as 5-Me-dCMP and AAI-dGMP, AAI-dAMP and AAII-dAMP.

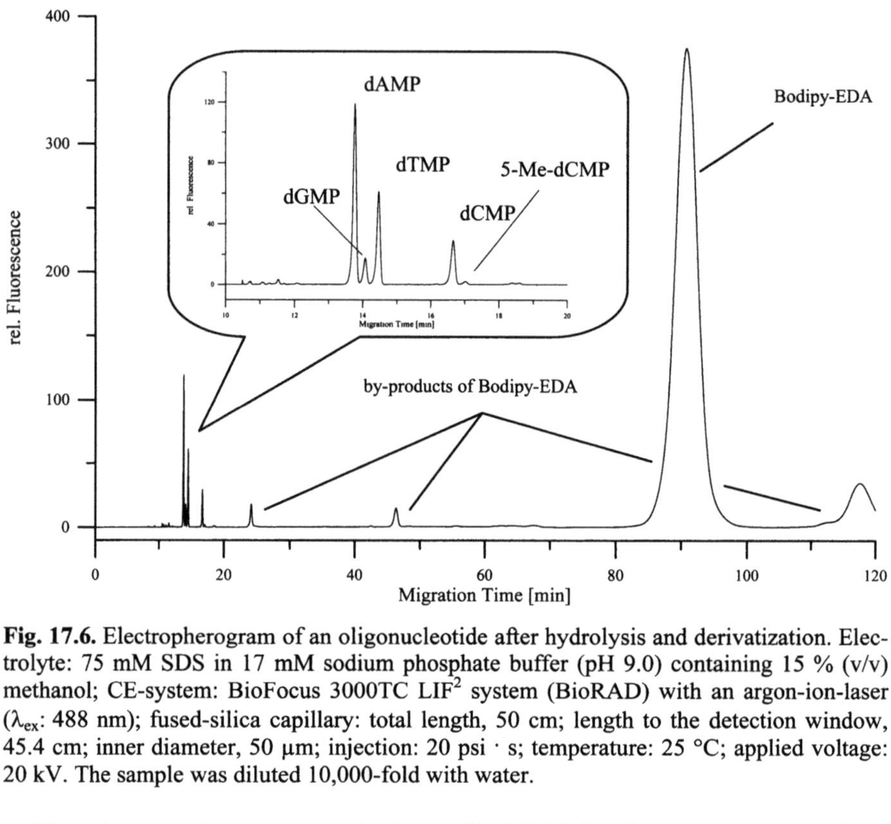

Fig. 17.6. Electropherogram of an oligonucleotide after hydrolysis and derivatization. Electrolyte: 75 mM SDS in 17 mM sodium phosphate buffer (pH 9.0) containing 15 % (v/v) methanol; CE-system: BioFocus 3000TC LIF^2 system (BioRAD) with an argon-ion-laser (λ_{ex}: 488 nm); fused-silica capillary: total length, 50 cm; length to the detection window, 45.4 cm; inner diameter, 50 µm; injection: 20 psi · s; temperature: 25 °C; applied voltage: 20 kV. The sample was diluted 10,000-fold with water.

There is a growing awareness in the medical field that the correct pattern of genomic methylation is essential for healthy cells and organs. If methylation patterns are not properly established or maintained, diverse disorders as mental retardation, immune deficiency, and sporadic or inherited cancers may follow. Through inappropriate silencing of growth regulating genes and simultaneous destabilisation of whole chromosomes, methylation defects may induce a chaotic state from which cancer cells evolve. Methylation defects are present in cells before the onset of obvious malignancy and therefore cannot be explained simply as a consequence of a deregulated cancer cell (Costello and Plass 2001). Therefore, the simultaneous determination of DNA adducts and 5-Me-dCMP is very interesting for investigations of the development of cancer.

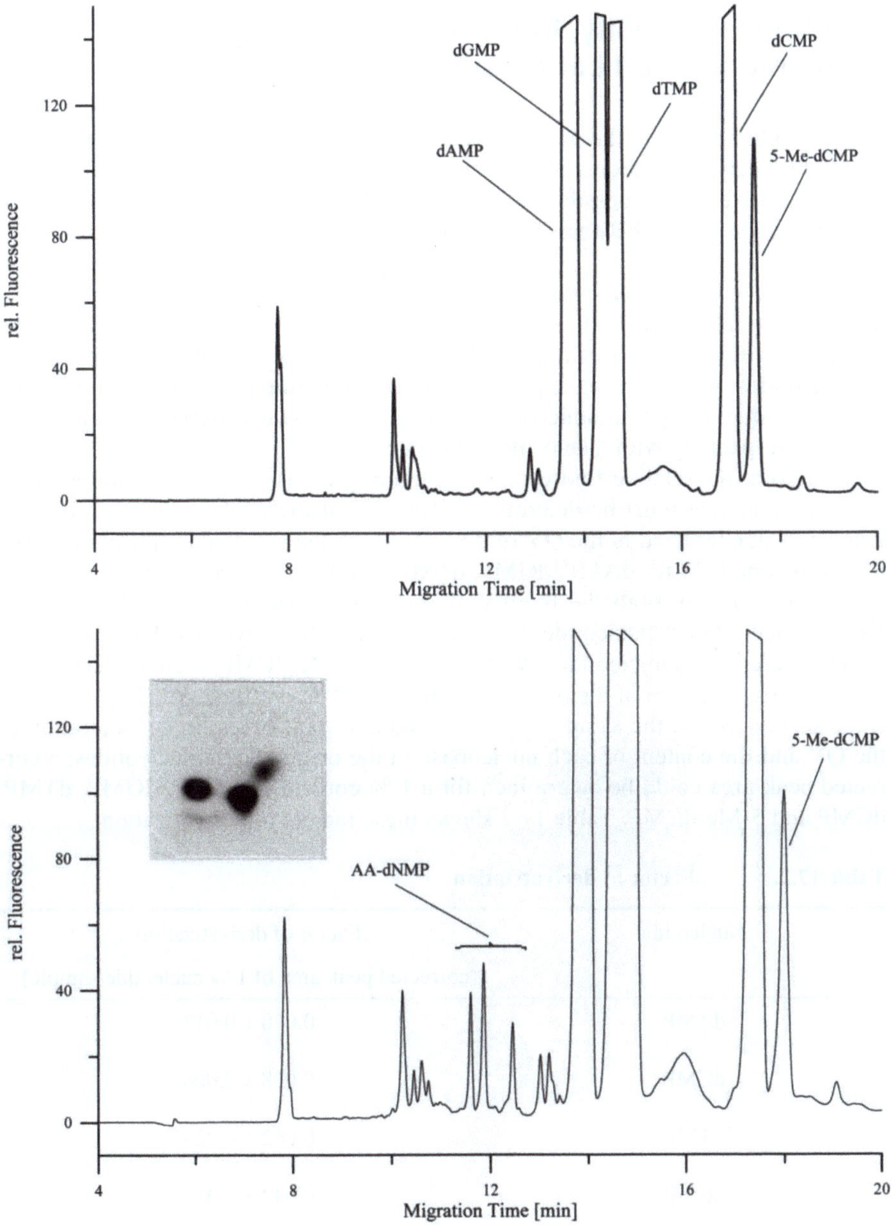

Fig. 17.7. Analysis of DNA adducts of aristolochic acid in CT-DNA. The samples were diluted 100-fold with water. The separation conditions were as in Figure 17.6.

17.3 Reproducibility, Fluorescence-Quenching Phenomenon and Labeling Efficiency

The reproducibility of the hydrolysis, derivatization and separation was determined by analysis of aliquoted oligonucleotides. The use of fluorescein as an internal standard made it possible to correct the peak areas and led to a relative standard deviation (RSD) between 5.6 and 8.4 % (n = 29) for the nucleotides of CT-DNA.

Previous work has reported that the fluorescence of some fluorescent dyes is quenched by an interaction between dyes and a nucleobase (Seidel et al. 1996; Lieberwirth et al. 1998; Torimura et al. 2001). The CE-LIF analysis of several oligonucleotides with known sequences showed that there is also fluorescence-quenching due to a photoinduced electron transfer between BODIPY and the nucleotides, especially with guanosine as the nucleobase.

Therefore, the relative fluorescence quantum yield (QY) of each fluorescence-labeled nucleotide must be determined. The QY of the nucleotides were investigated and standardized to the QY of dAMP, which led to quantum factors of 1.00, 0.55, 0.97 and 0.98 for dAMP, dGMP, dTMP and dCMP, respectively.

For quantitation, both the relative fluorescence quantum yield (QY) of each fluorescence-labeled nucleotide and the labeling efficiency must be determined. Eight different oligonucleotides (four containing 5-Me-dCMP) were divided into 3 to 5 aliquots, and each of them was hydrolyzed and derivatized. After addition of a fluorescein solution the samples were analyzed several times. In consideration of the QY and the content of each nucleobase in the original oligonucleotides, a corrected peak area could be determined for a 1 % content of dAMP, dGMP, dTMP, dCMP and 5-Me-dCMP. Table 17.2 shows these factors of derivatization.

Table 17.2. **Factor of derivatization**

Nucleotide	Factor of derivatization (corrected peak area of 1 % nucleotide/sample)
dAMP	0.070 ± 0.012
dGMP	0.038 ± 0.006
dTMP	0.062 ± 0.008
dCMP	0.042 ± 0.007
5-Me-dCMP	0.046 ± 0.007

With the use of these factors the analysis of CT-DNA (Figure 17.8) showed that 2.95 ± 0.18 % (RSD = 6.1 %; n = 7) of all cytosines were methylated. Since no standard of 5-Me-dCMP was available, the QY of dCMP was used for unmodified and modified cytosine.

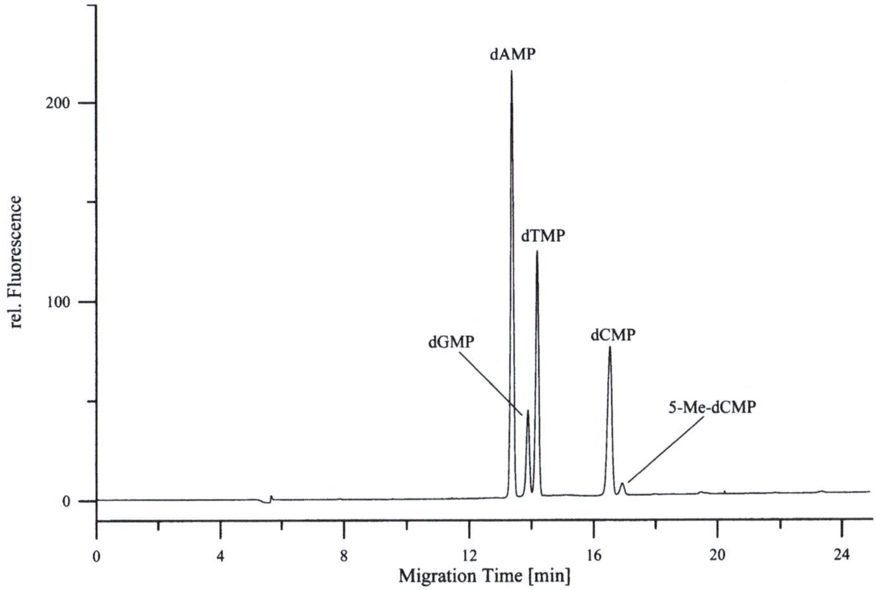

Fig. 17.8. Analysis of the content of 5-Me-dCMP in CT-DNA. The sample was diluted 10,000-fold with water. The separation conditions were as in Figure 17.6.

17.4 Sensitivity

The strength of the ^{32}P-postlabeling procedure is its high sensitivity which is made possible by the enrichment of modified nucleotides by butanol extraction or digestion with nuclease P1. The strength of the CE-LIF method described here is its excellent selectivity. The samples must be diluted 100 fold because of the high salt concentration and the great surplus of unmodified nucleotides. This leads to a relatively high detection limit of 850 pM (determined for dAMP), which in turn means a relative adduct labeling of 2 DNA adducts in 10^6 nucleotides. Electrostacking with reversed field (Chien and Burgi 1992) as a focussing step improved the sensitivity to 50 pM or an RAL of 1.4 DNA adducts in 10^7 nucleotides. The high detection limit, compared with ^{32}P-postlabeling, is nevertheless sensitive enough for the detection of many endogenous DNA adducts, the exact quantitation of DNA methylation (5-Me-dCMP) in tumor cells and surrounding normal tissue.

References

Al-Deen AN, Cecchini DC, Abdel-Baky S, Abdel Moneam NM, Giese RW (1990) Preparation of ethylenediaminephosphoramidates of nucleotides and derivatization with fluorescein isothiocyanate. J. Chromatogr. 512:409-414

Apruzzese WA, Vouros P (1998) Analysis of DNA adducts by capillary methods coupled to mass spectrometry: A perspective. J. Chromatogr. A 794:97-108

Baylin SB, Herman JG (2000) DNA hypermethylation in tumorigenesis: epigenetics joins genetics. Trends Genet. 16:168-174

Chien RL, Burgi DS (1992) Sample stacking of an extremely large injection volume in high-performance capillary electrophoresis. Anal. Chem. 64:1046-1050

Chu BCF, Wahl GM, Orgel LE (1983) Derivatisation of unprotected polynucleotides. Nucl. Acids Res. 11:6513-6529

Costello JF, Plass C (2001) Methylation matters. J. Med. Genet. 38:285-303

Esmans EL, Broes D, Hoes I, Lemiere F, Vanhoutte K (1998) Liquid chromatography mass spectrometry in nucleoside, nucleotide and modified nucleotide characterization. J. Chromatogr. A 794:109-127

Gupta RC, Reddy MV, Randerath K (1982) ^{32}P-postlabelling analysis of non-radioactive aromatic carcinogen-DNA adducts. Carcinogenesis 3:1081-1092

Hemminki K, Koskinen M, Rajaniemi H, Zhao C (2000) DNA adducts, mutations, and cancer 2000. Regul. Toxicol. Pharmacol. 32:264-275

Jones PA, Buckley JD (1990) The role of DNA methylation in cancer. Adv. Cancer Res. 54:1-23

Kelman DJ, Kenneth TL, Sharma M (1988) Synthesis and application of fluorescent labeled nucleotides to assay DNA damage. Chem. Biol. Interact. 66:85-100

Lan ZH, Wang P, Giese RW (1999a) Matrix-assisted laser desorption/ionisation mass spectrometry of deoxynucleotides labeled with an IMI dye. Rapid Commun. Mass Spectrom. 13:1454-1457

Lan ZH, Qian X, Giese RW (1999b) Preparation of an IMI dye (imidazole functional group) containing a 4-(N,N-dimethylaminosulfonyl)-2,1,3-benzoxadiazole fluorophore for labeling of phosphormonoesters. J. Chromatogr. A 831:325-330

Lee T, Yeung ES, Sharma M (1991) Micellar electrokinetic capillary chromatographic separation and laser-induced fluorescence detection of 2'-deoxynucleoside-5'-monophosphates of normal and modified bases. J. Chromatogr. 565:197-206

Li W, Moussa A, Giese RW (1992) Capillary electrophoresis of fluorescein-ethylendiamine-5'-deoxynucleotides. J. Chromatogr. 608:171-174

Lieberwirth U, Arden-Jakob J, Drexhage KH, Herten DP, Müller R, Neumann M, Schulz A, Siebert S, Sagner G, Klingel S, Sauer M, Wolfrum J (1998) Multiplex dye DNA-sequencing in capillary gel-electrophoresis by diode laser-based time-resolved fluorescence detection. Anal. Chem. 70:4771-4779

Nortier JL, Muniz Martinez MC, Schmeiser HH, Arlt VM, Bieler CA, Petein M, Depierreux MF, de Pauw L, Abramowicz D, Vereerstraeten P, Vanherweghem JL (2000) Urothelial carcinoma associated with the use of a chinese herb (*Aristolochia Fangchi*). N. Engl. J. Med. 342:1686-1692

Phillips DH, Castegnaro M (1999) Standardization and validation of DNA adduct postlabelling methods: report of interlaboratory trials and production of recommended protocols. Mutagenesis 14:301-315

Phillips DH, Farmer PB, Beland FA, Nath RG, Poirier MC, Reddy MV, Turteltaub KW (2000) Methods of DNA adduct determination and their application to testing compounds for genotoxicity. Environ. Mol. Mutagen. 35:222-233

Poirier MC (1993) Antisera specific for carcinogen-DNA adducts and carcinogen-modified DNA: Application for detection of xenobiotics in biological samples. Mutat. Res. 288:31-38

Poirier MC, Santella RM, Weston A (2000) Carcinogen macromolecular adducts and their measurement. Carcinogenesis 21:353-359

Randerath K, Reddy MV, Gupta RC (1981) ^{32}P-labelling test for DNA damage. PNAS 78:6126-6129

Reddy MV, Gupta RC, Randerath E, Randerath K (1984) ^{32}P-postlabeling test for covalent DNA binding of chemicals in vivo. Carcinogenesis 5:231-243

Reddy MV (2000) Methods for testing compounds for DNA adduct formation. Regul. Toxicol. Pharmacol. 32:256-263

Schmeiser HH, Pool BL, Wießler M (1994) Identification and mutagenicity of metabolites of aristolochia acid formed by rat liver. Carcinogenesis 7:59-63

Schmitz OJ, Wörth CCT, Stach D, Wießler M (2002) Capillary electrophoresis analysis of DNA adducts as biomarkers for carcinogenesis. Angew. Chem. Int. Ed. 41:445-448

Seidel CAM, Schulz A, Sauer M (1996) Nucleobase-specific quenching of fluorescent dye: 1. Nucleobase one-electron redox potentials and their correlation with static and dynamic quenching efficiencies. J. Phys. Chem. 100:5541-5553

Sharma M, Jain R, Isac TV (1991) A novel technique to assay adducts of DNA induced by anticancer agent cis-diaminedichloroplatinum (II). Bioconjug. Chem. 2:403-406

Soglia JR, Turesky RJ, Paehler A, Vouros P (2001) Quantification of the heterocyclic aromatic amine DNA adduct N-(deoxyguanosin-8-yl)-2-amino-3-methylimidazo[4,5-f]quinoline in livers of rats using capillary liquid chromatography/microelectrospray mass spectrometry: a dose-response study. Anal. Chem. 73:2819-2827

Stiborovà M, Frei E, Bieler CA, Schmeiser HH (1998) ^{32}P-postlabelling: A sensitive technique for the detection of DNA adducts. Chem. Listy 92:661-668

Sugimura T, Ushijima T (2000) Genetic and epigenetic alterations in carcinogenesis. Mutat. Res. 462:235-246

Sweetman GMA, Shuker DEG, Glover RP, Farmer PB (1998) Mass spectrometry in carcinogenesis research. Adv. Mass Spectrom. 14:343-376

Torimura M, Kurata S, Yamada K, Yokomaku T, Kamagata Y, Kanagawa T, Kurane R (2001) Fluorescence-quenching phenomenon by photoinduced electron transfer between a fluorescent dye and a nucleotide base. Analytical Sciences 17:155-160

Vanhaelen M, Vanhaelen-Fastre R, But P, Vanherweghem JL (1994) Identification of aristolochic acid in Chinese herbs. The Lancet 343:174

Vanherweghem JL, Depierreux M, Tielemans C, Abramowicz D, Dratwa M, Jadoul M, Richard C, Vandervelde D, Verbeelen D, Vanhaelen-Fastre R, Vanhaelen M (1993) Rapidly progressive interstitial renal fibrosis in young women: association with slimming regimen including Chinese herbs. The Lancet 341:387-391

Wang P, Giese RW (1993) Phosphate-specific fluorescence labeling under aqueous conditions. Anal. Chem. 65:3518-3520

Wang P, Giese RW (1998) Phosphate-specific fluorescence labeling with BO-IMI: reaction details. J. Chromatogr. A 809:211-218

Index

SacherLasertechnik^{Group}

Sacher Lasertechnik GmbH
Hannah-Arendt-Str. 3-7,
D-35037 Marburg, Germany
Tel.: +49 6421 30 52 90, Fax.: +49 6421 30 52 99
eMail: contact@sacher-laser.com Web: http://www.sacher-laser.com

Sacher Lasertechnik is leading manufacturer of tunable external cavity diode lasers (ECDLs) with more than 10 years of experience. The product range includes ECDLs in Littrow and in Littman/Metcalf configuration as well as driver electronics for the LD and sophisticated measuring electronics. ECDLs are small-sized and suitable for electronic high frequency modulation which results in desirable light sources for sensor application [1].

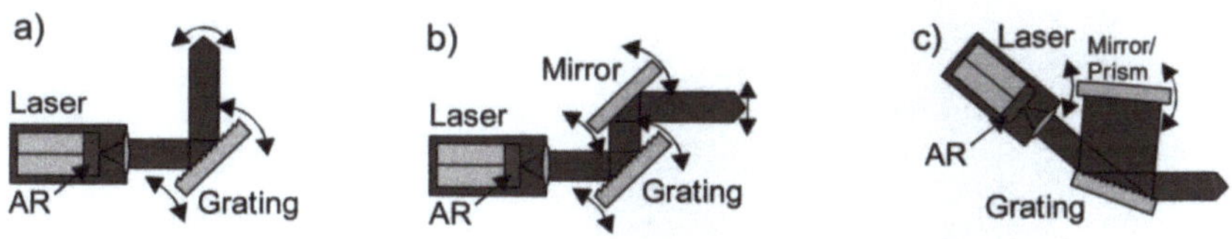

Fig.1: Design of the resonator in Littrow and in Littmann configuration

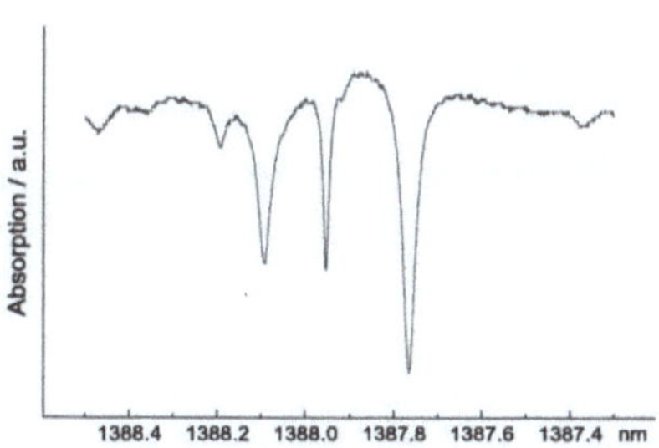

An overview is given on our web site: http://www.Sacher-Laser.com. ECDLs are easy to align, cost effective and offer large output power (see table). Major applications are Raman spectroscopy [2], atom absorption spectroscopy [3] or optical water concentration measurements shown in figure 2.

Fig.2: Spectrum of H_2O in ambient air.

wavelength	Littrow cw-power	Littman cw-power
400 - 420 nm	15 - 30 mW	3 - 8 mW
625 - 700 nm	8 - 15 mW	2 - 5 mW
730 - 1100 nm	50 - 200 mW	10 - 40 mW
1230 - 1660 nm	5 - 15 mW	2 - 5 mW

Please refer to our web page for detailed information about specifications.

[1] C.E. Wieman, L. Hollberg: Using diode lasers for atomic physics. Rev. Sci. Instrum. 62 (1), pp. 1-19, 1991.
[2] http://data.sacher.de/RAMAN/gas.pdf, http://data.sacher.de/RAMAN/milkfat.pdf or http://data.sacher.de/RAMAN/headache.pdf
[3] http://data.sacher.de/publications/ao2003.pdf

MIX
Papier aus verantwortungsvollen Quellen
Paper from responsible sources
FSC® C105338

If you have any concerns about our products,
you can contact us on
ProductSafety@springernature.com

In case Publisher is established outside the EU,
the EU authorized representative is:
Springer Nature Customer Service Center GmbH
Europaplatz 3, 69115 Heidelberg, Germany

Printed by Libri Plureos GmbH
in Hamburg, Germany